POPE, PRINT, AND MEANING

POPE, PRINT AND MEANING

JAMES McLAVERTY

OXFORD
UNIVERSITY PRESS

OXFORD
UNIVERSITY PRESS

Great Clarendon Street, Oxford OX2 6DP
Oxford University Press is a department of the University of Oxford.
It furthers the University's objective of excellence in research, scholarship,
and education by publishing worldwide in

Oxford New York
Athens Auckland Bangkok Bogotá Buenos Aires Cape Town
Chennai Dar es Salaam Delhi Florence Hong Kong Istanbul Karachi
Kolkata Kuala Lumpur Madrid Melbourne Mexico City Mumbai Nairobi
Paris São Paulo Shanghai Singapore Taipei Tokyo Toronto Warsaw
with associated companies in Berlin Ibadan

Oxford is a registered trade mark of Oxford University Press
in the UK and in certain other countries

Published in the United States
By Oxford University Press Inc., New York

© James McLaverty 2001

The moral rights of the author have been asserted
Database right Oxford University Press (maker)

First published 2001

All rights reserved. No part of this publication may be reproduced,
stored in a retrieval system, or transmitted, in any form or by any means,
without the prior permission in writing of Oxford University Press,
or as expressly permitted by law, or under terms agreed with the appropriate
reprographics rights organization. Enquiries concerning reproduction
outside the scope of the above should be sent to the Rights Department,
Oxford University Press, at the address above

You must not circulate this book in any other binding or cover
and you must impose this same condition on any acquirer

British Library Cataloguing in Publication Data
Data available

Library of Congress Cataloging in Publication Data
McLaverty, J.
Pope, print, and meaning / James McLaverty.
p. cm.
Includes bibliographical references (p.) and index.
1. Pope, Alexander, 1688–1744—Criticism, Textual. 2. Pope, Alexander, 1688–1744—
Knowledge—Book industries and trade. 3. Pope, Alexander, 1688–1744—Contributions
in book design. 4. Authors and publishers—England—History—18th century. 5. Literature
publishing—England—History—18th century. 6. Printing—England—History—18th century.
7. Pope, Alexander, 1688–1744—Bibliography. 8. Meaning (Philosophy) in literature. I. Title.
PR3635 .L38 2001 821'.5—dc21 2001033967
ISBN 0-19-818497-2

1 3 5 7 9 10 8 6 4 2

Typeset in Ehrhardt MT
by Best-set Typesetter Ltd., Hong Kong
Printed in Great Britain
on acid-free paper by
T.J. International Ltd, Padstow, Cornwall

Acknowledgements

Acknowledgements are versions of pastoral; the fruits of scholarship reach down into the worker's hand. In acknowledgement land, every librarian is communicative, every colleague genial, every partner enthusiastic, and babes in arms are of invaluable assistance in compiling the index. I have tried writing the academic equivalent of *The Village*, but after addressing myself to 'every care that reigns | O'er youthful peasants and declining swains', I soon found myself lapsing into the embraces of the cloud-compelling queen. The material is stale, and in any case, I have debts to pay and promises to keep. So, with an apology to Crabbe, I amble into the land of content.

This is one of several books I should have written on Pope. The first, under the influence of David Fleeman and David Foxon, would have been more strictly bibliographical than this. From the first of the Davids I absorbed a liking for examining books, from the second a love of puzzles, but not unfortunately his knack of solving them. I nevertheless hope this book forms a useful supplement to *Pope and the Early Eighteenth-Century Book Trade*. The second book would have been more sociological. Gordon Fyfe first encouraged me to think about the problems of being a businessman and an artist, and his influence was reinforced by that of my brother Peter, who also did his best to rescue me from drowning in detail. The third book would have been concerned with the theory of textual criticism, drawing on some of the work of Nelson Goodman and John Searle. These ideas still lurk somewhere in the background of this study, but they have resisted most of my attempts to tease them forward. I am grateful to Jonathan Dancy for his willingness to discuss some associated problems with me—and still more for his general intellectual hygiene.

This book was written around the editing of David Fleeman's Johnson bibliography, during relief from teaching in 1993, 1997, and 1999–2000. I am grateful to Keele University for supporting these periods of leave and to the Arts and Humanities Research Board for the Award that enabled me to take the third. The additional burdens that fell on my colleagues in the English Department were shouldered without complaint and with disturbing ease. I am particularly grateful to Anthea Trodd, Ed Larrissy, and Simon Bainbridge for organization, encouragement, and advice. In 1997 I was elected a Fellow of the Centre for the Book at the British Library, and Richard Price, Mervyn Jannetta, Richard Goulden, and others in the ESTC office helped me out in the race to get round the stacks and locate books before they were shipped off to St Pancras. I am also deeply grateful for the kindness and patience of the staff of the Bodleian Library, and for the help I have received from the university libraries of Keele (especially from Martin Phillips), Cambridge, and London, and from the New York Public Library.

I am indebted to those scholars who have tried to clarify my ideas on Pope and correct my mistakes over the last few years. The three original readers of my manuscript for OUP were encouraging and perceptive, and I have tried to follow their advice. A fourth adviser read the whole book with painstaking generosity, made numerous valuable suggestions, and saved me from many errors. Michael Suarez gave Chapter 5 a detailed and perceptive reading, and David Amigoni did the same for an early draft of Chapter 7. Sally Gray corrected mistakes in my Latin and helped tidy an early draft of Chapter 6. A rough version of Chapter 3 was read at Birmingham University at the invitation of Mark Storey, and I hope I profited from the subsequent discussion. I am particularly grateful to Tony Davies for drawing my attention to important parallels with Erasmus. At Keele, John Bowen has borne years of haphazard discussion with remarkable stamina and acuity, the late John Goode made me rethink approaches to Bakhtin, and Alison Sharrock helped me with some classical puzzles. At the Press Jason Freeman was encouraging, Sophie Goldsworthy patient, Jackie Pritchard thorough, and Frances Whistler, as always, shrewd. In New York, Eric and Wendy Rose were generous in their hospitality. Of other scholars who have helped me, Bruce Redford, Valerie Rumbold, Charles Swann, and David Vander Meulen have been steadfast counsellors and sources of information and encouragement.

Finally I remember with gratitude R. B. Eldred, with whom I first read Pope many years ago at Derby School, and Richard Rouse, whose occasional 'Haven't you finished that book yet?' has done much to ensure its completion.

J. McL.

Contents

List of Illustrations	viii
Short Titles	ix
1. Introduction	1
2. *The Rape of the Lock*: From Miscellany Endpiece to Illustrated Independence	14
3. The *Works* of 1717: Building a Monument	46
4. *The Dunciad Variorum*: The Limits of Dialogue	82
5. *An Essay on Man* and Harte's *An Essay on Reason*: Title Pages and Implied Authorship	107
6. *The First and Second Satires of the Second Book of Horace* (1733–1734): Parallel Texts	142
7. *To Arbuthnot* and *Sober Advice*: Textual Variation, Sexuality, and the Public Sphere	175
8. The *Works* of 1735–1736: Pope's Notes	209
Works Cited	242
Index	251

List of Illustrations

The illustrations are reproduced by kind permission of the Bodleian Library, Oxford University.

1. Frontispiece to *The Rape of the Lock*, 1714 (Bodl. Don. e. 115)	27
2. Plate to canto 2 of *The Rape of the Lock*, 1714 (Bodl. Don. e. 115)	32
3. Frontispiece to *Works*, 1717 (Bodl. Vet. A4 d. 140)	62
4. First page of *An Essay on Criticism*, *Works*, 1717 (Bodl. Vet. A4 d. 140)	65
5. First page of *The Rape of the Lock*, *Works*, 1717 (Bodl. Vet. A4 d. 140)	72
6. First page of *The Dunciad Variorum*, 1729 (Bodl. CC 76 (1) Art)	83
7. Final-page advertisement from *An Epistle to a Lady*, 1735 (Bodl. Fol. Δ 760)	114
8. Title pages of *An Epistle to Dr Arbuthnot* and *An Essay on Reason*, 1735 (Bodl. Vet. A4 c. 289 (6 and 13))	115
9. Parallel of the Characters of Dryden and Pope, *The Dunciad Variorum*, 1729 (Bodl. Vet. A4 d.128 (2))	150
10. First page of *The First Satire of the Second Book of Horace*, 1733 (Bodl. Fol. Δ 696)	156

Short Titles

Anecdotes	John Nichols, *Literary Anecdotes of the Eighteenth Century*, 9 vols. (London, 1812–15)
Correspondence	*The Correspondence of Alexander Pope*, ed. George Sherburn, 5 vols. (Oxford: Clarendon Press, 1956)
Elwin–Courthope	*The Works of Alexander Pope*, ed. W. Elwin and W. J. Courthope, 10 vols. (London, 1871–89)
Foxon	David F. Foxon, *English Verse 1701–1750*, 2 vols. (Cambridge: Cambridge University Press, 1975)
Foxon, *Pope and the Book Trade*	David F. Foxon, *Pope and the Early Eighteenth-Century Book Trade* (Oxford: Clarendon Press, 1991)
Griffith	Reginald Harvey Griffith, *Alexander Pope: A Bibliography*, 2 vols. (Austin: University of Texas Press, 1922–7)
Mack, *Collected in Himself*	Maynard Mack, *Collected in Himself* (Newark: University of Delaware Press, 1982)
Mack, *Last and Greatest Art*	Maynard Mack, *The Last and Greatest Art: Some Unpublished Poetical Manuscripts of Alexander Pope* (Newark: University of Delaware Press, 1984)
Mack, *Life*	Maynard Mack, *Alexander Pope: A Life* (New Haven: Yale University Press, 1985)
Spence, *Anecdotes*	Joseph Spence, *Observations, Anecdotes, and Characters of Books and Men*, ed. James M. Osborn, 2 vols. (Oxford: Clarendon Press, 1966)
Twickenham	*The Twickenham Edition of the Poems of Alexander Pope*, ed. John Butt et al., 11 vols. (London: Methuen, 1939–69)

1. *Introduction*

Pope was fascinated by print throughout his life. In his youth he penned manuscript pages that looked like print; in early manhood he designed elaborate editions of the *Iliad* and *Odyssey*; and in middle age he ran his own printing and publishing business by proxy. Print was for him both a serious vocation and an elaborate game. This study approaches Pope's major poems through their printing and publication, each chapter being stimulated by some new edition, elaboration of apparatus, or detail of typography. Although interpretation sometimes ranges widely in search of an explanation that will satisfy an initial curiosity, the materiality of the texts is always kept in mind and the discussion informed by Pope's printing practice. The thesis of this book is that, in reading Pope, print matters.

An introduction to Pope, print, and meaning must face up to the two Popes: the Pope who loved print and the Pope who hated it. The Pope who hated print loathed the great mass of printed matter: Grub Street scandal, party pamphlets, weekly journals, scholarly editions, boring poems, most plays, critics, and booksellers. He also hated attacks on Pope, though he collected them. This Pope is well known. He featured in Aubrey L. Williams's *Pope's 'Dunciad'* in the fifties, was taken up and re-educated in Marshall McLuhan's *The Gutenberg Galaxy* in the sixties, and was revived by Alvin Kernan as a starting point for his *Printing Technology, Letters & Samuel Johnson* in the late eighties.[1] But the other Pope, the one I believe is important for the interpretation of the poetry, was fixated on print. He loved the look of print: dropped heads, italics, black letter, caps. and smalls; fine paper, wide margins, and good ink; headpieces, tailpieces, initials, and plates. This Pope is a better-kept secret. He features in Maynard Mack's edition of Pope's manuscripts and in David Foxon's Lyell lectures on Pope and the book trade.[2] Whether as lover or hater of print, Pope understood the layers of its culture so well that he could move subtly within them, adopting a variety of roles, denying his agency when necessary or even implying it when it was not there. By 'print' in this study I have in mind primarily products, such as books, pages, and type-areas, which have a variety of symbolic values, but I am also interested in the processes by which these objects are created and later distributed. 'Print' is a difficult, polysemous word, and I use it broadly. It means lettering,

---

[1] *Pope's 'Dunciad': A Study of its Meaning* (London: Methuen, 1955); *The Gutenberg Galaxy: The Making of Typographic Man* (London: Routledge & Kegan Paul, 1962); *Printing Technology, Letters & Samuel Johnson* (Princeton: Princeton University Press, 1987). An exception to the general run of studies is Brean Hammond's *Professional Imaginative Writing in England, 1670–1740: Hackney for Bread* (Oxford: Clarendon Press, 1997), which contains a characteristically shrewd assessment of Pope's business commitments and skills.

[2] *The Last and Greatest Art* (1984); *Pope and the Book Trade* (1991).

typography, the impress of type (as in 'Can you read small print?'); or it means an impression, a run; or, in an older sense given by the *OED*, 'the work of the press, the process of printing'. 'In print' means 'in a printed state', 'in a printed form', but it also now means 'available for sale'; that the book has been printed, and in sufficient quantity, is the condition of its availability for a mass market. Bibliographers sometimes talk of the cost of print and paper, thinking of print as both what has to be added to the paper to make the book and the process (composition and press-work) that is charged for. Pope was engaged by all of these. If he never actually set type, as Virginia Woolf did, he made up for this omission by his commitment to advertisement, distribution, and price-fixing.[3] Pope and print also involves Pope and publishing.

The conflict between Pope the lover and Pope the hater of print is on the face of it easy to resolve. The step from loving beautiful manuscripts to loving beautiful printed books is a short one, and it need not involve acceptance of mass production or the other cultural developments associated with print. Pope might have loved print only in so far as it could be adapted to producing elegant volumes which served as substitute manuscripts for elite readerships. But the detail of Pope's literary career shows this was not so; his career was worked out, both practically and conceptually, through a constant engagement with print. Even a beautiful manuscript like that of the *Pastorals* seems a stage on the road to publication rather than an artefact in itself. Admittedly the manuscript inverts the relationship between roman and italic, but it otherwise keeps to the conventions of print, with the text written out in italic and roman used for proper nouns and for emphasis, the lines are justified, and footnotes cued to the text. More remarkably, the text begins with a dropped head so closely imitating type that it might be mistaken for the real thing, with elegant details such as a hollowing out of the apex to 'A' and a slanting serif to 'L'. When we get to the manuscript of *An Essay on Criticism* a few years later, Pope is to be found counting the lines per page in preparation for the printed edition.[4] And although there were rather grand books throughout Pope's career, often for subscribers, there were also small ones for the general public on which he seems to have lavished no less attention. The marked-up copy of a volume of an octavo *Works* II in the British Library (C.122.e.31) shows a characteristic care for the smallest details of wording, capitalization, and punctuation. Yet Pope's love of print was never pure, uncontaminated by self-consciousness and suspicion. Often it was antagonistic or ironic. Parodies of reviews (of Philips in the *Guardian*), of journals (the *Grub-street Journal*), and of scholarly editions (*The Dunciad* and *Sober Advice*) are signs of an unease that also manifests itself in the defensive preface to the *Works* of 1717 and the covert publication of the

---

[3] Virginia Woolf was the Hogarth Press's compositor and Leonard the press-man. See Hermione Lee's discussion of the setting up of the press and the influence on Woolf's art, *Virginia Woolf* (London: Chatto & Windus, 1996), 362–76. See also Jerome McGann's essay on 'Modernism and the Renaissance of Printing' in his *Black Riders: The Visible Language of Modernism* (Princeton: Princeton University Press, 1993), 3–41. Pope was much more in contact with the commercial London trade than the modernists were.

[4] See Mack, *Last and Greatest Art*, 24, and Foxon, *Pope and the Book Trade*, 162–7.

Letters. Although Pope used his imitation of Horace's *First Satire of the Second Book* to proclaim his duty to publish, his usual stance was that of someone who would rather not—and was looking for someone else to take the responsibility. As he grew more antagonistic both to the London book trade and to the court culture that represented its alternative, Pope found print an essential form of self-expression but one involving a necessary deformation.[5]

The first commentator on Pope and print was Martinus Scriblerus, and, as his was the major influence on Williams, McLuhan, and Kernan, his view and his poem must be disposed of before we can take a more positive view of Pope's relation to print.[6] In his 'Of the Poem' in *The Dunciad Variorum* Scriblerus describes the cultural crisis to which he thinks Pope is responding.

> We shall next declare the occasion and the cause which moved our Poet to this particular work. He lived in those days, when (after providence had permitted the Invention of Printing as a scourge for the Sins of the learned) Paper also became so cheap, and printers so numerous, that a deluge of authors cover'd the land: Whereby not only the peace of the honest unwriting subject was daily molested, but unmerciful demands were made of his applause, yea of his money, by such as would neither earn the one, or deserve the other: At the same time, the Liberty of the Press was so unlimited, that it grew dangerous to refuse them either: For they would forthwith publish slanders unpunish'd, the authors being anonymous; nay the immediate publishers thereof lay sculking under the wings of an Act of Parliament, assuredly intended for better purposes.[7]

The causal chain is laid out: invention of printing, cheap paper, numerous printers, deluge of authors, demand for praise and money, absence of regulation, slanders. The passage seems grounded in a fear of writing itself, with its delightful invocation of the 'honest unwriting subject' (the equivalent of the law-abiding citizen), but it also points to the transition from a manuscript culture to a print one. For Kernan this is the central preoccupation of *The Dunciad*:

> print in his [Pope's] understanding is a dullish, mechanical, undiscriminating, repetitive, mass medium, a true instrument of Dulness that gives extraordinary opportunities to those already inclined that way, greedy booksellers, vain, dull gentlemen, poor scribblers, pedantic schoolmasters. . . . the old aristocratic society and its system of letters was also being distorted and beginning to disintegrate under pressures from new print-fostered kinds of party-politics, market-place economics, rationalistic philosophy, and machine technology. Print was both the image and the instrument of these new ways of thinking and doing, and in Pope's apocalyptic vision, a flood of printer's ink was a darkness that spread across the land, staining . . . the white

---

[5] Self-expression seems to me a much more plausible objective than it does to Helen Deutsch in her stimulating *Resemblance & Disgrace: Alexander Pope and the Deformation of Culture* (Cambridge, Mass.: Harvard University Press, 1996).

[6] I take up the question of the characterization of Scriblerus again in Chapter 4. Although there are dangers in attempting too radical a separation of Scriblerus and Pope, it seems to me the lesser of two evils. Williams identifies Scriblerus and Pope, *Pope's 'Dunciad'*, 9.

[7] *Twickenham*, 5, ed. James Sutherland (1943), 49. The passage is virtually unchanged in 1743, though there 'Liberty' is changed to 'licence'. Subsequent quotations from *The Dunciad* in this chapter are from the *Twickenham* edition.

page, darkening the minds of the people and their rulers, obliterating polite letters, and finally extinguishing all light, to leave the land in ancient night and ignorance. (15–16)

This is very well put, a wonderfully comprehensive vision of the poem, but it is somehow not quite right. The hesitation comes in with that image of printer's ink flooding the world: Pope did not use it. *The Dunciad* would have been a different poem if it had portrayed technology as the root of social evil. I do not believe the historical situation Scriblerus sketches is pictured in the poem or particularly engaged by it, and *The Dunciad*'s attention to print, in any specific sense, is wavering.

It is worth looking at the passages in *The Dunciad* that take a detailed interest in print in order to assess whether Pope's position coincides with Scriblerus'. An interest in books surfaces at several key points in the poem. Our first encounter with them as objects occurs in book 1, while Bays sits supperless in his study. Significantly Pope's criticism in this episode is not of the press's capacity to produce in bulk for the popular market but of the contrast between appearance and content. Bays's books are chosen because they make a good show, with large paper, illustration, and gilding—books in fact dangerously like Pope's own translations and *Works*. The library also includes some very old books, the work of the first English printers, Caxton and Wynkyn de Worde, but the emphasis is not on the invention of printing but on the state of learning in the period: these are 'The Classicks of an Age that heard of none'. At the end of the book Dulness shows her own works, which complement Bays's library:

> Here to her Chosen all her works she shews;
> Prose swell'd to verse, verse loit'ring into prose:
> How random thoughts now meaning chance to find,
> Now leave all memory of sense behind:
> How Prologues into Prefaces decay,
> And these to Notes are fritter'd quite away:
> How Index-learning turns no student pale,
> Yet holds the eel of science by the tail . . . (*Four Book Dunciad*, 1. 273–80)

This passage does portray a physical progress reflecting an artistic deterioration: kinds and forms dissolve as the parts of the book crumble into one another; everything gravitates to the fragmentation of notes and index-learning. But what is displayed is a subversion of the codes of the book, rather than creativity being determined by those codes.

The three later books similarly fail to present the Scriblerian perspective. Book 2 is the most concerned with the book trade, with its degrading games involving the booksellers Lintot, Curl, and Chetwood/Osborne, and it might be expected to tell us a good deal about Pope and print; but there is no sign of technological change or influence. At the end of the book, there are readings from Blackmore and Henley, but the scene is one associated with an early literate culture rather than with the explosion of print:

> The pond'rous books two gentle readers bring;
> The heroes sit, the vulgar form a ring. (2. 383–4)

Even in book 3, with its historical perspective, Pope misses his opportunities. Chi Ho-am-ti, Caliph Omar I, and the Roman synods are condemned for the destruction of books. Wormius (Thomas Hearne) is ridiculed for indiscriminate preservation of old learning. But there is no culminating criticism of a modern mass market. Sir Thomas Hanmer's appearance in book 4 leads to an attack on scholarly editing, but his Shakespeare is mocked because it resembles the grand volumes of Bays's library. This is not the market in operation at all, but private publication, distribution by the University to its commoners. The new scholarship is thought of as fancy bookmaking and, simultaneously, destruction of classic authors. The appearance later of Bentley, with his boast of discovering a single letter, the digamma, seems to confirm the attack on pedantry, but it does nothing to advance an attack on print.

An examination of *The Dunciad* for its interest in books and print shows up its lack of interest in technological change: no invention of printing; no mass production; no marketing. There is principally an attack on scholarly editing, with supplementary ridicule of low authors and scandalous journals. *The Dunciad* is not guilty of technological determinism. Of course, it could be claimed that my examination is inadequate and literal-minded, and that technological change, though not directly addressed, is of pervasive importance. That I am willing to accept as long as Pope's response to the technology is allowed to be discriminating and his response to the growth of the market ambivalent. Scriblerus is a pedant, a scholar with no literary culture or critical sense. He can be expected to be opposed to the democratization entailed by print. Pope is much more conscious of a failure of court culture that makes a return to the past undesirable.

Whatever Pope's reservations about the mass market, he showed a career-long determination to involve himself in the printing of his work. A note sent to William Bowyer about his collected *Works* in 1717 (he was only 29) is typical: 'I desire, for fear of mistakes, that you will cause the space for the initial letter to the Dedication to the Rape of the Lock to be made of the size of those in Trapp's Prælectiones. Only a small ornament at the top of that leaf, not so large as four lines breadth. The rest as I told you before.'[8] The combination of minute instructions on the printing of the book and mistrust of the printer's willingness to carry them out is common in Pope's dealings with the book trade. The second paragraph of this letter begins 'I hope they will not neglect'. The William Bowyers, father (1663–1737) and son (1699–1777), printed most of Pope's work before the 1728 *Dunciad* (John Watts was responsible for the rest), and he turned to them again at the end of his career.[9] They were the most distinguished London printers of their day, but they still seem not to have enjoyed Pope's confidence. The letters to the younger Bowyer, 'the most learned Printer of the Eighteenth Century',[10] over the printing of *The Dunciad in Four Books*

---

[8] *Correspondence*, 1. 394. The letter is to Broome, but the instructions are clearly to be passed on to Bowyer.

[9] Foxon, *Pope and the Book Trade* has tables listing Pope's printers and other information for each poem.

[10] *Anecdotes*, 1. 2. Bowyer's learning is one of the justifications for the nine-volume collection.

and the epistles at the end of Pope's life show some warmth, but even then there are problems: Bowyer 'does not quite answer my Impetuosity for getting this poem out of my hands'; 'they have still not separated it [*The Dunciad*] right'; and Warburton's corrections have not been included in one of the sheets (*Correspondence*, 4. 428, 477, 514).

The Bowyers were not singled out for criticism in any special way. When Tonson's printer John Watts, the Bowyers' chief rival for pre-eminence early in the century, was chosen for printing the *Odyssey*, Pope wrote a letter to Tonson calling into question Watts's competence:

I must desire a favor of you . . . which is also to redound to the credit of Mr Lintot. I mean in regard to the beauty of the Impression, that you will use your interest with Mr Watts, to cause them to work off the Sheets more carefully than they usually do: & to preserve the blackness of the Letter, by good working, as well as by the best Ink. The sheets I've seen since the first Proof, are not so well in this respect as the first. I beg your Recommendation as to this particular, There's nothing so mu[ch] contributes to the Beauty & credit of a Book, which would be Equally a reputation to Mr Lintot & to me. (*Correspondence*, 2. 217)

A concern about illustration and composition would not be surprising, but an interest in presswork is exceptional and reveals a professional's sensitivity about the quality of the printed product. The letter is surprising, because, as David Foxon shows, Watts was an exceptionally fine printer by English standards. Pope seems to feel that he alone has a proper commitment to his book, and that he, as much as the bookseller and more than the printer, will be judged by its quality. Any sense that the printer is a potential collaborator with a pride in his craft is missing.

It seems unlikely that Pope supervised John Wright, who became his printer from 1728 onwards, any less intently than he did Bowyer and Watts. Wright's seems to have been a small shop, possibly worked by him and his son, and most of his printing was for Pope or Pope's friends, though he would have been engaged in small-scale jobbing work as well.[11] He must have been very dependent on Pope's goodwill, which may not have been easy to maintain. John Dennis provides an entertaining, if hostile, glimpse of the printing of the *Dunciad Variorum*:

Does not half the Town know, that honest *J. W.* was the only Dunce that was persecuted and plagu'd by this Impression? that Twenty times the Rhapsodist alter'd every thing that he gave the Printer? and that Twenty times, *W.* in Rage and in Fury, threaten'd to turn the *Rhapsody* back upon the Rhapsodist's Hands?[12]

---

[11] There are no records from Wright's shop, so the inferences about his business are speculative. He did not use press-figures, which suggests he was using only one press, or at least did not need a system of identification for his press-men. His ornaments do not lead to identification of a large number of publications, but, of course, he may not always have used them (see James McLaverty, *Pope's Printer, John Wright*, Oxford Bibliographical Society Occasional Publication 11 (Oxford: Bibliographical Society, 1977)). He did small printing jobs for Christ's Hospital, and his son succeeded him (Minutes of the Committee of Almoners, Guildhall Library, MS 12,811/9 and 12).

[12] 'Remarks upon the *Dunciad*', *The Critical Works of John Dennis*, ed. Edward Niles Hooker, 2 vols. (Baltimore: Johns Hopkins University Press, 1939–1943), 2. 356. The 1728 *Dunciad* was printed by James Bettenham; I do not know why.

The incessant revision, particularly of works as complicated as the *Variorum*, where text and two varieties of notes had to be juggled on the same page, must have strained the relationship, though a lot would depend on the system of payment. Pope seems to have rewritten in proof a great deal, and Wright presumably charged extra for this work.

This brief survey of Pope's relations with his printers, based on little more than scraps of documentation, nevertheless suggests the sort of interest he took in his books, and provides a context for my discussion of particular poems and collections. The chief point that emerges is that Pope took immense care over the text of his poetry, but that his interest did not stop there, ranging instead over every aspect of design and production. The second paragraph of the letter to Bowyer about the 1717 *Works* may serve as representative:

I hope they will not neglect to add at the bottom of the page in the Essay on Criticism, where are the lines 'Such was the Muse whose rules,' &c., a note thus: 'Essay on Poetry, by the present Duke of Buckingham,' and to print the line 'Nature's chief masterpiece' in italic. Be pleased also to let the second verse of the Rape of the Lock be thus,

What mighty contests rise from trivial things.[13]

This paragraph shows the usual enthusiasm for revision, but also the concern for how the text is to be mediated to the reader. The line in the 1714 *Rape of the Lock* had been 'What mighty Quarrels rise from trivial Things,' and Pope simply increases the mock-epic grandeur of his line by elevating a tiff into a contest. Simple revision of the couplet in *An Essay on Criticism*, however, is not enough. Pope must also acknowledge his debt to the Duke of Buckingham, and in doing so highlight the friendship with him that was originally to begin and end the *Works*. Pope's continuing care for his text and for its presentation is the necessary background to the interpretative focus of this book.

Although I would reject any simple separation between looking at a book and reading it—looking is a necessary but not sufficient condition for reading, while reading is a sufficient but not necessary condition for looking—one of the interests of this study is in looking at Pope's books while reading them and in the consequences that approach has for meaning and interpretation. My understanding of meaning is a thoroughly intentionalist one, and I read and look at the books in an attempt to be clearer about Pope's intentions. I am happy to agree with John Searle that meanings are 'in the head' and with H. P. Grice that the object of my attention is an utterer's intended meaning.[14] The search is not merely for some general institutional significance, but for what Grice calls a non-natural meaning: when a person means something by an utterance. An example from Grice explains the

---

[13] *Correspondence*, I. 394. The changes to the *Essay on Criticism* are illustrated in *Pope and the Book Trade*, fig. 43.

[14] John R. Searle, *Intentionality: An Essay in the Philosophy of Mind* (Cambridge: Cambridge University Press, 1983), 197–230. H. P. Grice, 'Meaning', *Philosophical Review*, 66 (1957), 377–88. I draw on Grice's thought in my 'Issues of Identity and Utterance', in Philip Cohen (ed.), *Devils and Angels: Textual Editing and Literary Theory* (Charlottesville: University Press of Virginia, 1991), 134–51.

distinction and draws us, I think, surprisingly close to Pope's world of ambiguity and implication:

> I have a very avaricious man in my room, and I want him to go; so I throw a pound note out of the window. Is there here any utterance with a meaning$_{NN}$[non-natural]? No, because in behaving as I did, I did not intend his recognition of my purpose to be in any way effective in getting him to go. . . . If on the other hand I had pointed to the door or given him a little push, then my behavior might well be held to constitute a meaningful$_{NN}$ utterance . . . (384)

There are potential problems of infinite regression with this as a theory of meaning (do you need to recognize this meaning intention as supported by a prior one?) but the stress on recognition of the intention seems right—and appropriate to Pope. Pope would have left you guessing whether he had thrown the pound note out of the window to tempt you to leave, and he would have opened up the possibility of interpreting the push as a pat on the back. The line, 'Better than lust for Boys, with *Pope* and *Turk*' (*Sober Advice*, 43) raises precisely the Gricean question, 'Does he intend me to recognize a reference to himself?'

Poetry tends to sport in the margins identified by philosophers of language, and two detailed examples from John Searle's work on speech acts throw light on Pope's practice and on the sort of curiosity it gives rise to. Building on the work of J. L. Austin, Searle points out in *Speech Acts* that illocutionary acts (doing something in saying something) can often be expressed in an illocutionary verb: for example, 'I promise', 'I warn', 'I command'. But in refining his account in *Expression and Meaning*, he explains that it does not follow that all verbs taking this form name an illocutionary act:

> There are many illocutionary verbs that are not restricted as to illocutionary point, that is, they can take a large range of illocutionary points, and thus they do not genuinely name an illocutionary force. 'Announce', 'hint', and 'insinuate', for example, do not name types of illocutionary acts, but rather the style or manner in which a rather large range of types can be performed . . . for example, hinting is not part of meaning in the sense that hinting is neither part of illocutionary force nor propositional content.[15]

I believe that 'emphasize' is another of these verbs with unrestricted illocutionary point and that this feature of utterance goes some way to explain Pope's freedom in changing italics and capitals from edition to edition of his poetry. The best example is *An Essay on Criticism*, where the first edition makes heavy use of italic for emphasis:

> Learn then what MORALS Criticks ought to show,
> For 'tis but *half a Judge's Task*, to *Know*.
> 'Tis not enough, Wit, Art, and Learning join;
> In all you speak, let Truth and Candor shine:
> That not alone what to your *Judgment*'s due,
> All may allow; but seek your *Friendship* too.[16]

---

[15] *Expression and Meaning: Studies in the Theory of Speech Acts* (Cambridge: Cambridge University Press, 1979), p. ix. I am not sure why these verbs should be called illocutionary verbs at all.

[16] *An Essay on Criticism (1711)* [facsimile], ed. David F. Foxon (Menston: Scolar Press, 1970), 32. Foxon illustrates and analyses the development of this passage in *Pope and the Book Trade*, 167–80.

I think Pope adopted this emphatic way of writing from Thomas Creech's translation of Horace's *Art of Poetry*. Creech uses italic elsewhere in his translations of Horace to mark directly reported speech, but in the *Art of Poetry* he diffuses the italic, perhaps in an attempt to capture a didactic tone. Pope follows Creech, highlighting key terms, as in this passage, or pointing up antitheses by placing nouns in italic. The practice continues into the second edition, fades away in the third, and is dropped in the *Works* of 1717. It does not subsequently reappear and no other Pope poem is written or printed in that way. This poses a problem for interpreters of Pope's printing practice. The claim that the meaning of a large number of passages has changed between the first and third editions is implausible, but we could not be satisfied either with the claim that the changes were insignificant. It seems that the typography compensates for the absence of the spoken voice by offering a binary (roman and italic) or ternary (if we include caps. and smalls) opposition that enables the writer to imitate a feature of speech acts, by changing the manner in which something is said without, strictly, changing the meaning. I have commented frequently on such features of typography in this study, but I have tried to resist the temptation to argue that the typography changes the meaning of the word. Its role is much more, I believe, to construct broad relationships with the reader within which specific poetic effects can function.[17]

The second element in Searle's detailed analysis that I have found helpful takes us back to Grice. In the example in which Grice pushes his visitor in the back, the agent is taken to mean something because he acts with the intention of producing an effect through recognition of his intention. But Searle explains that one can mean without intending to produce an effect:

We need to have a clear distinction between representation and communication. Characteristically a man who makes a statement both intends to represent some fact or state of affairs and intends to communicate this representation to his hearers. But his representing intention is not the same as his communication intention. . . . One can make a statement without intending to produce conviction or belief in one's hearers or without intending to get them to believe that the speaker believes what he says or indeed without even intending to get them to understand at all. There are, therefore, two aspects to meaning intentions, the intention to represent and the intention to communicate. (*Intentionality*, 165–6)

There is an interesting connection here with the peculiar status of hinting, which, as we have seen, is not an illocutionary act, though it looks like one, and works by withholding from the receiver a full appreciation of the utterer's communication intention. Emphasizing, on the other hand, makes the communication intention clear. One might imagine that Pope's poetry worked through full disclosure of its communication intentions, but the opposite is often the case: the representation is

---

[17] Literary criticism is perhaps over-anxious to call anything important meaning, but I admit that typography can change reference, both by convention (Othello or *Othello*) and by stipulation. For a view different from mine, see Tania Rideout, 'The Reasoning Eye: Alexander Pope's Typographic Vision' (unpublished doctoral thesis, University of Cambridge, 1994), and 'The Reasoning Eye: Alexander Pope's Typographic Vision in the *Essay on Man*', *Journal of the Warburg and Courtauld Institutes*, 55 (1992), 249–62.

clear, the communication is in doubt. The potential for Jacobite allusion is present in *The Rape of the Lock*, but the question whether Pope intends those allusions to be communicated to his readership is an open one. Such ambiguities lend importance to Pope's self-expression through print.

Although this study deals very widely in Pope's utilization of print, touching on format, general typography, illustration, italics, capitals, footnotes, endnotes, prefaces, and advertising, it does not do so systematically, and its exploration of the poems is not confined by attention to these features. Each discussion starts with attention to print, but explanation, as I believe it must, deals with many different aspects of the poems and their context. There is no implicit claim that Pope consistently uses typography as a cipher for occult messages. Print is one stimulus to interpretation among many. It certainly would have been possible to write chapters on capitals and italics, for example, tracing developments chronologically through revisions and from text to text, but such an approach necessarily confines itself to issues of immediate communication intentions, hints and emphases, and avoids major issues of interpretation. The full significance of typographical features comes from the evidence they provide for the author's intentionality in the context of the work as a whole. Consequently I have organized this study by particular poem or book, and I have involved myself in controversial questions and tried to pursue them to a conclusion. Each chapter starts with a particularly important feature of print and uses it to raise critical questions. The principal feature—revised edition, annotation, title page, parallel text—starts off the debate, which then develops freely. The two chapters on Pope's *Works* are exceptions because volumes of *Works* create new units of interpretation, different from the solitary text that is usually the prime focus of literary study, and I believe they consequently merit special attention.

Pope's primary interest in printing his poetry was in the substantives of his poetic text, and I have tried not to neglect them. Chapters 2 (*Rape of the Lock*) and 7 (*To Arbuthnot* and *Sober Advice*) are particularly concerned with changes of conception that required extensive revision and new editions; I even suggest that in the case of *To Arbuthnot* Pope had built textual revision into his programme of publication—a measure of textual instability was a deliberate part of his plan. But although Pope's primary attention was to his poetry, he also cared for his notes and for paratext in general. One of my emphases is on the importance Pope attached to notes, prefaces, appendices, and other forms of mediation between his poetry and his public, though like all textual thresholds they can constitute a barrier as well as an invitation to the reader. Chapters 4 (*Dunciad Variorum*) and 8 (*Works*, 1735–6) focus on Pope's notes, the former considering them as part of a complex text (poetry and prose) in some ways analogous to the novel, and the latter investigating them as an attempt to cast Pope as a historical figure in order to court a new public and guide it through the intricacies of the poetic texts. One of Pope's aims in annotation was to clarify the relation of his work to other texts, and a preliminary exercise in dealing with parallel texts is discussed in Chapter 3 (*Messiah* in *Works*, 1717). An even more complex deployment of typography in the service of textual comparison appears, however, in the Horatian imitations, and Chapter 6 attempts a detailed analysis of all aspects of that apparatus. Pope's interest in other writers was not, of course, confined to classical texts, and in discussing his use of critical apparatus in Chapters 4 and 8, I detail his debts to the Geneva edition of Boileau's *Works* and some of his engagement with his contemporaries. From 1728 onwards Pope appears to have run his own mini printing and publishing business, and in Chapter 5 I discuss the relation of his own *Essay on Man* with Walter Harte's *Essay on Reason*, whose title page shows it to be a product of that business. I claim that Pope intended Harte's poem to be taken for his own in the hope that he might profit from its orthodoxy, but then suggest that Harte's orthodox intentions were not always realized in his writing. Pope's correspondence shows a keen interest in the decoration of his books, and in Chapters 2 (*Rape of the Lock*) and 3 (*Works*, 1717) I discuss the illustrations and their interaction with the text, suggesting their important role in expressing Pope's politics and in fashioning the poet's self. I have been attentive to issues of format throughout, and in the final chapter I attempt a brief view of the progress of Pope's annotation through different editions in 1735 and 1736.

I have tried to show the relevance of bibliography to literary criticism by moving from bibliographical evidence to critical positions that are potentially refutable. I argue that the revision and illustration of *The Rape of the Lock* rob Arabella Fermor of her experience, attack a decadent court culture, and imply that prominent Catholic families are aping its corruption; that the *Dunciad Variorum* resembles a novel in its plurality of voices, but uses the codes of the scholarly edition to subordinate and discipline voices that rival the poet's; that the *Essay on Man*'s twin publication, *An Essay on Reason*, reveals both the comparatively low value Pope attached to human reason and the possibility that he regarded his poem as entirely compatible with Christian orthodoxy; that the parallel texts of the imitations of Horace declare Pope's identity with Horace; and that the revision of *To Arbuthnot* reveals the complexity of the parallelism between Hervey and Pope. I also claim that Pope's career shows ever-increasing attempts to shape the response to his work, and that these attempts involve a form of narcissism, putting his private life and reputation at risk.

In comparison with other accounts, for example, Maynard Mack's *Life* (1985), which deservedly holds a central place in Pope studies, my exploration of Pope pushes him into the margins of his society in terms of social class and sexuality. The pre-Warburton Pope I discuss is a more thoroughly alienated figure than the usual one. With his father's background in trade and his own skilled dealing in books, Pope should have been the great bourgeois poet of the eighteenth century, but he lacked a developed class-consciousness, and indeed such a consciousness was probably unavailable to him. Paul Langford writes well about class in *A Polite and Commercial People*, capturing the difficulties facing someone like Pope. He notes that the language of class emerged during the eighteenth century, partially displacing terms such as 'order', 'rank', and, Pope's favourite, 'quality'. The term 'the middle class' was still a novel expression in 1784. One consequence was that the middle class 'was united in nothing more than in its members' determination to make themselves gentlemen and ladies, thereby identifying themselves with the upper class'.[18] Herein lay Pope's problem. Although his satiric impulse, and a sense of merit inadequately recognized, drove him towards an anti-aristocratic stance after 1729, he identified himself with a small group of alienated aristocrats and with writers of the past rather than with any contemporary social grouping. Nevertheless, like Langford, I have used the term 'class' very freely and regarded class-antagonism as central to the poetry of the 1730s. The attacks on *To Burlington* express a powerful class-consciousness and a contempt for the professional man; and that same class-consciousness speaks powerfully in the attack on Pope's birth in *Verses to the Imitator of Horace*. Pope's response was only partially based on the claim to gentility found in the notes to *To Arbuthnot*. He also responded by drawing attention to his unpropertied independence and by basing his critique on claims to an exceptional personal probity. The aristocratic attacks of the 1730s account in part for Pope's intensification of his own self-portrayal.

From 1717 onwards Pope was printing versions of himself. Print gave him the distance necessary for a free shaping of his image; and his relations with the book trade gave him the power to realize his intentions. Printing his poetry and letters seems to have served as a kind of surrogate narcissism. Illustration, annotation, and revision were increasingly directed at projecting an image of the poet. There were, I think, restrictions on what Pope thought it proper to do. Though he never regarded himself as on oath before his public, he did accept limits on what he claimed, and even the doctored edition of his letters was constructed out of letters actually sent, though redirected in print. From 1731 onwards his main dangers came from the powerful social position of his enemies and from the degree to which he felt it necessary to stake his personal integrity on the outcome of his quarrels; he risked cannibalizing himself in order to dictate the terms of the debate. I suggest in Chapter 7 that Pope's attack on Lord Hervey involved reflections on his own sexuality in *Sober Advice from Horace*, with hints of pederasty and onanism fulfilling his duty to represent his own 'spots' to the public. In doing so, I am slightly embarrassed to find myself publicizing charges against Pope that even his most virulent opponents neglected to make, but I think Pope wanted to give them an airing himself. Fortunately, he was rescued from further introspection by the *Essay on Man* controversy and William Warburton. This study stops short of Pope's joint productions with Warburton, though I discuss the detail of Warburton's editing at various points in my analyses. In spite of some valuable work, we still lack the understanding of Warburton necessary for a proper investigation of his collaboration with Pope. Some scholar with a relish for *The Divine Legation of Moses* is needed to complete the account of Pope's literary career.

Some general positions are implicit in this study and I should like it to be clear that I have not wandered into them inadvertently. The account is a realist and even materialist one. I am committed to the view that there is a world independent of our

---

[18] *A Polite and Commercial People: England, 1727–1783* (Oxford: Clarendon Press, 1989), 652–5, 752–4 (653).

representations of it. Though knowledge of this world is not easy, I am also committed to some correspondence theory of truth. I believe there can be no meaning without intentionality, and that there is both collective and individual intentionality. Pope loves to play on the borders of collective and individual intentionality, but I hope to retrieve his particular intentionality and meanings. I believe that the features I discuss, such as capitals, italics, prefaces, footnotes, endnotes, are part of the work of art, which exists through its material tokens; but ultimately that question may not be of great moment. Even if these features are not part of the work, they are important clues to the writer's intentionality and tease us with their potential significance. My main concern is that I may not have been materialist enough. I suspect there are clues I have missed, and that further attention to publication-dates, paper, and type has yet more to tell us about the games Pope played with meanings.

2. The Rape of the Lock: *From Miscellany Endpiece to Illustrated Independence*

The Rape of the Lock was Pope's first major experiment in textual revision and showed his power to utilize the resources of print to transform his materials. The poem was published in 1712 and 1714 in two quite different forms. In 1712 it appeared at the end of *Miscellaneous Poems and Translations*, a collection of pieces by Pope and his friends, as a twenty-four-page semi-independent afterthought; in 1714, expanded to forty-eight pages, it was issued as a separate octavo volume with six engraved plates, and engraved headpieces and initials, partly as a trial run for the subscription *Iliad*. The expansion of the poem marked a crucial stage in Pope's progress from being what Ian Jack has called a 'court-poet', a writer working face-to-face with his readers or audience (and for whom publication was a secondary matter), to being a professional writer aiming at a market and large readership.[1] This expansion, which produced Pope's first popular and commercial success, with 3,000 copies sold in four days, illustrates Pope's harmonizing of the competing interests that shaped his career. There were generic requirements to be met, the tradition which suggested a mock-epic without a developed machinery was defective; there was a relationship with differing readerships to be negotiated (Addison had already praised *The Rape of the Lock* in the *Spectator*[2] but Arabella Fermor was uneasy); there was the desire to make beautiful books that symbolized artistic achievement; and there was making money, which underpinned and engaged the other three without ever quite determining them.

The anxiety generated by these conflicting interests focused, as so often in Pope, on the margins of the public and the private. In freeing his poem from its *Miscellany* context, Pope recognized his debt to its essentially private origins, but he wanted to take full possession of it and to explore its potential public significance. Commercial publication brought its dangers because it could be associated with vulgarizing advertisement, servile dedication, shoddy printing, crude illustration, indifferent context, and hostile reception, but Pope's response, as throughout his career, was to occupy all these potentially hostile zones himself. By drafting the advertisements, remodelling the dedication, supervising the printing, designing the illustrations, publishing his sources, and writing his own criticism, he could protect his poem and

---

[1] Ian Jack, *The Poet and his Audience* (Cambridge: Cambridge University Press, 1984), 1–59.
[2] *The Spectator*, ed. Donald F. Bond, 5 vols. (Oxford: Clarendon Press, 1965), 4. 361–4 (No. 523). Addison's essay moves on to a discussion of heathen mythology in mock-heroic, a discussion which may have stimulated Pope's revision. The close relation of the world of the *Rape* to that of the *Spectator* is made clear in Bond's notes to his edition.

enhance its powers of expression. What seems most external to the author's work—its material and commercial character—is made to express the poet, while what seems to be most private—the original rape and Pope's relation to a particular circle—is first severed from its context and then reinstated in changed form. This invagination, folding the sheath inside out, extends to the structure of the revised poem, where the sylphs are both external to the action and internal to Belinda, and teasing allusions to politics are encountered as eruptions of an inner meaning and yet remain obstinately matters of the surface.[3] *The Rape of the Lock* invites readings on different levels—the political, the private, the epic, the religious, the bawdy—but the levels refuse to operate independently or allegorically; the poem insists on its own fusion of different elements. The real rapist was Alexander Pope himself: he had gained metaphorical possession of Arabella Fermor's lock of hair, and he was careful that no one should exercise the same power over his poem as he had exercised over her lock.

LINTOT'S *MISCELLANEOUS POEMS AND TRANSLATIONS* (1712)

In recollecting the first publication of *The Rape of the Lock* in conversation with Spence, Pope made an interesting slip by locating it in one of Tonson's miscellanies.[4] Pope's first publications, 'January and May', 'The Episode of Sarpedon', and *The Pastorals*, had appeared in the sixth and final volume of Tonson's *Poetical Miscellanies* in 1709, but *The Rape of the Lock* first appeared in *Miscellaneous Poems and Translations, by Various Hands*, published by Tonson's great rival Bernard Lintot on 20 May 1712, and it marked the beginning of a partnership between the poet and the ambitious bookseller. Tonson's miscellanies, first appearing in 1684, had provided one solution to the problem, later solved by the magazines, of finding a market for short and occasional poems, and providing a vehicle for young writers starting their careers. By balancing important translations by Dryden against lesser pieces, Tonson showed how to form a collection that was not based on a particular event (the death of a prince or the visit of a sovereign), and in doing so he established the 'miscellany' or 'miscellanies' as a term and something like a literary form.[5] Pope made fun of the practice in a letter to Wycherley shortly after his own appearance in a Tonson miscellany, 'This modern Custom of appearing in Miscellanies, is very

---

[3] The term is appropriated from Derrida's 'Living on. Border Lines', in *Deconstruction and Criticism*, ed. Harold Bloom et al. (New York: Seabury Press, 1979), reprinted in *A Derrida Reader: Between the Blinds*, ed. Peggy Kamuf (Hemel Hempstead: Harvester Wheatsheaf, 1991), 265–8. I use it without endorsing the philosophy that generates it, though Derrida's acknowledgement of the sexual signification of his term could lead back again into the world of the *Rape of the Lock* (see his interview with Christie V. McDonald, *A Derrida Reader*, 453–4). David C. Greetham comments on invagination in his important essay '[Textual] Criticism and Deconstruction', *Studies in Bibliography*, 44 (1991), 1–30 (23–4).

[4] Spence, *Anecdotes*, No. 105. Anecdotes 104–7 concern *The Rape of the Lock*.

[5] *Correspondence*, 1. 60; 20 May 1709. For a list of miscellanies, see A. E. Case, *A Bibliography of English Poetical Miscellanies* (Oxford: Bibliographical Society, 1935). David Foxon discusses the relation of Lintot's *Miscellany* to Tonson's practice in *Pope and the Book Trade*, 18–38. G. F. Papali has some discussion of miscellanies in his account of Tonson and Dryden in *Jacob Tonson, Publisher* (Auckland: Tonson, 1968), 24–8.

16   The Rape of the Lock

useful to the Poets, who, like other Thieves, escape by getting into a Crowd, and herd together like *Banditti*, safe only in their Multitude' (*Correspondence*, 1. 60), but he seems to have become fully involved with Lintot's volume three years later.

Miscellaneous Poems is one of the most cheerful and optimistic episodes in Pope's literary career. At this point his own ambitions and Lintot's seemed to coincide. Lintot wanted to be a major literary publisher; Pope wanted to be a major poet. The project included not only Pope, but also, perhaps through his agency, his close friends Gay and Rowe, and his later collaborators on the *Odyssey*, Fenton and Broome.[6] In addition to their major items, Gay and Pope both decided to contribute poems providing comic definitions of the project. These verses might have been used as initial commendatory poems, but Lintot, probably thinking them too undignified, sandwiched them in the middle of the volume, at the beginning of the satiric section. Gay's 'On a Miscellany of Poems. To Bernard Lintott' is wonderfully clear about both the trade and literary aims of the collection. The trade aim is to 'rival *Jacob* [Tonson]'s mighty Name'; the literary aim is to have all the major kinds represented: lyric, heroic, elegy, love poetry (not too much), and satire. Vinton Dearing says that this edition 'corresponds almost exactly to Gay's prescription', a consequence, doubtless, of his close involvement with the whole project.[7] At the end of the poem, Gay promises that if Lintot follows his formula, he will gain book-trade success:

> Then, while Calves-leather Binding bears the Sway,
> And Sheep-skin to its sleeker gloss gives way;
> While neat old *Elzevir* is reckon'd better
> Than *Pirate Hill*'s brown Sheets, and scurvy Letter;
> While Print Admirers careful *Aldus* chuse
> Before *John Morphew*, or the weekly News:
> So long shall live thy Praise in Books of Fame,
> And *Tonson* yield to *Lintott*'s lofty Name. (*Miscellaneous Poems*, 173–4)

The poem provides a rare insight into Gay's (and, as it turns out, Pope's) sense of taking a place in the history of fine books. Its understanding of the contemporary book trade is rich and detailed: sheepskin was the cheapest of leather bindings, and the Bodleian copy of Lintot's *Miscellaneous Poems* is bound attractively in calf, with blind panelling, but I doubt this is a trade binding. Henry Hills was a notorious pirate and John Morphew was a leading London publisher and distributor of newspapers. These commonplace details of the contemporary trade are contrasted with the great names of the past; the promise of immortality is tongue in cheek but it is connected to genuine aspirations. Pope was stimulated by Gay's poem into writing a companion piece, combining learning and informality:

---

[6] See the discussion of Pope's relations with Lintot in James McLaverty, 'The Contract for Pope's Translation of Homer's *Iliad*: An Introduction and Transcription', *Library*, 6th ser. 15 (1993), 206–25.

[7] *John Gay: Poetry and Prose*, ed. Vinton A. Dearing, 2 vols. (Oxford: Clarendon Press, 1974), 2. 488.

> Some *Colinæus* praise, some *Bleau*
> Others account them but so, so;
> Some *Stephens* to the rest prefer,
> And some esteem old *Elzevir*:
> Others with *Aldus* would besot us;
> I, for my part, admire *Lintottus*. (*Miscellaneous Poems*, 174)

Publication involved a place in the history of print as well as in the history of literature, and the history of print was potentially an important and dignified one.

Miscellaneous Poems and Translations was a significant stage in Pope's professionalization as a writer, but I am not confident that Norman Ault is correct in casting him as the collection's editor.[8] Tonson's miscellanies had sometimes been advertised as 'edited by Mr Dryden', and Pope certainly had the equivalent central place in this collection. He was also very busy in collaborating and planning, but he was not paid for editorial work, as Lintot's accounts make clear, and the emphasis is on a circle of friends rather than on professional editorial skills. The contributors named in addition to Pope are Prior, Bate, Gay, Broome, Southcott, Cromwell, Edmund Smith, Fenton, Betterton, Barret. All but two of these (Smith and Barret) were closely connected with Pope. After Pope himself, the major contributor in lines was his fellow Catholic Thomas Betterton, the actor, who had died two years earlier. A letter from John Caryll to Pope shows that friends saw publication as a personal kindness to Betterton and to his widow: 'I am very glad for the sake of the Widow and for the credit of the deceas'd, that *Betterton*'s remains are fallen into such hands as may render 'em reputable to the one and beneficial to the other.'[9] Pope was paid five guineas (£5 7s. 6d.) for Betterton's modernization of 'The Miller's Tale' and he received a grand total of £26 19s. 0d. for his own contributions:

1712, Feb. 19. Statius, First Book, Vertumnus and Pomona £16 2 6
1712, Mar. 21. First Edition of the Rape £7 0 0
1712, 9 April. To a Lady Presenting Voiture, Upon Silence, To the Author of a Poem called *Successio* £3 16 6

I have wondered whether at this stage Pope was paid so much per line: fourpence for translations (the first group), fivepence for long poems (the second group), and sixpence for short poems (the third group). The total would have been rounded up or down to a respectable sum in guineas or pounds.[10] The amount Pope was paid for all these contributions came close to the sums paid for his major poems published in folio, *Windsor-Forest* and *The Temple of Fame*, which earned thirty guineas (£32 5s.) each, but there is no doubt the first *Rape of the Lock* came to Lintot cheap; the

---

[8] *New Light on Pope* (London: Methuen, 1949; repr. Hamden, Conn.: Archon, 1967), 27–48. The arguments in Ault's *Pope's Own Miscellany* (London: Nonesuch, 1935) are considerably stronger; I suspect there was some deal about the *Works* that supported the *Poems on Several Occasions* (1717).

[9] *Correspondence*, I. 142. In the same letter, of 23 May 1712, Caryll wonders whether Pope has published or suppressed 'the Lock'.

[10] The first group averages 3.9d. per line, the second 5.0d., and the third 6.2d. Betterton's work was not so well rewarded at 1.4d. per line.

following year he paid £15 for the 'Ode on St. Cecilia's Day', which was only eight pages long.[11]

Pope's leading role in *Miscellaneous Poems and Translations* is clear from the make-up of the book. The first poem, 'The First Book of Statius his Thebais', is his, and it has its own section title and headlines. Roughly halfway through the volume, between pages 168 and 175, come Gay's 'On a Miscellany of Poems' and Pope's 'Verses design'd to be prefix'd to Mr. Lintott's *Miscellany*'. They come after translations, and paraphrases of scripture, and before the humorous pieces. Betterton's adaptations of Chaucer, which had been revised by Pope, stand out from the rest by having their own section title (with a mini-imprint) and headlines. In this they set the pattern for *The Rape of the Locke* which undoubtedly became the pride of the collection.

The Rape of the Locke sits quite comfortably in *Miscellaneous Poems and Translations* but there are already signs of its potential detachment. The collational formula and pagination of the volume (A^4 B–X^8 Aa–Bb^8; [1–5], 6–320, [353] 354–76) captures its status perfectly.[12] *The Rape* fits physically at the end of the book but only after gaps that indicate its quasi-autonomy: it begins a new alphabet, with the omission of two signatures, and it takes up the pagination after a gap of thirty-two pages. More important, it has a modest title page of its own, not merely the customary section title but the name of the poem and an abbreviated imprint, 'Printed for Bernard Lintott. 1712.' There is no author's name; the poem appeared anonymously. This quasi-independence is reflected in the running head, which announces '*The Rape of the Locke*' rather than the 'Miscellaneous POEMS and TRANSLATIONS' of most of the rest of the volume. At this stage Pope and Lintot were probably not looking ahead towards independent publication but backward to manuscript circulation. The task of writing *The Rape of the Lock* came to Pope as a kind of private commission from John Caryll, who was looking for ways of reconciling the Fermor and Petre families, at war over Robert Petre's snipping off of a lock of Arabella Fermor's hair. Pope claimed in the dedication to the 1714 volume that the poem 'was intended only to divert a few young Ladies, who have good Sense and good Humour enough, to laugh not only at their Sex's little unguarded Follies, but at their own' (*Twickenham*, 2. 142), and this stage in publication seems designed to retain contact with the original manuscript circulation and to keep open the route to the poem's social origins.[13] Pope was trying to fuse market publication with private circulation. *The Rape of the Locke* made an attractive pamphlet that could be distributed to his circle. His correspondence shows him making special arrangements to ensure that Robert Petre, Arabella Fermor, and John Caryll got pre-publication copies; and friends like the Blount sisters would also have got them.

The poem has the same ambivalent relation to its companion poems as its pages do to the collation. Its genre relates it to the classical pieces in the volume, but the

---

[11] *Anecdotes*, 8. 299–300.
[12] David Foxon considers the possible explanations for the gap in *Pope and the Book Trade*, 34–8.
[13] For these social origins, see *Twickenham*, 2, ed. Geoffrey Tillotson, 81–7 and 142, and Valerie Rumbold, *Women's Place in Pope's World* (Cambridge: Cambridge University Press, 1989), 48–82.

more important connections are with the occasional verse, whose conventions and attitudes the *Rape of the Locke* displays and distances itself from. The prevailing tone of the shorter, amatory verses in the collection is one of sophisticated gallantry, oscillating between hyperbolic compliment and earthy cynicism: the woman is a goddess, she rivals the sun, but she will encounter age and death; the mirror which shows her beauty should also remind her of mortality.[14] As Geoffrey Tillotson shows in his Twickenham edition of the poem, Arabella Fermor belonged to this world and had taken her place in the verse of the period before Pope recorded her lost lock in *Miscellaneous Poems and Translations*. She was complimented in *St James's Park: A Satyr* (1708) and *The Mall: or, the Reigning Beauties* (1709), and she appears in *The Celebrated Beauties* in the same Tonson *Miscellany* (1709) in which Pope's *Pastorals* were first published (*Twickenham*, 2. 373-4). 'Belinda' is the usual name for the loved or admired one in the Pope–Lintot collection. In 'On a Flower which Belinda gave me from her Bosom', Broome emphasizes not the whiteness of the breast (a feature that links Arabella Fermor and Pope's Belinda) but its scent:

> A Store of such a rich Perfume
> Must from *Belinda*'s Bosom come;
> Thence, thence such Sweets are spread abroad
> As might be Incense for a God. (*Miscellaneous Poems*, 116)

The 'Incense for a God' recalls Pope's heroine's sparkling cross. But if this Belinda is associated with divinity, she also shares the rose's mortality: the rose will decay, its beauties fade; Belinda, as in a glass, will descry 'How Youth, and how soft Beauty die'. In 'Rapin Imitated, in a Pastoral Sent to *Belinda* upon her leaving *Hattley*' Fenton also issues a warning. Urging Rosalind's return, Florus advises,

> forbear to urge your homeward Way,
> While *Phœbus* riots in Excess of Day;
> Least while his Beams infest the sultry Air,
> They shou'd your brighter Charms, O *Rosalind*, impair. (*Miscellaneous Poems*, 229)

Pope's Belinda also rivals the sun's beams; her eyes are as powerful but as undiscriminating. Other poems are more exclusively concerned with woman's weakness. In 'To a Lady Sitting before her Glass' Fenton warns Chloris of the example of Narcissus and the danger that she may find:

> No longer let your Glass supply
>   Too just an Emblem of your Breast;
> Where oft to my deluded Eye
>   *Love*'s Image has appear'd imprest;
> But play'd so lightly on your Mind,
> It left no lasting Print behind. (*Miscellaneous Poems*, 178)

[14] Many of these reflections on women follow patterns set out by Felicity Nussbaum in *The Brink of All We Hate: English Satires on Women, 1660–1750* (Lexington: University Press of Kentucky, 1984), see particularly her chapter on Pope, 'The Glory, Jest, and Riddle of the Town', 137–58, and by Ellen Pollak, *The Poetics of Sexual Myth: Gender and Ideology in the Verse of Swift and Pope* (Chicago: University of Chicago Press, 1985), 77–107.

The unfixedness of woman is emphasized and also the danger that she may be her own victim, 'Yourself the Captive, and the Snare', even as Belinda is both snare and captive. A few pages later R.F. epigrammatically tells the story of Chloe, 'a Coquet in her Prime', who marries at 45 and, compared to an old weathercock, is rusted into fixity (181). The most surprising link with Pope's poem, however, comes in a contribution that lacks any air of gallantry. In Fenton's 'The Fair Nun. A Tale' a fair nymph leaves 'Balls and Play' for the convent; her lover joins her in disguise; she becomes pregnant, but she signs a deed with the devil so that they can both escape. When the devil returns for his bargain, he is told that the covenant is void unless he can straighten a hair 'curl'd like a Bottle-Scrue'. Fenton has Popeian fun with the question of the hair's origins, 'whence it came I cannot swear', and the devil fails in his task (210–21).

THE FIRST VERSION OF *THE RAPE OF THE LOCK*

The Rape of the Locke has an evident kinship with the other poems in the miscellany. As a poem responding to a particular occasion and to a particular person, 'This Verse to C—l, Muse! is due', it sits comfortably alongside 'On the Birth-Day of Mr. *Robert Trefusis*; Being Three Years Old', or the poems to the memory of Dryden or John Philips. Like them, it presents itself as having a primary social function to which publication is a mere adjunct; writing verse is an ordinary part of social intercourse. It shares with several of the poems a fetishizing of female beauty that expresses wonder but is deeply troubled by materiality. *The Rape of the Locke* differs from these other occasional pieces not only in the sustained ironic self-awareness with which it deploys its motifs and in the elegance of the structure that integrates them, but in the air of sympathetic inwardness with its heroine that characterizes the narrative. The first version of *The Rape of the Lock* is a strikingly well-balanced and integrated piece. Its economical narration—no sylphs, no toilet, no cards, no cave of spleen— does not raise the same questions of misogyny and of skewed attention that are forced on modern readers by the final version. Without the sylphs there is no extended, displaced, externalized exploration of female psychology; this poem is concerned with the display of behaviour, the behaviour of men as well as women— the curiosity of a well-bred lord assaulting a gentle belle as well as of a gentle belle rejecting a lord. The balancing of the self-dramatizing protagonists is reinforced by the spelling 'locke', which is associated with them. This would-be archaic form is not used consistently in the poem, nor is it integrated into a general archaizing pattern. The narrator uses 'lock' (1. 36, 45, 116) except when he is influenced by the Baron's thoughts (1. 100) or the general mood (2. 149). The Baron uses 'locke' (2. 51) and so does Belinda (2. 86, 148). Pope is tentatively using typography to display the distorted consciousness of his two main figures.

The poem is organized through the progress of a single day. The action is bounded by the movement of the sun, whose power is consistently compared with that of Belinda's eyes, thereby preparing the final emphasis on the transcendence of mutability. Belinda is woken by the sun:

> *Sol* thro' white Curtains did his Beams display,
> And op'd those Eyes which brighter shine than they . . . (1. 13–14)

She rises promptly at ten and travels to Hampton Court, her eyes in her progress 'Bright as the Sun'. A little after noon ('The Sun obliquely shoots his burning Ray'), they drink coffee, the Baron severs the lock, and the first canto comes to an end. Throughout the canto interest in Belinda is balanced by attention to the Baron. The thirty-two lines describing Belinda's rising are offset by twenty describing the Baron, who has taken an unfair advantage by making his sacrifice 'e'er *Phœbus* rose'. The Baron's eighteen-line speech of triumph at the end of the canto places the emphasis, and ridicule, squarely on rape and the rapist. The second canto is to be given over to the counteraction and the victim.

Before allowing Belinda to speak Pope generalizes the conflict through Thalestris's speech of twenty-six lines and Sir Plume's of three and a half. The contrast is significant, for the woman is articulate and sophisticated while the man is inarticulate and incoherent. Finally the Baron is given a short (eight-line) expression of repulsive, unyielding superiority. The first version of the poem, then, permits Belinda a genuine, though comic and tightly controlled, pathos in her speech of reproach and regret that she ever left the country for Hampton Court:

> What mov'd my Mind with youthful Lords to rome?
> O had I stay'd, and said my Pray'rs at home! (2. 76–7)

The speech, an imitation of Achilles' lament for Patroclus, is the self-dramatization of inexperience, the contrary of the Young Lady Leaving the Town after the Coronation; it combines the homely, 'my best, my fav'rite Curl', with the self-adulatory, 'thy sacrilegious Hands'. This speaker is entitled to her self-pity; she has not just herself been the exulting victor at a game of cards. The consequent battle, which occupies the major portion of the second canto, is also even-handed in its treatment of the antagonists; if the Baron has the advantage in range of innuendo, Belinda has the physical victory with the charge of snuff. The poem ends with night and the poet's triumph. Belinda's lock is seen by the Muse as a comet; those suns which are Belinda's eyes shall set but her lock shall have an everlasting fame. Pope now openly claims that he has taken what the Baron stole; what he can achieve for Belinda is not restoration but translation, a translation enacted by the poem. The first version conveys a wry amusement at an eccentric pattern of behaviour. The epic colouring flatters with attention even as it ridicules passionate egoism.

Although *The Rape of the Locke* and its author received important critical praise from Addison in *Spectator* 523,[15] neither the collection nor the poem was in other respects successful. *Miscellaneous Poems and Translations* was not a commercial success and had to be reissued with a cancel title page in 1714, with *An Essay on Criticism* and *Windsor-Forest* added to strengthen Pope's contribution. The published poem seems not to have been a complete social success either. Pope's

---

[15] 'I have read over, with great Pleasure, the late Miscellany published by Mr. Pope, in which there are many Excellent Compositions of that ingenious Gentleman.' *Spectator*, 523, 30 Oct. 1712 (4. 361).

report to Spence says that the poem was well received and 'had its effect in the two families', but as Petre had married someone else, Catherine Warmsley, on 1 March 1712 (before the *Miscellaneous Poems* was published) and was to die of smallpox on 22 March 1713, the original quarrel had lost its power. The Fermors, moreover, were displeased:

> Sir Plume blusters, I hear; nay, the celebrated lady herself is offended, and, which is stranger, not at herself, but me. . . . Is not this enough . . . to make a writer never be tender of another's character or fame? as in Belinda's [case]; to act with more reserve and write with less.[16]

The anger expressed at Arabella Fermor here may have fed into the development of Belinda in the revised poem; certainly Pope seems to entertain thoughts of appeasing the woman and avenging himself on, or perhaps through, the character. Tillotson's conjectures on the reasons for Arabella Fermor's unhappiness, that they concern her marriageability, have received some recent confirmation from Valerie Rumbold's research on the Blount circle.[17] All Arabella Fermor is recorded as saying is that fools had talked and fools had heard them, but Pope's defensive reaction suggests that publication, even anonymous, miscellany publication, was at the heart of the matter. It was one thing to laugh at a poem circulating in manuscript; it was another to be the public focus of jests about honour, integrity, and intimacy.

THE DEDICATION TO THE REVISED *RAPE OF THE LOCK*

The reaction of the Fermors to the miscellany publication influenced the task of revision or supplementation; for it justified a decision to break off the poem from its particular social context. Pope himself encouraged the idea that the expansion of the poem was a formal matter, the fulfilment of a pre-existing plan, and in the dedication to the revised poem he explained that publication had come 'before I had executed half my Design' (*Twickenham*, 2. 142); but the revision to form a five-canto illustrated book was in reality a much more radical matter. In devising his machinery Pope reinterpreted his poem. He took advantage of the resources available to him, poetic and bibliographic, to increase the density of his satire. What had been merely glanced at before—underlying patterns of social and sexual behaviour and their politics—could now be fully engaged by the poem. The whole poem becomes more complex, as Pope explores contemporary sexual mores and their relation to the politics of the late Stuart court.

The break from the immediate social context is announced in the dedication to the new edition, a fine example of Pope's deploying the resources of the book in order to achieve his aim. He begins with a ground-clearing account of the first publication:

> You may bear me Witness, it was intended only to divert a few young Ladies, who have good Sense and good Humour enough, to laugh not only at their Sex's little unguarded Follies, but at their own. But as it was communicated with the Air of a Secret, it soon found its Way into

---

[16] *Correspondence*, 1. 151; Pope to Caryll, 8 Nov. 1712. The omissions concern Mr Weston.
[17] *Women's Place in Pope's World*, 69–74.

the World. An imperfect Copy having been offer'd to a Bookseller, You had the Good-Nature for my Sake to consent to the Publication of one more correct: This I was forc'd to before I had executed half my Design, for the *Machinery* was entirely wanting to compleat it. (*Twickenham*, 2. 142)

This is a beguiling picture of a poem first drifting from the private into the public sphere and then having its full private intentionality restored to it in public. Without predatory booksellers there would be no problem; the poet belongs to society not the market. The ruse about an imperfect copy and an eager bookseller is too common to gain credence; Johnson writes ironically, 'with the usual process of literary transactions, the author, dreading a surreptitious edition, was forced to publish it'.[18] Pope was particularly fond of this device and its required pose as a private individual, though he found it difficult to keep a straight face, operating with a strong sense of its conventionality and potential transparency. A similar case is the 'Publisher to the Reader' in the first *Dunciad*, where the publisher declines to explain how she came in possession of the manuscript, but claims, 'If it [this edition] provoke the Author to give us a more perfect edition, I have my end.'[19] Pope's own purposes emerge because he allows them to infect the publisher's motives; if she had stayed in character she would have aimed at profit regardless of quality. The dedication to *The Rape of the Lock* presents a subtle form of the same manœuvre. The disjunction between private and public is not complete: Pope admits he might always have chosen to publish when the poem was complete; the book trade has merely forced precipitate publication. But the result of the manœuvre is the usual one: the author has no choice but to bring about an ideal reconciliation of public and private by publishing an authorized version.[20]

The dedication to *The Rape of the Lock* is equivocal in another important respect—its attitude to Arabella Fermor. Pope wanted to keep some relation between the poem and its original incident—a poem with no specific social basis would have less authority—but he wanted to be free of obligation to the original participants. In particular, Pope recognized a duty to oblige Arabella Fermor, even as the urge to blacken Belinda in the poem grew stronger. As Tillotson shows, Pope gave her the choice between a preface and the dedication: 'A preface which salved the lady's honour, without affixing her name, was also prepared, but by herself superseded in

---

[18] Samuel Johnson, *Lives of the English Poets*, ed. George Birkbeck Hill, 3 vols. (Oxford: Clarendon Press, 1905), 3. 102.

[19] *Twickenham*, 5. 203–4. See also *Pope's Dunciad of 1728: A History and Facsimile*, ed. David L. Vander Meulen (Charlottesville: University Press of Virginia for Bibliographical Society of University of Virginia and New York Public Library, 1991), 16–23, and p. v of facsimile.

[20] One of the consequences of the stern moral tone adopted towards Pope's conduct by friend and foe is that little attention has been paid to his skill in equivocation. In this case he does not say that he was not the one offering an imperfect copy to a bookseller. On equivocation, see Albert R. Jonsen and Stephen Toulmin, *The Abuse of Casuistry: A History of Moral Reasoning* (Berkeley and Los Angeles: University of California Press, 1988), 195–215, and Johann P. Sommerville, 'The "New Art of Lying": Equivocation, Mental Reservation, and Casuistry', in Edmund Leites (ed.), *Conscience and Casuistry in Early Modern Europe* (Cambridge: Cambridge University Press, 1988). I am grateful to Jonathan Dancy for introducing me to this topic.

favour of the dedication' (*Correspondence*, 1. 207). Pope took great care over the dedication, even to the extent of consulting the Earl of Oxford:

> As to the *Rape of the Lock*, I believe I have managed the dedication so nicely that it can neither hurt the lady, nor the author. I writ it very lately, and upon great deliberation; the young lady approves of it; and the best advice in the kingdom, of the men of sense has been made use of in it, even to the Treasurer's. (*Correspondence*, 1. 207)

The dedication physically symbolizes Pope's manœuvre to take control of the poem. Arabella Fermor is moved from the central role in the narrative, which rightly seems to be hers, to the margins of the book. In achieving this, Pope indulges in some of that play with reference which was to characterize the second half of his career. The poem is and is not about Arabella Fermor. She is certainly present, her name appears in italic capitals stretching right across the first page after the title page, 'To Mrs. *ARABELLA FERMOR*', but the main purpose of this presence is to legitimize her disengagement from the poem. Pope's aim is to bring about the smooth transfer of power over the central experience from Arabella Fermor to himself. The dedication, therefore, acknowledges her significance, and even her authority, while putting her outside the work itself; she becomes more of a patron than a participant. First she is called on as a witness to the poem's publication history and as the authority permitting the original publication. Then she is relegated to the role of pupil, as Pope explains his machinery to her, a machinery that owes nothing to her. He then turns to the new poem, an amalgam, one might have expected, of what was hers and what is his:

> As to the following Canto's, all the Passages of them are as Fabulous, as the Vision at the Beginning, or the Transformation at the End; (except the Loss of your Hair, which I always mention with Reverence.) The Human Persons are as Fictitious as the Airy ones; and the Character of *Belinda*, as it is now manag'd, resembles You in nothing but in Beauty.
> If this Poem had as many Graces as there are in Your Person, or in Your Mind, yet I could never hope it should pass thro' the World half so Uncensured as You have done. (*Twickenham*, 2. 143)

Both these paragraphs have their oddities. The first version of the poem seems to be judged a representation of the participants in the original quarrel; it is the character of Belinda as it is 'now' managed that does not resemble Mrs Fermor. But much of the original was transferred without change into the revision, and, if all the passages in the new version are fabulous, how can the rape of the lock still be 'the Loss of your Hair', which, if it is mentioned in the poem, is mentioned without reverence? And how far and how justly has Mrs Fermor been censured? Pope's aim seems to be to retain a generalized reference but to position himself to deny any particular application. The dedication to the expanded *Rape of the Lock* hints at the increased censoriousness in the poem and the much rougher handling of Belinda in particular; it was important that readers should have these Catholic families in mind, but not that the details of the poem should be taken to represent details of their behaviour.

ILLUSTRATIONS

The style of the authorized edition of *The Rape of the Lock* is apparent from the advertisements:

> In a few Days will be publish'd, The Rape of the Lock; an Heroick-Comical Poem; by Mr. POPE now first publish'd complete in 5 Cantos; with 6 Copper Plates: Price 1s. The Tragedy of Jane Shore; by Mr. Row: Price 1s. 6d. There will be a small Number of each printed on fine Paper; those who are willing to have These, are desired to send in their Names to Bernard Lintott between the two Temple Gates: No more being to be thus printed than are bespoke. (*Post-Boy*, 23–6 Jan. 1714)

The expansion of the poem, from two cantos to five, goes with a change in its bibliographical status. The poem is now independent, important enough to warrant an extended advertising campaign.[21] There is no vestige of anonymity: Pope's name appears in capitals as a major selling point. More subtly, the relation to the readership has changed. *Miscellaneous Poems and Translations* envisaged two readerships, the ordinary purchasers and the recipients of the offprints; now there are still two readerships, but elite readers are distinguished by their willingness to pay, not by their membership of a particular circle. A residual delicacy in the advertisement obscures the reality of the situation by omitting the price of the fine-paper copies. In many respects this edition of the *Rape* served as a dummy-run for the *Iliad* translation that Pope and Lintot were planning and finally signed a contract for on 23 March.[22] Each volume of the *Iliad* was to be marketed initially as a subscription for Pope's benefit and then put on sale to the general public. Subscribers paid one guinea a volume for a quarto with engraved headpieces, tailpieces, and initials; wealthy clients outside the subscription circle paid one guinea for a large, but unillustrated, folio; while others paid 10s. for a small folio. At a later stage a duodecimo was made available at 2s. 6d. a bound volume. *The Rape of the Lock* was a simple project by comparison: by becoming a subscriber you received the same illustrated book but on better-quality paper. It seems possible that Pope got the money from this mini-subscription himself: on 20 February 1714 he was paid £15 for 'Additions to the Rape', which would represent 200 copies sold at a subscription price of 1s. 6d. Alternatively, Lintot may have printed a larger number of subscription copies and used the profit on them to pay for the illustrations. Whatever the arrangements, Pope was paid twice as much for the revision of the *Rape* as for the original poem.

The illustrations to the independent *Rape of the Lock* are the most striking signal of its development, flecking it with political satire, but they are missing from modern editions of the poem and this omission helps disguise the expansion of the poem's satiric range. The headpieces, initials, and tailpiece, as David Foxon has explained, originated with Joseph Trapp's *Praelectiones Poeticae*, but the six plates (a

---

[21] The first advertisement I have found is in the *Post-Boy* for 23–6 Jan. 1714. Publication was on 4 Mar.
[22] See 'The Contract for Pope's Translation of Homer's *Iliad*', 206–25.

frontispiece and one for the beginning of each canto) were specially engraved for this edition.[23] It is very likely that they derived from sketches by Pope himself, but all the plates are credited in the same way, 'Lud. Du Guernier inv.' and 'C. Du Bosc sculp.', and the evidence of Pope's involvement is cumulative and inferential rather than documentary. That he was keenly interested in the visual arts, and even spent time with his friend Jervas learning to paint, is well attested, not least in the *Epistle to Mr Jervas* that celebrates their time together.[24] The *Iliad* contract, signed only a month after the *Rape of the Lock* payments, specified that for that large and expensive undertaking the copperplates were to be engraved 'in such manner and by such Graver as the said Alexander Pope shall direct and appoint'. The *Odyssey* contract, signed on 18 February 1723, specified that Pope was to pay for the engravings and that, though Lintot was to keep them, Pope could use them in any other book he wished.[25] He took advantage of this provision and used them in the quartos of his 1735 *Works*. This evidence of Pope's interest in the illustration of his work is supported by analysis of the plates themselves. Elias F. Mengel's interpretation of the *Dunciad* engravings, detailed and wholly convincing, shows that they cannot be the work of a mere illustrator, because they do not merely illustrate.[26] The same proves to be true of the *Rape of the Lock*. Robert Halsband, who analysed the illustrations of the poem from 1714 to 1896, contrasts the poor technical quality of the *Rape of the Lock* illustrations with superior work by Du Guernier elsewhere.[27] He suggests the responsibility may lie with Du Bosc's engraving, but as Du Bosc was also capable of good work it is necessary to look for an explanation elsewhere. It is quite likely that Pope himself was responsible for the design of the illustrations and indirectly for their quality. If he provided the sketches, Du Guernier may have felt hampered in the scope of his own invention.[28]

The frontispiece to the 1714 *Rape of the Lock* (Fig. 1) makes a dramatic effect and declares the broadening of the poem's scope. The plate represents no particular scene in the poem—it is not, as might be expected, concerned with the rape of the title—but it elaborates the new machinery and it emphasizes the setting at Hampton Court. Three of the new episodes, involving the sylphs, are alluded to. First, the

---

[23] See *Pope and the Book Trade*, 42–6, and the facsimile edition *Alexander Pope: The Rape of the Lock*, ed. D. F. Foxon (Menston: Scolar Press, 1969).

[24] See Foxon, *Pope and the Book Trade*, 67–71; Maurice R. Brownell, *Alexander Pope and the Arts of Georgian England* (Oxford: Clarendon Press, 1978), 10–26; W. K. Wimsatt, *The Portraits of Alexander Pope* (New Haven: Yale University Press, 1965), 11–12.

[25] The *Iliad* contract is transcribed 'The Contract for Pope's Translation of Homer's *Iliad*', 220–3; the Odyssey contract in George Sherburn, *The Early Career of Alexander Pope* (Oxford: Clarendon Press, 1934), 313–16.

[26] Elias F. Mengel, jun., 'The *Dunciad* Illustrations', *Eighteenth-Century Studies*, 7 (1973–4), 161–78.

[27] Robert Halsband, *'The Rape of the Lock' and its Illustrations, 1714–1896* (Oxford: Clarendon Press, 1980), 9–23; see also Clarence Tracy, *The Rape Observ'd* (Toronto: University of Toronto Press, 1974).

[28] Du Guernier and Gribelin are both linked to Dorigny. Du Guernier came to London from Paris to work with him in 1708, while Gribelin was employed by him to work on the engravings of the Raphael cartoons at Hampton Court. This might provide an alternative explanation of the prominence of Hampton Court in the engravings, but whereas Gribelin is accurate, Du Guernier seems to have a weak grasp of the building. For the interrelations, see Hans Hammelmann, *Book Illustrators in Eighteenth-Century England*, ed. and completed by T. S. R. Boase (New Haven: Yale University Press, 1975).

Fig. 1. Frontispiece to *The Rape of the Lock*, 1714 (Bodl. Don. e. 115; 214 × 131 mm).

game of Ombre is represented by the playing cards that are being scattered over the group below. Second, the putto standing behind the mirror seems to represent Umbriel, returned with his phial from his visit to the Cave of Spleen. And third, the focus of the design is the young woman before the mirror: Belinda at her toilet, assisted by the sylphs and seeing a purer blush arise. In this plate the relation of the sylphs to the other figures and to the action differs from that in the poem and the other plates. The poem suggests an otherworldly intervention in the ministrations of the sylphs, an otherworldliness caught in their aerial delicacy but disguised by their ability to assume 'what Sexes and what Shapes they please' (1. 70). In this first engraving the distinction between the sylphs and the human characters is unclear. All the figures are of the same plump putto-like sort. The four airborne creatures are marked out as sylphs by their wings, but, except for the figure tentatively identified as Umbriel, none of the others has wings. The satyr, bottom right, half-removing his mask, has a physique similar to the others, and Belinda is not sharply distinguished from the rest. She is larger, but putto-like, and fits comfortably into her surrounding group. We might be observing children at play. Whereas the poem portrays Belinda's toilet as a religious ritual and the card game as a heroic struggle, both laying the gravest responsibilities on the sylphs, the frontispiece evokes a frivolity, a carelessness, a wantonness, encouraged by the sylphs. Halsband is right to emphasize the erotic presentation of Belinda, her gown open to reveal her thighs, and its relation to the iconography of Venus. The frontispiece puts aside a mask of irony to reveal naughty children playing sexual games. The frankness of the frontispiece is connected with the appearance there of the satyr half-removing his mask. The symbol of the satyr is a common one for representing the genius of satire, and it was important to Pope. A striking example of it is to be found in the British Museum Department of Prints and Drawings collection of political satires, where a mocking, lascivious face emerges from behind its mask (No. 1414).[29] The same motif appears again, though this time there are two satyrs, adult and dangerous, with not one but two masks each, in the headpiece to the *Rape of the Lock* in the 1717 *Works*. This iconography is related to Renaissance theories of satire. As André Dacier, who disapproved of the etymology, explains, some writers decided on the basis of the spelling 'satyr' and its equivalents that 'the Divinities of the Groves, which the *Grecians* call'd *Satyrs*, the *Romans Fauns* gave their Names to these Pieces; and that of the Word *Satyrus* they had made *Satyra*, and that these Satyrs had a great affinity with the Satyrick Pieces of the Greeks'.[30] The spelling makes a link with Greek phallic songs and satyr plays. Puttenham presents a view in line with Pope's representation, as Robert C. Elliott explains:

Puttenham, for example, wrote in the *Arte of English Poesie* (1588) that the ancients had a kind of poem called satyre, a bitter invective against vice and vicious men, named after the Satyrs, 'those terrene and base gods being conuersant with mans affairs and spiers out of their secret faults . . .'[31]

[29] The face is circled by 'ROTAVELTRAGOPANN'. George William Reid's *Catalogue of Prints and Drawings in the British Museum*, division 1, volume 2: *June 1689 to 1733* (London: Printed by Order of the Trustees, 1873), remains an invaluable guide to these materials. Most of the prints are now available on the microfilms published by Chadwyck-Healey (Cambridge, 1992). For possible links between sylphs and satyrs, see *The Diverting History of the Count de Gabalis*, 2nd edn. (London: Lintott and Curll, 1714), 62–3.
[30] *Essay upon Satyr* in René Le Bossu, *Treatise of the Epick Poem*, 2 vols. (London, 1719), 2. 309.
[31] Robert C. Elliott, *The Power of Satire: Magic, Ritual, Art* (Princeton: Princeton University Press, 1960), 102; the Puttenham quotation is from the edition of the *Arte of English Poesie* by Gladys D. Willcock and Alice Walker (Cambridge: Cambridge University Press, 1936), 31.

This account catches the figures in the *Works* illustration perfectly. These Pan-like creatures present a spirit of lasciviousness but they do not participate in the scenes in which they appear. They embody what they pleasurably observe. Howard Weinbrot points to a long tradition of associating satire with masking in book illustration. For example, a Juvenal of 1685 shows 'a satyr holding a long rope on which are placed eleven different masks—including the old, the young, the grave, the gay, the arch, the innocent'. He separates this from the tradition which shows 'satire not putting on but stripping off a mask, commonly one of hypocrisy', and describes an illustration in a Juvenal and Persius of 1671: 'a bare-breasted two-faced woman whose skirt raised to her calf reveals a large clawed foot and a tail. She holds two burning hearts in her left hand and a mask, presumably removed by the satires within the book, in her right hand.'[32] Similarly the plate to the first canto of *The Rape of the Lock* shows a hairy and hoofed leg emerging from beneath Belinda's bedroom table. The unmasking satyr, then, represents a complex set of ideas: the lustful man or woman; the animal lurking behind a civilized façade; the antagonistic spiers-out of human weakness; the self-disguising satirist; the satirist unmasking the hypocrisy of others. The reflexive nature of the relations between observer and observed undoubtedly appealed to Pope; the unmasking of others became connected to a masking and unmasking of the self.

Two prominent features of the frontispiece are difficult to explain in relation to the poem itself: the east front of Hampton Court Palace, which overshadows the action, and the putto in the left foreground, who is drawn to our attention by the pointing finger of one of his fellows. Hampton Court is present in both versions of the poem as the scene of the rape, but the illustrations in the 1714 book make it much more significant. The frontispiece provides the most powerful image of the palace, but the plate to canto 2 shows a distant prospect of the east front, misleadingly suggesting it can be approached by the river, while the plates to cantos 3 and 5 use grand but unidentifiable parts of a palace as a backdrop.[33] Whether Arabella Fermor actually had her lock cut off at Hampton Court cannot be determined. Tillotson quotes a letter from the historian Edward Yates which identifies the rooms in which the rape took place, but Tillotson himself regards the location as improbable (396 and 81) and I think it much more likely that Hampton Court was chosen for its value as a symbol. Nearly twenty years after the publication of the poem, with the Hanoverians on the throne and new patterns of royal residence, Hampton Court seems to have become associated with sexual

---

[32] *Eighteenth-Century Satire: Essays on Text and Context from Dryden to Peter Pindar* (Cambridge: Cambridge University Press, 1988), 37–8.

[33] The river flows past the south front; the canal constructed under Charles II's direction leads to the east front. There is an excellent picture of the south front in *Britannia Illustrata or Views of Several of the Queens Palaces as Also of the Principal Seats of the Nobility and Gentry of Great Britain Curiously Engraven on 80 Copper Plates*, drawn by L. Knyff and engraved by I. Kyp (London, 1709), No. 6. This engraving is reproduced in Charles Harding Firth's edition of Macaulay's *The History of England from the Accession of James the Second*, 6 vols. (London: Macmillan, 1913), 3. 1361, which contains a wealth of illustrations from the period, including Sutton Nicholls's prospect of the east front, which it dates 'about 1695'. Kyp's (or Kip's) engraving of the east front is reproduced in Ernest Law, *The History of Hampton Court Palace*, 3 vols. (London: George Bell, 1885–91), 3. 178–9.

misdemeanours, with *The Hampton-Court, Richmond, and Kensington Miscellany* (1733) containing items such as 'A Review of Alexis's Seraglio', but in 1714 Hampton Court's chief association was with the reign of William and Mary and the court of William's successor, Anne. The importance the illustrations give to this royal palace is especially interesting in the context of the publication of *Windsor-Forest* in 1713, between the two-canto and five-canto versions of *The Rape of the Lock*. In *Windsor-Forest* the castle and park are seen as a microcosm of the nation, whose history can be traced through its welfare. Pope presents what is essentially a Tory history from the Normans to Queen Anne, drawing parallels between the Norman invasion and the revolution of 1688. Hampton Court itself had a history to rival even Windsor's, and from the beginning of the reign of William and Mary it had been the chief royal palace. As Kerry Downes says in his book on Wren, 'From the gardens it is the chief architectural monument of the reign of William III.'[34] One of William's first actions as monarch was to remove his residence from Whitehall to Hampton Court, which he believed was better for his health. The subsequent schemes for the rebuilding of the palace and the eventual scheme for partial demolition and extension by Wren were doubly criticized: the King was extravagant, foolishly attempting to rival Versailles and to duplicate his own palace at Het Loo, and he was destroying the English court.[35] Ernest Law, in his history of Hampton Court, summarizes the case against William:

great dissatisfaction was already beginning to be expressed, in various quarters, at the King's spending so little of his time in London. Even his ardent supporter, Bishop Burnet, is constrained to admit the justice of the complaint. 'The King,' he says, 'a very few days after he was set on the throne, went out to Hampton Court, and from that palace came into town only on council days: so that the face of a court and the rendezvous, usual in the public rooms, was now quite broken. This gave an early and general disgust. The gaiety and diversions of a court disappeared.' The founding of an English Versailles was, in fact, an idea in every way repugnant to the ordinary Londoner; 'and,' as the Bishop adds, 'the entering so soon on so expensive a building afforded matter of censure to those who were disposed enough to entertain it.'[36]

Pope's hostility to William is apparent in *Windsor-Forest*, where the poem's attack on the Norman kings carries modern reverberations and the account of the death of

---

[34] *The Architecture of Wren*, 2nd edn. (Reading: Redhedge, 1988), 95.
[35] Bevil Higgons in *The Mourners* presents the parallel charge, 'Mourn for the Statues, and the Tapestry too, | From Windsor, gutted to aggrandize Loo', in G. de F. Lord et al. (eds.), *Poems on Affairs of State*, 7 vols. (New Haven: Yale University Press, 1963–75), 6. 362. Pope identifies Higgons as the author in his own copy, see Mack, *Collected in Himself*, 436. Higgons wrote a poem in praise of Pope, published by Ault in *Pope's Own Miscellany*, 81.
[36] Law, *Hampton Court Palace*, 3. 11. Gilbert Burnet, *History of his Own Time*, 2nd edn., 6 vols. (Oxford: Oxford University Press, 1833), 4. 3, adds immediately, 'And this spread a universal discontent in the city of London. And these small and almost indiscernible beginnings and seeds of ill humour, have ever since gone on in a very visible increase and progress.' A poem that expresses that discontent is 'A Description of Hampton Court Life', *Poems on Affairs of State*, 5. 56. Pope's satire on court life, *The Impertinent*, published in 1733 but written earlier, seems to be set at Hampton Court (*Twickenham*, 4. 48–9).

William Rufus in the forest evokes the death of William III, whose horse stumbled while he was out hunting in the area of Hampton Court.

Since Hampton Court is associated with the 1688 revolution and the subsequent settlement, it is worthwhile reading the frontispiece to *The Rape of the Lock* in relation to contemporary political iconography, even though, I believe, such a reading must ultimately be discarded. Certainly, divorced from the poem, the picture would invite such interpretation. If this were a print of the 1690s the figure absorbed in her reflection in the mirror would be Mary, and the putto being pointed at in the foreground, dressed in Belinda's shoes and absorbed by the casket of trinkets, would be William. William was represented by Jacobite propaganda as homosexual and effeminate; he was also small, his mother's nickname for him being 'Piccinino'.[37] The casket is not out of place in a political satire. The British Museum Department of Prints and Drawings has a satirical print representing a sword, a carafe, and a harlot's scent box; in the text they discuss which has inflicted the most damage on the enemy. A political interpretation might also be elaborated by reference to the star at the top left of the picture, drawn to our attention by a pointing sylph (rather as the putto bottom left is). On one level the star is the lock, the only episode from the original poem represented in the frontispiece:

> A sudden Star, it shot thro' liquid Air,
> And drew behind a radiant *Trail of Hair*. (5. 127–8)

But the dead William was also commonly represented by a star. The British Museum has two particularly strong examples: 'Europe in Mourning' in which 'The soul of William III. ascends to a temple which, enshrining a radiant star, stands on rocks in the distance'; and 'Truth above all things' in which 'The etching represents a genius holding a radiant star of great magnitude, on which is "W III".' At other times the symbolism is rendered ambiguous, as in 'To the eternal memory of the victorious year 1702' in which Deceit holds an *ignis fatuus*.[38] Although these are Dutch engravings, it is likely that the symbolism was common currency, for William's death produced a flurry of political writing, both Whig and Tory.

The plate that introduces canto 3 (Fig. 2) also teases the viewer with the potential for political readings. The boat carrying Belinda to Hampton Court corresponds neither to the vessel in the poem nor to the craft depicted on the river in contemporary illustrations. There are sails with sylphs hovering around them, but there are also oarsmen. *Britannia Illustrata* shows several barges on the river and a variety of traffic, but none of the vessels uses both oars and sails. *The Rape of the Lock* oarsmen look angry and disaffected—a nice contrast with the fair nymphs and well-dressed youths—and there may be an allusion to the contrast represented in the British

---

[37] See Nesca A. Robb, *William of Orange*, 2 vols. (London: Heinemann, 1962–6), 2. 398–406, 448–50; *Poems of Affairs of State*, 5. 37–8; 6. 18, 244; and Paul Kléber Monod, *Jacobitism and the English People, 1688–1788* (Cambridge: Cambridge University Press, 1989), 55–6.

[38] British Museum, Political and Personal Satires, 1362 'De Ongerse Pallasch, de Boere Caraffa, en Hoerensmeerdoos', 1420 'Europe in Rouw', 1421 'De Waarheid Boven Al', and 1423 'Ter Eeuwige Gedachtenisse van het Zeeghaftig Jaar 1702'.

32   The Rape of the Lock

Lud. Du Guernier inv.   C. Du Bosc sculp.

FIG. 2. Plate to canto 2 of *The Rape of the Lock*, 1714 (Bodl. Don. e. 115; 214 × 131 mm).

Museum, Political and Personal Satires 1532 'A Dialogue betwixt Whig and Tory' of 1710: slavery is represented by a galley rowed by slaves; liberty by a boat with sails. Anne's reign is a compromise between the two, a state of affairs in which the actors of *The Rape of the Lock* are trapped.

These interpretations hover uncomfortably around these plates, but I do not think they can be endorsed. Political readings of aspects of the poem and its illustrations are easy to generate but they are difficult to validate. Howard Erskine-Hill has raised many of the issues in a vital but circumspect way in his articles on the politics of the *Rape*.[39] He argues that Pope's poem draws on the rhetoric of Jacobitism: rape was the image for William's conquest of the kingdom; the rape in the poem is preceded by a card game, a form of struggle frequently allegorized in political literature; it is possible to identify some of the cards with political figures, especially, for example, Pam with Marlborough and 'Queen of Hearts' with Anne. In his essay on 'Literature and the Jacobite Cause', Erskine-Hill concluded, 'That the poem is also, by allusion, an heroi-comical reworking of 1688, I am tempted to think' (54), and he has recently developed his analyses to place these particular allusions in a broader moral context, without suggesting that he finds any tidy political allegory. There is little doubt that interpretative traps for the modern reader lie in the limited range of motifs available at any historical time; they form a pool that writers with quite different purposes draw on. Just as it is possible for the same locution to be used in different illocutions, so it is possible for a symbol to be put to separate uses.[40] The major controversy of the immediate period of *The Rape of the Lock* surrounded Henry Sacheverell and his challenge to the 1688 settlement. Sacheverell became a major public figure, more important perhaps as a symbol than for the particulars of his views.[41] His cult status is evidenced by a pack of playing cards which records the history of his case, each card representing a new stage of his progress towards his final triumph. The playing cards invite a rereading of the game of Ombre, and look as though they might be the clue Erskine-Hill fears he may have missed. But I have not been able to work out such a reading, and I doubt whether one is available.[42] The teasing promise of political interpretation is further strengthened by a contemporary illustration which touches on the poem with surprising directness. In the frontispiece to Ned Ward's *The Fifth and Last Part of Vulgus Britannicus* (1710), a bishop kneels before the Queen, surrounded by other clergy, while a corner of his mitre is being cut off from behind by shears, even as Belinda's lock was cut off by the Baron; the picture might have been entitled 'The Rape of the Mitre'. If this were taken as a guide to the poem, analogies might easily be drawn between Belinda and the Church, particularly if the emphasis were satirical: the stance towards the Church would be generally

---

[39] Howard Erskine-Hill, 'Literature and the Jacobite Cause: Was There a Rhetoric of Jacobitism?', in Eveline Cruickshanks (ed.), *Ideology and Conspiracy: Aspects of Jacobitism, 1689–1759* (Edinburgh: John Donald, 1982), 49–69; 'The Satirical Game at Cards in Pope and Wordsworth', in Claude Rawson and Jenny Mezciems (eds.), *English Satire and the Satiric Tradition* (Oxford: Basil Blackwell, 1984), 183–95; *Poetry of Opposition and Revolution: Dryden to Wordsworth* (Oxford: Clarendon Press, 1996), 72–93. The last anticipates my interest in *New Atalantis* and the political colouring of *Rape of the Lock*.

[40] See J. L. Austin, *How to Do Things with Words*, 2nd edn. (Oxford: Oxford University Press, 1975) and John R. Searle, *Speech Acts: An Essay in the Philosophy of Language* (Cambridge: Cambridge University Press, 1969). I assume with them that exclusions and distinctions work for particular purposes.

[41] Geoffrey Holmes gives a full account in *The Trial of Dr Sacheverell* (London: Eyre Methuen, 1973); George Macaulay Trevelyan gives a good Whig narrative in *England under Queen Anne*, 3 vols. (London: Longmans Green, 1930–4), 3. 47–60, while Howard Erskine-Hill presents an alternative perspective in *Poetry of Opposition and Revolution*, 65–8.

[42] The playing cards are *British Museum, Political and Personal Satires* 1546; the rape of the mitre is 1540.

sympathetic, but a concern for appearances and for social position, an abuse of the cross, and a triumphalism would make her vulnerable; the Baron would represent Godolphin, the false brother by whom the Church was imperilled. The trouble with such readings is not only that they are incompatible with a necessary attention to the poem's surface; they are also unenlightening. The only pleasure that might be given is coterie recognition of familiar materials in an unfamiliar context; but this is where we might possibly gain an insight into Pope's practice. The reference to Pam in the *Rape of the Lock* card game, like some of the allusions in *The Dunciad*, may be the equivalent of a masonic handshake, inviting a response of gratified surprise in a Catholic, Tory, or Jacobite reader. Such an effect is local—the impulse to make the poem one long handshake should be resisted—but in *The Rape of the Lock* it may alert readers to the colouring of the general social vision. Certainly I believe that the poem adopts a contemporary party-political reading of sexual mores, and criticizes the Catholic Fermors and Petres by implicating them in those mores.

NEW ATALANTIS, HAMPTON COURT, SEX, AND POLITICS

When Hampton Court appears in the first version of the poem, it is with a characteristic playing with syllepsis:

> Here *Britain*'s Statesmen oft the Fall foredoom
> Of Foreign Tyrants, and of Nymphs at home;
> Here Thou, great *Anna!* whom three Realms obey,
> Dost sometimes Counsel take—and sometimes *Tea*. (1. 69–72)

The connection between statesmen and the fall of nymphs is not a casual one. It had been fully established in the Tory propaganda of Delarivier Manley's *New Atalantis* in 1709. The book is referred to explicitly in both versions of the poem. The Baron directs us to it, at the point of supreme appropriateness, as he contemplates his triumph over his nymph and predicts its duration 'As long as *Atalantis* shall be read'. *New Atalantis* is composed of multiple stories of seduction, and for that reason it is often referred to as pornographic or semi-pornographic, but recent critical treatment, particularly from Rosalind Ballaster and Catherine Gallagher, has refocused attention on the journalistic skill with which it combines the sexual and the political. The exploiters of distressed young maidens, the adulterers, and the seducers of their wards are great Whig lords.[43] The work is set in 1702 at the death of William III. Astrea (Justice) returns to earth, and in the company of her mother, Virtue, and with the help of Intelligence surveys the contemporary scene. In the ensuing survey William himself is not spared—Bentinck's nursing of him in his illness is represented as a matter of sharing his bed—but few powerful Whigs escape criticism in these chronicles of scandal. Manley had early in her life moved in the circles she

---

[43] Delarivier Manley, *The New Atalantis*, ed. Rosalind Ballaster (London: Penguin Books, 1991); Rosalind Ballaster, *Seductive Forms: Women's Amatory Fiction, 1684–1740* (Oxford: Clarendon Press, 1992); Catherine Gallagher, *Nobody's Story: The Vanishing Acts of Women Writers in the Market Place, 1670–1820* (Oxford: Clarendon Press, 1994).

describes, but her literary life had been with the Tories, particularly as Swift's successor as writer of the *Examiner* and as the companion of the printer John Barber. It is difficult to estimate the impact of *New Atalantis* on the politics of the day. Gallagher sees the book as a significant critique of the 'civil humanism' associated with the Whigs and a defence of an older conception of virtue (88–144), while Trevelyan gives it an important role, subsidiary only to the Sacheverell trial, in bringing down Godolphin's ministry:

> Indeed the publication that did most harm to the Ministry that year [1709] was a book of the lowest order, the *New Atlantis*, wherein Mrs. Manley, a woman of no character, regaled the public with brutal stories, for the most part entirely false, about public men and their wives, especially Whigs and above all the Marlboroughs . . . Fancy names thinly veiled the persons libelled. The book soon had a second volume and passed through many editions.[44]

Consequently Swift thought Manley might be entitled to some reward for her services in the cause, 'by writing her Atlantis and prosecution [for it]'.[45] It is to this political atmosphere that *The Rape of the Lock* belongs.

The Tories attacked the Whigs on the ground that their scandalous sexual conduct showed them unfitted for power. The point of Pope's satire of the Roman Catholic Fermor and Petre families—the unifying criticism in *The Rape of the Lock*—is that they betray their religion and their politics by behaving in the same way. William B. Warner points out that Pope's satire 'exploits the gender strife, disguised intentions, and amorous battle found in the novels of amorous intrigue', while Valerie Rumbold draws attention to the vital importance of intermarriage in the Roman Catholic aristocracy and gentry. Only by careful alliances could their estates be kept of reasonable size; the circle of eligible partners was restricted and the need for rational choice was paramount.[46] When in the expanded version of his poem Pope presents a diagnosis of the failures that led to conflict between the two great Catholic families, it is of their being tainted by the world of the Whigs, the world of *New Atalantis*, rather than opposed to it. Pope's engagement with contemporary politics is not, therefore, at some level wholly remote from social activities, and the poem is not functioning on two separate levels. Consequently neither the poem nor the card game works consistently as an allegory, but the metaphor of rape does call up the events of 1688 and card games are related to political conflict and negotiation.[47] Pope needs to draw these matters into the reader's field of vision if the link between sexuality and politics is to be grasped; the two realms have to be brought together.

---

[44] Trevelyan, *England under Queen Anne*, 3. 38–9. Trevelyan's account itself brutalizes a narrative that is not without sensitivity and sympathy.

[45] *DNB* 12. 922. The best evidence of the importance of *New Atalantis* is its continued role in discussions of Marlborough's reputation. See Trevelyan, *England under Queen Anne*, 3, pp. x–xiii, for a dispute with Winston Churchill over whether Macaulay had a debt to Mrs Manley.

[46] William B. Warner, *Licensing Entertainment: The Elevation of Novel Reading in Britain, 1684–1750* (Berkeley and Los Angeles: University of California Press, 1998), 147; Rumbold, *Women's Place in Pope's World*, 58–67. Tillotson shows the care Caryll took of his own son, *Twickenham*, 2. 91.

[47] Erskine-Hill's 'The Satirical Game at Cards in Pope and Wordsworth', 183–95, can be supplemented by surviving political cards in the British Museum Department of Prints and Drawings collection: *Political and Personal Satires* 1435, 1461, 1491, 1492, 1543, 1544, 1546.

But the allegorical reading is always frustrated, folded back in again to a literal interpretation of events. When Pope produced his own reading of the poem, an elaborate *Key to the Lock*, the proposed interpretation was manifestly ridiculous. The *Key*'s interpretation is based on the idea that Pope meant 'something farther' and that something is discovered by taking the lock as the representation of the Barrier Treaty. The *Key* attacks malicious interpretation rather as *Gulliver's Travels* does, but there is also a small-scale double bluff. The politically sensitive reader knows that the political significance of *The Rape of the Lock* is not hidden away like the kernel of a nut; it is on the surface if the correct light is shone on it. The *Key to the Lock* draws attention to a way of reading that will not work; but this way of reading must nevertheless persist in our minds as an extension or exaggeration of the right one.

Manley's use of topography in her work throws light on the role of Hampton Court in *The Rape of the Lock*. *New Atalantis* is constructed around a tour of London and its environs, different places introducing us to different characters and stories. St James's Park and Hyde Park figure prominently, as do the royal palaces, Buckingham Palace, Parliament, and Salisbury Cathedral. Buildings represent their owners. Even if the foundations of Blenheim are inadequate the Duke of Marlborough will 'by superficial raisings and outside ornaments (as are most of his) fit it for show, since not for use' (161). The parallel with *To Burlington* ('pompous buildings once were things of Use' (24)) is striking. The contrast is provided by the Duke of Beaufort's 'goodly pile, which, with its proud eminence aspires almost above human sight' (230). The first royal palace to be considered in *New Atalantis* is either Hampton Court or Kensington Palace, though it is rather more likely to be the latter. The palace is associated with the whole system of favouritism, political and sexual, attacked by the novel; it is also associated with continuity between reigns. When Astrea asks Intelligence what will become of the old favourites, she replies, 'Why, they will be favourites still' (47). *New Atalantis* is unillustrated, but the *Memoirs of Mrs Manley* have a frontispiece depicting Old Somerset House, the antithesis of Hampton Court, as a stately backdrop, showing the potential for attaching symbols of aristocratic dignity, either positively or negatively, to scandalous narrative.[48] Buildings often feature as backdrops also in political prints. Several Dutch attacks on Louis XIV show Saint-Germain or Les Invalides as a background; in one of the caricatures by Romeyn de Hooghe, Louis is torn between his mistresses, while a background palace suggests neglected responsibilities.[49] The role of Hampton Court in *The Rape of the Lock* is to suggest a similar neglect of duties, but the neglect is twofold: the initial neglect by those empowered after 1688 and a mimicking neglect by these Catholic families we are currently concerned with. Reference to William and Mary lingers over the first plate and the poem like a ghost, but the chief point, the presence of vanity and frivolity at the seat of government, declares itself openly.

---

[48] British Museum, *Political and Personal Satires* 1264.
[49] Ibid. 1353–5, 1372. War tents similar to the one Belinda is shown sleeping in are common in these engravings.

REVISION: *LE COMTE DE GABALIS* AND THE ADDITION OF THE SYLPHS

The implicit political satire of the illustrations of the *Rape of the Lock* is linked to the predatory male behaviour which characterized Anne's court, but it also relates to female behaviour. The exploration of female psychology that came about with the introduction of the sylphs also has its socio-political importance. In creating the sylphs as machinery Pope greatly extended the intellectual horizon of his poem. As he explains in the dedication in 1714, his source was 'a French Book call'd *Le Comte de Gabalis*, which both in its Title and Size is so like a *Novel*, that many of the Fair Sex have read it for one by Mistake'. Even in jest Pope is aware that the appearance of a book influences the way it is read; the members of the fair sex reading Montfaucon de Villars's intellectual *jeu d'esprit* as if it were a novel are patterns for those misinterpreting *The Rape of the Lock* and its illustrations. *Le Comte de Gabalis* itself was soon made readily available as companion to the *Rape of the Lock*. There had been a translation in 1680, but in 1714 Lintot and Curll issued one jointly, proclaiming in the half-title that it was 'Very necessary for the Readers of Mr. Pope's Rape of the Lock'. It is to this translation I shall refer, taking it as part of the same publishing enterprise as *The Rape of the Lock*. The chief difficulty in understanding Pope's relation to his source, in spite of the extensive commentary on it, comes from the absence of any stable critical understanding of *Le Comte de Gabalis* itself. Villars's little book has been taken as a straight guide to Rosicrucianism, as an instance of Catholic pietism, and as an early expression of Enlightenment scepticism.[50] Registered on 28 November 1670, only ten months after publication of the Port Royal edition of Pascal's *Pensées*, which I strongly suspect it attacks, *Le Comte de Gabalis* in some ways resembles *A Tale of a Tub*, its ironic energy leading to a criticism more wide-ranging and destabilizing than its ostensible target would seem to allow. The apparent object of satire is Rosicrucianism; the main object, I suspect, Pascal and Jansenism.[51] The justification of the satire is a defence of Roman Catholic orthodoxy, to which both Jansenism and Rosicrucianism were opposed, but the energy of the satire endangers that orthodoxy itself. The tone of the narrator is

---

[50] See the edition 'Published by The Brothers' (London: Old Bourne Press, 1913), 'Let him whose quest is the gratification of a selfish intellectualism beware its pages'; Patricia Brückmann, 'Virgins Visited by Angel Powers', in G. S. Rousseau and Pat Rogers (eds.), *The Enduring Legacy: Alexander Pope Tercentenary Essays* (Cambridge: Cambridge University Press, 1988), 3–20; and the excellent edition by Roger Laufer, *Le Comte de Gabalis ou Entretien sur les sciences secretes. La critique de Bérénice* (Paris: A. G. Nizet, 1963). Easily the best recent treatment of the work as a source for Pope is Donna Scarboro, '"Thy Own Importance Know": The Influence of *Le Comte de Gabalis* on *The Rape of the Lock*', *Studies in Eighteenth-Century Culture*, 14 (1985), 231–41. My own impressions coincide with the judgement of Dominique Descotes that Villars is in intention more pro-Jesuit than he is modern; see her edition of *La Première Critique des 'Pensées'* (Lyons: Centre National de la Recherche Scientifique, 1982).

[51] For material Villars satirizes, see Blaise Pascal, *Pensées sur la religion et sur quelques autres sujets*, introd. Louis Lafuma, 3 vols. (Paris: Éditions du Luxembourg, 1951), i. 6; Pascal's *Pensées*, trans. and introd. A. J. Krailsheimer (Harmondsworth: Penguin Books, 1966), Pensées 269, 413, 713 (Brunschvicg 692, 162, 923). See also the chapter on 'The Other-Worldliness of Pascal' in F. T. H. Fletcher, *Pascal and the Mystical Tradition* (Oxford: Blackwell, 1954), 61–72.

sardonic throughout; he approaches his informant, the Comte de Gabalis ('so wise and so weak; so admirable and so ridiculous', 49), with a thoroughgoing scepticism and astonishment at the extravagance of his claims. At the centre of these is the condition that must be met in order to become an adept: '*You must*, (added he, whispering in my Ear) *you must renounce all carnal Commerce with Women*' (12). This is the joke which renders the Rosicrucians and, behind them, the Jansenists ridiculous, but it also implicitly carries an assertion of the power of ordinary sexuality that, together with a levity in treating scripture and the church fathers, lost Villars his licence to preach. Because Adam and Eve had carnal knowledge of one another, the Comte explains, human beings became unable to see creatures all around them—sylphs; undines or nymphs; gnomes; salamanders—whose souls are mortal. The only hope of immortality for these creatures is through sexual congress with human beings, and 'God has given leave to all these People to make use of all the innocent Artifices they can think of to converse with Men without their Privity' (75–6). The myths of female congress with animals represent what is really congress with sylphs. Once the sylph has established a relationship, a perfect purity is required from the partner. The sylphs, therefore, combine strict rules of chastity with animal passion. The first aligns them with propriety, but the second links them to the satyrs of Pope's illustrations. The 'companions of that sort of Satyr, which appear'd to St. Anthony' were called sylphs (62–3), and the satyrs in the first two illustrations may also be sylphs.

An interest in Rosicrucianism was something Pope shared with the *Spectator*. It is mentioned in numbers 198, 379, and 574, and that may have prompted his interest in *Le Comte de Gabalis*; he was, after all, in search of a new machinery. But the broader implications of Villars's satire did not pass Pope by as he expanded his poem. What he found in Villars was the possibility of providing an explanation for the behaviour that had simply been wondered at in the first version of the poem. The conclusion to the opening of the poem takes on a quite different force in the revised version:

> Say what strange Motive, Goddess! cou'd compel
> A well-bred *Lord* t'assault a gentle *Belle*?
> Oh say what stranger Cause, yet unexplor'd,
> Cou'd make a gentle *Belle* reject a *Lord*? (1. 7–10)

Although there is only a change in italicization of question marks between 1712 and 1714, the relation of the invocation to the rest of the poem has changed. In the first version, there is no real answer to Pope's question; there is merely the implication that the dispute is unintelligible. The revised poem does provide an explanation. Dennis claimed that the sylphs made nothing happen: 'For what he calls his *Machinery* has no Manner of Influence upon what he calls his *Poem*, not in the least promoting, or preventing, or retarding the Action of it.'[52] But, of course, to the informed understanding, they make it all happen; Dennis's interpretation is too

---

[52] *The Critical Works of John Dennis*, 2. 328; the point is developed on 337–9.

dependent on the contrast between the versions. The sylphs make Belinda beautiful; they determine her attitude to men; their abandonment of her permits her rape; and the gnomes shape her grief and anger. The whole drama stems from Belinda's remoteness from what is taken to be ordinary sexual life, a remoteness as ridiculous and culpable as that of the Comte de Gabalis.

In Gabalis's system, Ariel would be Belinda's lover, attaining immortality through their intercourse and anxious to protect the purity of his partner. Pope retains some of this idea but he modifies it in the interests of decency and of his own project. Pope's sylphs do give sexual comfort to their charges, but of an unspecified sort:

> Know farther yet; Whoever fair and chaste
> Rejects Mankind, is by some *Sylph* embrac'd . . . (1. 67–8)

And a vestige of the Gabalis system remains in Ariel's final withdrawal:

> Sudden he view'd, in spite of all her Art,
> An Earthly Lover lurking at her Heart.
> Amaz'd, confus'd, he found his Pow'r expir'd,
> Resign'd to Fate, and with a Sigh retir'd. (3. 143–6)

But the Ariel who deserts Belinda is not her lover. Pope's major departure from his source is to give his sylphs a history. They are no longer wholly separate beings; they are society ladies enjoying a life after death:

> As now your own, our Beings were of old,
> And once inclos'd in Woman's beauteous Mold;
> Thence, by a soft Transition, we repair
> From earthly Vehicles to these of Air. (1. 47–50)

The sylphs, then, do not represent an alternative masculinity, as Gabalis's do. They do not represent another world but an aspect of the present world reasserted and reinstated; the women who have fought the battle of the sexes when alive now fight on another plane. Tradition and influence are now dramatized as a quasi-spiritual intervention. Pope's world is, then, tightly bounded as Gabalis's is not. There is no escape from social custom and social pressure, but the social world is contested and it is possible to adhere to one party rather than another.

This appropriation of Villars allows Pope to present an expanded sexual politics but with the result that the presentation of male assault and female rejection becomes unbalanced. In the first version of the poem a balance between Belinda and the Baron was maintained throughout, and the Baron's assault was specifically related to the world of *New Atalantis*. In the revision this balance disappears and the emphasis falls on the world of women. Nothing in *Le Comte de Gabalis* requires this—the sylphs interact with men and women—and Pope's emphasis comes, I suspect, from a more specific connection he wanted to make between his poem and *New Atalantis*. Towards the end of Manley's fiction we are introduced to a group that may be forerunners of the sylphs and their allies, the 'Cabal' of women:

Oh how laudable! how extraordinary! how wonderful! is the uncommon happiness of the Cabal? They have wisely excluded that rapacious sex who, making a prey of the honour of ladies, find their greatest satisfaction (some few excepted) in boasting of their good fortune. (154)

Manley seems to anticipate Pope's appropriation of cabbala to the secrecy of female rejection of men. Certainly the Baron is an embodiment of 'that rapacious sex who, making a prey of the honour of ladies, find their greatest satisfaction . . . in boasting of their good fortune'. These women, who have gathered together to the exclusion of men, are regarded as participants in a mystery cult; all of them appear to have been court ladies and supporters of the Whigs. Astrea pronounces judgement on the phenomenon:

It is something so new and uncommon, so laudable and blameable, that we don't know how to determine, especially wanting light even to guess at what you call the mysteries of the Cabal. If only tender friendship, inviolable and sincere, be the regard, what can be more meritorious or a truer emblem of their happiness above? 'Tis by imitation, the nearest approach they can make, a feint, a distant landshape of immortal joys. But if they carry it a length beyond what nature designed and fortify themselves by these new-formed amities against the hymenial union, or give their husbands but a second place in their affections and cares, 'tis wrong and to be blamed. (161)

The predatory nature of the Whig lords is balanced by the caballing of the Whig ladies. Pope fuses the sylphs and the cabal, and in doing so sets women against men and the 'hymenial union'. In this context Hampton Court takes on another significance because of the ribald criticism of Queen Anne and her relations first with Sarah, Duchess of Marlborough, and later with Abigail Masham, which had come to prominence with the quarrel between the Queen and the Duchess between 1707 and 1708.[53] There had long been a tendency to see the Queen as having no mind of her own, her life shaped by her attendants, who were almost exclusively female. The Duchess shared this view: 'Q. has no Original Thoughts on any Subject; is neither good nor bad, but as put into' (Bucholz, *The Augustan Court*, 158). In her anger at the Queen's affection for Mrs Masham she fuelled innuendo that would formerly have been at her own expense. Aided by her secretary Arthur Maynwaring, she presented the relation as one of sexual infatuation. In a letter to the Queen of 26 July 1708, she says,

I remember you said . . . of all things in this world, you valued most your reputation, which I confess surpris'd me very much, that your Majesty should so soon mention that word after having discover'd so great a passion for such a woman, for sure there can bee noe great reputation in a thing so strange & unaccountable . . . *nor can I think the having noe inclenation for any*

---

[53] See, for example, Arthur Maynwaring or Mainwaring, *A Dialogue between Madame de Maintenon and Madam Masham* (1709). Edward Gregg gives an account of attacks on the Queen in *Queen Anne* (London: Routledge, 1980), 275–76, 295. Gregg's study is supplemented by R. O. Bucholz, *The Augustan Court: Queen Anne and the Decline of Court Culture* (Stanford, Calif.: Stanford University Press, 1993). Bucholz says the charge of lesbianism is 'ludicrous', 331 n. 100, but I am not sure how he knows.

but of one's own sex is enough to maintain such a charecter as I wish may still bee yours. (Gregg, *Queen Anne*, 275–6)

The letter was private, but Maynwaring provided public attacks. For example, 'A New Ballad to the Tune of Fair Rosemond' says of the Queen:

> Besides the Church, she dearly lov'd
> A Dirty Chamber-Maid . . .
> Her Secretary she was not,
> Because she could not write;
> But had the Conduct and the Care
> Of some dark Deeds at Night.[54]

And in *The Rival Dutchess*, Mrs Masham is portrayed saying, 'Especially at Court I was taken for a more modish lady, was rather addicted to another Sort of passion, of having too great a Regard for my own Sex, insomuch that few People thought I would ever have married' (Gregg, *Queen Anne*, 276). These are Whig complaints against the Queen, but *New Atalantis* shows how readily they could be adopted in the service of Tory propaganda. Pope could not have written a satire on women's efforts to exclude men, the 'rapacious sex', and have set it at Hampton Court without alluding obliquely to these rumours about the Queen and her court. This is the second prong of the attack on court mores. It explains why *The Key to the Lock* was written. The politics it deals in, chiefly the Great Barrier Treaty, are unmediated by socio-sexual behaviour. In that respect, those are right who have seen it as a double bluff.

THE APPROPRIATION OF ARABELLA FERMOR'S EXPERIENCE

Although the dedication and the illustrations are important, the chief way in which Pope lays claim to authority over the central experiences of the poem and their context is through the sylphs. As a glance at the list at the end of the chapter suggests, the sheer volume of material on the sylphs modifies the debt to the originating incident. In the revised version of the poem Belinda is emptied of internal life; she is cut off from others and from the hymeneal bond; her life ceases to be her own. Pope hands her over to his machinery—the sylphs contrive it all. Pope's creation of his machinery enables him to assert his own power over Belinda's experience.

Ariel's speeches are taken up a good deal with explaining the sylphs, to Belinda and to themselves. Their duties are not confined to the external protection of women; they are invasive to such an extent that Ariel provides an all-sufficient explanation of female behaviour. Whereas in the first edition Belinda is endowed with intentionality and feeling analogous to the Baron's, in 1714 she is denied it. Women—Nymphs, the Fair—have no inner life of their own. They are subject, we are to understand, to appetite and appetite only. They are, therefore, subject to temptation, 'When kind Occasion prompts their warm Desires' (1. 75). But reason and a moral sense are lacking, even when behaviour is good:

---

[54] *Poems on Affairs of State*, 7. 309; quoted by Gregg, *Queen Anne*, 275.

42   The Rape of the Lock

> 'Tis but their *Sylph*, the wise Celestials know,
> Tho' *Honour* is the Word with Men below. (1. 77–8)

Women are led by appetite, either directly in sensory experience,

> When *Florio* speaks, what Virgin could withstand,
> If gentle *Damon* did not squeeze her Hand? (1. 97–8)

or indirectly, when gnomes crowd the brain with gay ideas. Imagination and creativity are allowed to women, but as the gift of the sylphs,

> Nay oft, in Dreams, Invention we bestow,
> To change a *Flounce*, or add a *Furbelo*. (2. 99–100)

Warburton notes in his 1751 edition of the *Rape* how careful Pope is to keep his machinery before the mind of the reader.[55] This is partly a matter of thorough rewriting, but it is also a matter of Pope's conception of his heroine. Belinda lacks power to act without the sylphs. Her beauty, which is her chief mode of social identification, is their product; she owes her victory at Ombre to their help and the intensity of her grief to Umbriel's vial; even the pinch of snuff which disables the Baron is directed by a gnome. The pervasive agency of the sylphs gives the drama of the poem a special twist. In an epic, gods and goddesses interact with human figures; there is conflict and resistance; Aeneas may or may not obey; Venus may or may not assist him. Because Pope's machinery fills the intentionality of his characters, no such drama is possible for most of *The Rape of the Lock*, but the exception proves the turning point of the revised poem. At first the sylphs protect Belinda from the Baron's scissors, but suddenly their power is lost when Ariel sees 'An Earthly Lover lurking at her Heart'. Belinda's ideas are here her own, not the sylphs' or the gnomes'. For 'An Earthly Lover' to lurk 'at her Heart' is for her to have given way to the temptation resisted in the case of Florio and Damon; it is for her to have shown an impermissible autonomy.[56] This autonomy is punished by the rape; in the original poem the rape was the Baron's fault, now it seems to be Belinda's.

Arabella Fermor could never, therefore, claim the Belinda of the revised *Rape of the Lock* as her own. Belinda was the possession of the sylphs, her one act of autonomy revealed as forbidden female desire. Pope also took steps in the detail of his revision to make it unlikely that Arabella Fermor, or anyone else, would want to claim Belinda as her own. The blackening begins in the bedroom. That Belinda now stays in bed too long need not be regarded as a serious blot, particularly as it enables her to hear Ariel's message, but in retrospect it can be seen as the first in a series of smudges. Her subsequent river journey is marked by a series of changes to her disadvantage. In 1712,

---

[55] *Rape of the Lock*, 5. 131 n. Warburton traces the changes to the poem very carefully.
[56] There are complications here stemming from Pope's adaptation of his source. A Gabalisian sylph would immediately punish infidelity with desertion, but Ariel has no claim to Belinda's conscious devotion, having introduced himself only that morning. He should have made her think of somebody else; but we are to assume the attraction of the earthly lover (perhaps a gnome?) is too powerful.

> A Train of well-drest Youths around her shone,
> And ev'ry Eye was fixed on her alone (1. 21–2)

but in 1714,

> Fair Nymphs, and well-drest Youths around her shone,
> But ev'ry Eye was fix'd on her alone. (2. 5–6)

Belinda's triumph is now at the expense of her rivals. In 1712 Belinda's beauty provoked a moral generosity:

> If to her share some Female Errors fall,
> Look on her Face, and you'll forgive 'em all. (1. 33–4)

In 1714 amorality had taken its place:

> Look on her Face, and you'll forget 'em all. (2. 18)

In 1712 the chatter at Hampton Court initially passed without censure, 'In various Talk the chearful hours they past' (1. 75), but in 1714 it is tinged with ironic criticism, 'In various Talk th' instructive hours they past' (3. 11); Hampton Court should be worth something better.

More important changes take place in the account of Belinda's behaviour after the rape. In 1712 Belinda feels rage, resentment, and despair at the assault upon her, but, in keeping with the aim of reconciling the families, Pope suggests that she might have been comforted when instead she was enraged by others:

> While her rackt Soul Repose and Peace requires,
> The fierce *Thalestris* fans the rising Fires. (2. 11–12)

But in 1714 Umbriel returns from the Cave of Spleen to find her prepared for his intervention:

> *Belinda* burns with more than mortal Ire,
> And fierce *Thalestris* fans the rising Fire. (4. 93–4)

The changed relation between Thalestris and Belinda furnishes Pope's most significant change in these details. In 1712 Thalestris maintains the code in which honour is reduced to a matter of appearances and Belinda is not implicated:

> Oh had the Youth but been content to seize
> Hairs less in sight—or any Hairs but these!
> Gods! shall the Ravisher display this Hair,
> While the Fops envy, and the Ladies stare!
> *Honour* forbid! at whose unrival'd Shrine
> Ease, Pleasure, Virtue, All, our Sex resign. (2. 19–24)

But in 1714 the first two lines are taken from Thalestris and given to Belinda to end her speech of remonstrance and bring down the curtain on the fourth canto:

> Oh hadst thou, Cruel! been content to seize
> Hairs less in sight, or any Hairs but these! (4. 175–6)

44   The Rape of the Lock

In thus committing herself to a world of appearances, Belinda may be innocent of the sexual significance of her speech (and the change from a dash to a comma before 'or any Hairs but these' seems to open up that possibility), but Pope permits a guffaw at her expense that would have been out of place in the first version.

The Rape of the Lock is a good example of how making a poem and making a book can complement each other. The first version of the poem found an appropriate place in Lintot's *Miscellany*, alongside classical translations and occasional verse. It would have made a satisfactory folio pamphlet, like *Windsor-Forest*, but it could not have been made into a book on its own. The revised poem finds an appropriate typographic form in its octavo booklet. Each of the five cantos can be properly marked by a new section of the book, with a fresh page and a dropped head, and the important preliminary matter, the epistle dedicating the poem to Arabella Fermor, can also take its place as a separate section. The six plates permitted by the format also allowed Pope to review and emphasize elements of his poem: the politics associated with Hampton Court; the sexuality of Belinda, the playfulness of the sylphs, and the power of his own art, represented by Pegasus in headpiece and initial. The independence of the booklet can be taken to represent a new independence established by the author, developing his own treatment of classical forms and escaping from the restrictions of the social circles in which the seeds of the poem originated. It was a good way to end the first part of his career and begin his 'ten years to comment and translate' (*Dunciad*, 3. 332).

APPENDIX: LIST OF MAJOR CHANGES IN *THE RAPE OF THE LOCK* FROM 1712 TO 1714

A square bracket separates 1712 and 1714. The list notes the major revisions, the 1714 additions, and the 1714 canto breaks. Additions are introduced by a line common to the two texts.

The Rape of the Locke] *The Rape of the Lock*
——] Written by Mr. *POPE*
——] Frontispiece and five plates
——] Dedication
Canto I] Canto I
[Following '*Belinda* rose, and 'midst attending Dames (1. 19)] *Belinda* still her downy Pillow prest,' (1. 19)]
——] 132 lines: Ariel introduced (7 lines); Ariel's speech (88 lines); Awakening by Shock (6 lines); Belinda's toilet (28 Lines)
——] [Canto II] Belinda and rising sun (3 lines)
Launch'd on the Bosom (1. 20)] Lanch'd on the Bosom (2. 4)
[Following 'The rest, the Winds dispers'd (1. 64)] The rest, the Winds dispers'd (2. 46)']
——] 96 lines: The painted vessel (6 lines); Attendant sylphs (20 lines); Ariel's speech (70 lines)
Close by those Meads (1. 65)] [Canto III] Close by those Meads (3. 1)

[Following 'And the long Labours of the *Toilette* (1. 88)] And the long Labours of the *Toilette* (3. 24)']
———] 80 lines: The game of Ombre
[Following 'While frequent Cups prolong (1. 96)] And frequent Cups prolong (3. 112)']
———] 4 lines: Sylphs protect from coffee
[Following 'As o'er the fragrant Steams (1. 114)] As o'er the fragrant Steams (3. 134)']
———] 12 lines: The sylphs try to intervene
[Following 'T'inclose the Lock; then joins it (1. 116)] T'inclose the Lock; now joins it (3. 148)']
———] 4 lines: Interposed sylph severed
Canto II] Canto IV
[Following 'As Thou, sad Virgin! (2. 10)] As Thou, sad Virgin! (4. 10)']
———] 82 lines: Visit to Cave of Spleen
Oh had the Youth but been content to seize | Hairs less in sight—or any Hairs but these! (2. 19–20)]] For this with Fillets strain'd your tender Head, | And bravely bore the double Loads of Lead? (4. 101–2)
[Following 'The long-contended Honours of her Head. (2. 58)] The long-contended Honours of her Head. (4. 140)']
———] 2 lines: Umbriel breaks the vial
[Following 'Nay, *Poll* sate mute (2. 81)] Nay, *Poll* sate mute (4. 164)']
———] 2 lines: A sylph had warned
[Following 'And tempts once more (2. 89)] And tempts once more (4. 174)']
———] Oh hadst thou, Cruel! been content to seize | Hairs less in sight, or any Hairs but these! (4. 175–6)
———] [Canto V]
[Following 'And the pale Ghosts start (2. 111)] And the pale Ghosts start (5. 52)']
———] 4 lines: Umbriel observes the fray
[Following 'A Charge of *Snuff* the wily Virgin (2. 137)] A Charge of *Snuff* the wily Virgin (5. 82)']
———] 2 lines: Gnomes direct snuff
[Following 'And drew a deadly *Bodkin* from her Side. (2. 141)] And drew a deadly *Bodkin* from her Side. (5. 88)']
———] 8 lines: History of the bodkin
[Following 'The Skies bespangling with dishevel'd (2. 175)] The Heav'ns bespangling with dishevel'd (5. 130)']
———] 2 lines: Sylphs see constellated hair

3. The Works of 1717: Building a Monument

THE MODERN CLASSIC

In many ways the *Works of Mr Alexander Pope*, published on 3 June 1717, dignified, decorated, and expensive, is the boldest as well as the most beautiful of Pope's books, proclaiming both the poet's youth and his achievement with a confident flourish. The *Works* gave Pope the status of a classic author, opened up new opportunities for authorial control and self-definition, and set a pattern of publication that persisted until his death. But in June 1717 Pope had only just passed his twenty-ninth birthday, and in publishing a volume of *Works* he was engaging in an act of self-promotion that any celebrated 79-year-old contemporary would have blenched at. It is true that his action was not altogether unprecedented. In 1616 Ben Jonson had taken a similar step, publishing *The Workes of Beniamin Jonson* in folio, with a monumental title page and with his plays presented in the manner of the 'earliest editions of Plautus, Terence, and Aristophanes'.[1] Jonson was 44 at the time of publication and had reached the climax of a distinguished career, but his chosen mode of publication still excited derision:

> Pray tell me *Ben*, where doth the mystery lurke,
> What others call a play you call a worke. (*Ben Jonson*, 9. 13).

In publishing a volume of *Works* during his lifetime, Jonson broke a taboo going back to the earliest days of printing. Even Erasmus, the pioneer in successful manipulation of the resources of print to advance image, influence, and reputation, stopped short of publishing his works, leaving it to Froben to publish an edition four years after his death.[2] Consequently Jonson's precedent went virtually ignored, his seventeenth-century successors lacking his sense of tradition or his confidence. Wycherley and Dryden are sometimes adduced as exceptions to this pattern of self-censorship because they both had volumes called *Works* published in their lifetime, but Wycherley's *Works* of 1713 omitted the poems that he had been revising with Pope's help (a source of subsequent controversy), and the volumes of Dryden's *Works* issued by Tonson were merely made-up volumes of old publications, cashing in on a ready market.

---

[1] David Riggs, *Ben Jonson: A Life* (Cambridge, Mass.: Harvard University Press, 1989), 221. This biography has a reproduction and analysis of the title page. See also *Ben Jonson*, ed. C. H. Herford, and Percy and Evelyn Simpson, 11 vols. (Oxford: Clarendon Press, 1925–50).

[2] See Lisa Jardine, *Erasmus, Man of Letters: The Construction of Charisma in Print* (Princeton: Princeton University Press, 1993), 49.

Pope's volume, in contrast to Tonson's *Dryden*, was an elegant companion to his *Iliad* translation and therefore self-evidently an edition of classical writing. The description of a section of Theobald's library in *The Dunciad* is apposite:

> Volumes, whose size the space exactly fill'd;
> Or which fond authors were so good to gild;
> Or where, by sculpture made for ever known,
> The page admires new beauties, not its own. (*Dunciad Variorum*, 1. 117–20)

The *Works* of 1717 is a grand volume of this type, but it was carefully planned and unified, probably the most successful of Pope's experiments in large-format bookmaking. It was issued alongside the third volume of the translation of the *Iliad* (by then an established critical success), and it was offered in the same formats as Homer's epic. There were illustrated quartos to sit alongside the *Iliad* on the shelves of Pope's subscribers, illustrated folios to be sold to Lintot's richer clients, and unillustrated pot folios to be sold to the general public.[3] Lintot's confidence in the new venture is evident from the size of the edition. At this stage of the *Iliad* publication, he was printing 660 illustrated quartos for Pope's subscribers (200 on superior writing royal paper), 250 illustrated folios, and 1,000 pot folios. He printed the same number of folio *Works*, in the same styles, and also 750 quartos (250 on fine paper, only 120 of which went to Pope).[4] The edition was a large one, given that these were large-format books, costing between 12*s*. and a guinea each; the print-run is much the same as for the octavo *Works* printed by Woodfall in the 1730s. The best-quality copies, the quartos, are on royal paper, and the typography is impressive, with wide margins and twenty lines to a page. The printing, by William Bowyer, is good, and the project was a well-disciplined one, with none of those bibliographical eccentricities that trouble some of Pope's later works. More important, the book is well illustrated. There is a large frontispiece engraving of Pope, by George Vertue after a portrait by Charles Jervas, and the major poems have attractive headpieces and initials, designed by Simon Gribelin, doubtless in collaboration with Pope. As David Foxon has remarked, the *Works* headpieces represent an advance over those of the *Iliad* by breaking up the long horizontal space into oval scenes encased in decorative but symbolic frames (Figs. 3 and 5), and thus making a success of this attempt to integrate type and illustration. At just over at 450 pages long, these books are impressive without being unwieldy. In size, sculpture, and format they are worthy companions of the *Iliad*.

That Pope should eventually arrange the publication of such a volume, given his interest in book production, may be unsurprising, but that he should publish at this stage of his career is perplexing. As we shall see, his preface shows some authorial nervousness at his temerity. The title 'Works' was not thrust on him by the need to find a title for a collection; an alternative title—*Poems on Several Occasions*—was well

---

[3] According to Griffith, the advertising campaign for 'the 3d Vol. of Mr. Pope's Homer, and all his Works, wherein are several Poems never before published' began in the *Evening Post* on 16–19 Mar. (1. 65).

[4] The arrangements for both the *Works* and the *Iliad* are outlined by David Foxon in *Pope and the Book Trade*, 47–8, 52–9.

established by the early years of the century. In this period we find *Poems on Several Occasions* by, among others, Prior, Fenton, Gay, Harte, Rowe, and Parnell. Representative writers having their *Works* published in the same period include Virgil, Anacreon and Sappho, Tibullus, Dryden, Boileau, Shakespeare, Spenser, and Chaucer—all authors of classic status, and dead.[5] Why, then, did Pope choose to publish a *Works* at all? Why not wait and give both the *Iliad* and the *Works* independent clear runs? Or why not wait until a larger body of material was available, and then incorporate the translation into it?[6]

POPE, LINTOT, AND THE *WORKS*

A possible solution to the problem of the premature publication of the *Works*—one that calls into question my emphasis on the author's agency—is presented by Maynard Mack. Mack shares the view that the volume is in some ways an expression of Pope's attitudes and personality, even going so far as to say that to some extent it is a 'monument to vanity', but he sees Lintot as the prime mover in the publication. His argument springs from two factors: David Foxon's calculations that the early stages of the *Iliad* provided disappointing profits for Lintot, and the details of the contract. The contract was not signed until 28 December 1717, six months after publication. Pope had already received his side of the bargain, 120 copies of the quarto on royal paper, and, in return for receiving these books free of all charges and the additional sum of 5s. on signing, he gave Lintot the rights to print and publish all his works, not only for the first fourteen years, but also for any other term that might become Pope's to grant.[7] Noting that Pope's financial interest was limited to the 120 royal quarto copies, Mack argues:

This suggests that, however welcome the undertaking may have been to a young man's ego, it was essentially an effort by the publisher with the poet's help, and with some new important pieces, to compensate for the somewhat less than hoped return that his [Lintot's] part in the Homer, the folio without plates, had brought in.[8]

I think that the balance of evidence is against this view. Large-format editions like the 1717 *Works* involved heavy outlays and slow returns; eventually they would return a profit, but they were chiefly an investment for the future, with the possibility of steady returns from subsequent small-format editions. The last way for Lintot to make a quick compensatory profit was through a large-format edition. It seems more likely that he saw Pope as a valuable property and was prepared to go to

---

[5] I owe these illustrations to Vincent Carretta's excellent essay '"Images Reflect from Art to Art": Alexander Pope's Collected Works of 1717', in Neil Fraistat (ed.), *Poems in their Place: The Intertextuality and Order of Poetic Collections* (Chapel Hill: University of North Carolina Press, 1986), 195–233. I draw on this essay again in my discussion of the illustrations.

[6] Pope and Lintot's decision to keep the Homer translations separate has had long-running consequences. Whereas Dryden's translation of the *Aeneid* is printed with his collected poems, Pope's *Iliad* and *Odyssey* most often are not.

[7] I discuss the contract in rather more detail in Foxon, *Pope and the Book Trade*, 239–41.

[8] *Life*, 333–4. Mack presents the best general view of the place of the *Works* in Pope's career.

great lengths, and some expense, to promote and retain him. The expenditure on the illustrations, and Pope's likely role in them, is the best indication of this attitude. The extravagant frontispiece to the *Works* is a deliberate attempt to build on the celebrity of the *Iliad* translator, and Lintot was willing to base the engraving on an original painting by Pope's friend Charles Jervas, and allow Jervas to supervise Vertue's work on the engraving and keep Pope informed.[9] Similarly, although no instructions for the illustrations survive, we do have notes from Pope to the printer that show him giving detailed and explicit direction about the typography (*Correspondence*, 1. 394), and he would have had similar control over the headpieces and tailpieces. But even more important evidence that this was Pope's book rather than Lintot's is the history of Pope's interest in publishing a collection of his work and the arrangements he made to facilitate it. His earliest publications (including the *Pastorals*) had been with the Jacob Tonsons, senior and junior, and the *Essay on Criticism* had been published by William Lewis; Lintot did not own the copyright to all Pope's work and would not have been in a position to publish a collected edition without his help. The first step was taken on 5 October 1713 (before Pope had even signed the contract for the *Iliad*), when Pope was paid for his contributions to Tonson junior's *Poetical Miscellanies* of 1714. In a memorandum of that date Tonson agreed that

> notwithstanding any consideration given him for any poems of his printed by me, the said Mr Pope shall have full liberty, whenever he thinks fit, to cause the said poems to be reprinted by any other bookseller in what volume he pleases, and that neither himself nor that bookseller shall suffer any hindrance or molestation from me, that bookseller allowing me books in proportion to the number of sheets the said poems amount to in such volume and in proportion to the impression.
>
> And I likewise engage my self to Mr Pope that in case he shall desire me to reprint all his poems in a collection, I will therein cause to be reprinted all such pieces of his as have been before printed by any other bookseller in what Volume and manner he shall appoint. (*Correspondence*, 1. 191–2)

The first paragraph sets out the arrangements that were eventually invoked for the *Works* of 1717; the second paragraph sets up the possibility that Pope would ask Tonson, rather than Lintot, to publish a collected poems. Notably the initiative is left with the poet. Pope also seems to have sorted out the problem of *An Essay on Criticism*. Either he had retained the copyright when the first edition was published, or he had sold it to Lewis for a short period, or he had bought it back, for Lintot paid him for it on 17 July 1716. Thanks to Pope, the copyright problems had been solved and the arrangements for a collected *Works* were in place.

The sum paid for the *Essay on Criticism*, only £15, throws some light on those 120 copies, which were Pope's payment for the new pieces in the *Works*: preface, 'Discourse on Pastoral Poetry', 'Fable of Dryope', 'Two Choruses', 'Verses to the Memory of an Unfortunate Young Lady', 'To the Same on Leaving the Town after

---

[9] 'I am just going to Vertue to give the last hand to that Enterprize which is our Concern. He has done the King from Kneller, but so wretchedly that I can scarcely imagin how bad the Picture must be from which the Artist has performed so poorly' (*Correspondence*, 1. 310).

the Coronation', 'On a Fan', 'Epitaph', 'Epilogue to *Jane Shore*', 'Occasion'd by some Verses of his Grace the Duke of Buckingham', *Eloisa to Abelard*. Some of these are important poems, but only *Eloisa to Abelard* is of any length; and in the case of *Eloisa* Lintot was missing potential profits because he had no opportunity to publish an independent first edition. In this light, the 120 copies, which would be worth 120 guineas or more to Pope, begin to look like a generous payment, which certainly compares favourably with £15 for *An Essay on Criticism*. I suspect they were a pro-rata payment, like Tonson's. Tonson took the proportion of the folios and quartos that roughly corresponded to his copyright holdings; as it turned out, he and Lintot agreed on a neat quarter. Pope's share in the copyright is difficult to estimate because many of the new pieces are short, but if Lintot allowed him the commendatory poems as part of his share, he would have had roughly $9\frac{1}{2}$ out of $58\frac{1}{2}$ sheets, the equivalent of something over 121 copies. Of course, Tonson had to pay Lintot for the cost of printing his share of the quartos and folios, but Pope was also selling Lintot the copyright of his poems and consequently his contract specifically exempts him from paying any share of the costs.

THE *WORKS* AND THE MODERN AUTHOR

It is clear, then, that as early as the end of 1713 Pope was planning a collected works, and that he subsequently made further arrangements with the book trade to bring the project to fruition. I suspect that in doing so he was giving his individual response to a turning point in the history of authorship: a point at which the author's person, personality, and responsibility were becoming matters of public interest as never before. There are competing definitions of the change taking place in authorship in the eighteenth century, and any understanding needs to take account of multiple and overlapping strands, but the strongest contemporary account, which comes from Martha Woodmansee and Mark Rose, has many merits, not least of which is the role that it allows to Pope as the precursor of later developments. Woodmansee sees the eighteenth century as the moment of the birth of the modern author.[10] Emphasizing the relation between romantic ideology and copyright, her history develops and refines the view that the eighteenth century marks the transition from patronage to the market. Her starting point is Foucault's essay 'What Is an Author?' and in her essay 'The Genius and the Copyright' she supplies some of the socio-historical analysis Foucault declines to give, and attempts to answer some of the vital questions he poses: how did the author become individualized? what status has he been given? when did studies of authenticity and attribution begin? how was the author evaluated? when did the lives of authors become more popular than the lives of heroes?

---

[10] Martha Woodmansee, 'The Genius and the Copyright: Economic and Legal Conditions of the Emergence of the "Author"', *Eighteenth-Century Studies*, 17 (1984) 425–48; Martha Woodmansee and Peter Jaszi (eds.), *The Construction of Authorship: Textual Appropriation in Law and Literature* (Durham, NC: Duke University Press, 1994); Mark Rose, *Authors and Owners: The Invention of Copyright* (Cambridge, Mass.: Harvard University Press, 1993).

when did 'the-man-and-his-work' criticism begin?[11] Woodmansee's answer is that the author became individualized in the late eighteenth century, between 1773 and 1794. German writers, anxious to break free of a system of patronage and to earn their living by their pens, needed copyright. They established a view of the author as original and creative and, therefore, the potential owner of a work. The problem of knowledge's being free, immediately communicated on publication, was solved by Fichte's development of Young's idea of organic creation. Some précis of thought might be common and already transmitted by publication, but each mind appropriated thought individually and made it its own. Forms of thought and expression are individual and can be copyrighted. In this history, therefore, the author is a creation of romanticism and its emphasis on interiority. Before romanticism, the writer was either a craftsman or an instrument of the muse or the Holy Spirit; the writer was not original, or, as Woodmansee puts it, personally responsible.

Woodmansee's account is valuable in demonstrating the importance of the relationship between romantic ideology and copyright legislation in Germany and the United States, but it is difficult to reconcile it with British and French literary history. Rose, in his very clever accounts, tries to cope with this problem. He suggests that in Britain the legislation actually preceded the institutions of authorship associated with it, but that the legislation was then in a position to influence ideology and institutions. And he makes a distinction between the work as act (and therefore subject to penalty) and the work as aesthetic object (and therefore subject to ownership). The late eighteenth century, he argues, marks this change, and with it the development of the modern notion of the author.[12]

The views of Woodmansee and Rose have important implications for Pope. In their strongest form they amount to a claim that the agitation for copyright in Germany between 1773 and 1794, and the coincident British case of *Donaldson* v. *Becket* in 1774, mark what Foucault calls 'The coming into being of the notion of "author" ... the privileged moment of *individualization* in the history of ideas, knowledge, literature, philosophy, and the sciences' ('What Is an Author?', 197). Pope's career can be seen both as a preparation for this moment and as a troubled response to the conflicts that generated it. Some evidence for the importance of this period in the individualization of the author, and of Pope's response, falls readily to hand. Pope's career began in the same year as the introduction of the first Copyright Act, 1709, and as Rose recognizes, he was prominent in invoking the new legislation. In his preface to the *Iliad*, he identifies Homer's greatest quality as 'Invention', which is 'the very foundation of poetry', and praises the creativity and originality of his author rather than his judgement. There are also stirrings of those developments that Foucault associates with modern authorship. If, for example, Johnson's *Lives of the Poets* (1779) has a vital founding role in 'the-man-and-his-work' school of

---

[11] Michel Foucault, 'What Is an Author?', in David Lodge (ed.), *Modern Criticism and Theory* (London: Longman, 1988), 196–210, esp. 197. Lodge's inclusion of this essay in his collection has done much to popularize it.
[12] See *Authors and Owners*, 13, and 'The Author in Court: Pope v. Curll (1741)', in Woodmansee and Jaszi (eds.), *The Construction of Authorship*, 213–15.

criticism (and was also a project initiated by booksellers in order to protect their copyright with Johnson's name), Pope's interest in Giles Jacob's *Lives*, his publication of his own *Letters*, and the biographical elements of *To Arbuthnot* all anticipate this development. Again, if an important element of Boswell's *Life* is the identification of Johnson's writing, with authenticity and attribution subsequently becoming vital issues for nineteenth-century editors, Pope in 1717 was already anxious to create a canon by separating those poems he was willing to acknowledge from those he was not, and yet he was also enough of a modern editor to keep the unacknowledged poems in print. All this seems to provide evidence for the claim that Pope, anticipating the developments identified by Woodmansee and Rose, was one of the first modern authors.

However, when we turn to the *Works* in search of the vindication of such a claim, we encounter a problem. *The Works of Mr Alexander Pope* surely suggests that there is an institution of authorship already well in place. This book has an author not merely incidentally but necessarily. It illustrates very persuasively Foucault's insight into the importance of authors' names. Foucault observes that if proper names are regarded as the equivalent of definite descriptions, then the names of authors function in a special way:

> If, for example, Pierre Dupont does not have blue eyes, or was not born in Paris, or is not a doctor, the name Pierre Dupont will still always refer to the same person; such things do not modify the link of designation. The problems raised by the author's name are much more complex, however. If I discover that Shakespeare was not born in the house that we visit today, this is a modification which, obviously, will not alter the functioning of the author's name. But if we proved that Shakespeare did not write those sonnets which pass for his, that would constitute a significant change and affect the manner in which the author's name functions.[13]

This argument brushes aside the problems of reconciling the identifying descriptions of Dupont too easily—Frege would argue that if we are using Dupont's name with different senses, we are using different languages—but it presents an interesting modification of John Searle's view of proper names. Searle argues that if we had a set of identifying descriptions of Dupont from all users of the name, no single description would be analytically true of Dupont but their disjunction would be. We could drop a particular belief about Dupont, without having to abandon the name. But the same applies to Aristotle or Shakespeare; we could still use the proper name Shakespeare if Shakespeare had not written the sonnets. What I think Foucault is really pointing to here is a different 'sense' of the word 'Shakespeare' that evolves from the proper name, but is closely related to it. Shakespeare might be identified by listing the texts he had written, but in turn those texts could be identified by

---

[13] 'What Is an Author?', 201. Foucault acknowledges a debt to John Searle here (*Speech Acts*, 162–74). Searle argues that neither Mill (who argues that proper names have denotation but not connotation) nor Frege (who argues that proper names have senses) is wholly right. He argues that proper names are logically connected with characteristics of the object they refer to, but in a loose sort of way (170). Looseness of identity criteria for proper names enables them to function in the happy way they do (172).

Shakespeare's name (Shakespeare's *Macbeth*, not Verdi's), or classified by that name, or designated as a group by that name ('Brush up your Shakespeare'). It is this second sense of 'Shakespeare', as a degenerate proper name, that has a specifiable meaning through the listing of Shakespeare's works, and the meaning of this word would be changed by new information about what the writer had written. The *Works of Mr Alexander Pope*, therefore, shows this Foucauldian function of the author in exemplary form: the name of the author is used to classify texts and becomes the name of texts, and this very practice of classification generates the book. Such books celebrate authors and authorship. This is Pope's book and *Pope*.[14]

Although Pope was exceptional in publishing his *Works* as a living author, he was drawing on established patterns of publishing *Works*. A general search for '*Works of*' in the British Library Catalogue in 1997 generated 11,981 hits (and, of course, omitted all *opera*, *œuvres*, and so on); this was reduced to 631 by filtering out dates after 1717; and still further to 431 by limiting the search to London imprints. Of these, around 10 per cent could be eliminated on the ground that they contained only a subsidiary reference to *Works*. Of the remaining volumes of *Works* published up to 1717, by far the largest number are the work of clergymen (Richard Hooker outstanding, with six editions), followed by books by lawyers and physicians. I suspect there are so many volumes of collected sermons and theology not only because there was a large market for such work, but also because *Works* could be expected to provide a sound and comprehensive body of doctrine and instruction. As Father Michael Suarez has pointed out to me, theologians' *Works* are published in order to demonstrate the systematic and comprehensive nature of their understanding and thus to establish their doctrinal orthodoxy. A theologian like Karl Rahner's collected papers are called, in their English translation, *Theological Investigations* precisely in order to suggest a departure from old scholastic systems. There are relatively few literary names in the list of *Works of* in the early and mid-seventeenth century, but there is a sharp increase at the end of the seventeenth and beginning of the eighteenth (Beaumont and Fletcher, Congreve, Cotton, Defoe, Dryden, Etherege, Lee, Otway, Rochester, Rowe, Sedley, Sidney, Suckling, Wycherley). This suggests, though the whole topic merits further research, that in the early years of the eighteenth century Tonson and Lintot started to publish the works of dead literary men, building on established publishing practice in relation to clergymen, lawyers, physicians, philosophers, and a few established literary figures. There was an established readership, a public, for various forms of professional writing and evolved institutions of authorship; parallel arrangements were now being developed for literature.

The established institutions of authorship that the booksellers and Pope could draw on in 1717 are linked to issues of responsibility. In this respect there is a revealing shift in definition in Woodmansee's work. In the essay on 'The Genius and the

---

[14] Such proper names are a problem for dictionaries. I note that the old *OED* defines Shakespearian as 'Of or pertaining to, or having the characteristics of William Shakespeare (1564–1616) or his dramatic and poetical productions.' But the term surely refers more often to the products than the person.

Copyright', she says, 'In contemporary usage an author is an individual who is solely responsible—and therefore exclusively deserving of credit—for the production of a unique work' (426). In *The Construction of Authorship*, a different definition is introduced: 'an author is the creator of unique, original works such as stories, plays, and poems.' The first explanation is much less romantic than the second and makes room for a greater range of writing, as I believe it should; the second definition unnecessarily (at least if we are thinking of England in the eighteenth century) narrows authorship to the romantic and literary. Authors could be said to be responsible for their work in three ways: they produce it; they are subject to legal remedies; and they are subject to critical appraisal. Production is not necessarily the same as creation. In particular, historians, clergymen, and lawyers might regard themselves as compilers of works and yet as authors fully entitled to copyright protection.[15] Authors provide works for the circumstances in which they are consumed. In the sort of face-to-face society preceding the market, they would deliver their sermons, histories, and poems directly to their congregations, colleagues, patrons, or friends; they had no control over further dissemination of the material, but neither did anybody else. Or the playwright would provide the players with a play that they would then perform. With the coming of print culture, responsibility for dissemination becomes shared between author and publisher, and authors have to insist on their rights as primary producers. Rose cites a decree by the Council of Ten in sixteenth-century Venice to the effect that printers should not publish works without their author's consent, and cases in sixteenth-century France that asserted the author's right to control of publication of his work. When dissemination becomes broader, the issue becomes more vital, and Pope's concern over canon formation in the *Works* is an important step in his growing concern over what are sometimes called the author's moral rights.[16] The second and third sorts of responsibility, to the law and to criticism, are closely linked because both require somebody to take the blame. The second sort, legal responsibility, has been treated by Rose,[17] who sees it as preceding aesthetic responsibility. Certainly nervousness of the authorities gave an increased importance to authorship in the seventeenth century. An edict of the Long Parliament, for example, insisted that title pages bore the name of the author, and this was precisely because they wanted someone to prosecute if the book was

---

[15] The Carte papers in the Bodleian provide a fascinating glimpse of the resentments of non-literary authors. Carte was secretary for the Society for the Encouragement of Learning, and in 1735 he was engaged in pressing for better rewards for authors. At one point he lists books that he thinks have not or will not cover the cost of their production. They include *Additions to Whitby on the New Testament*; *Nichols on the Common Prayer*; *Nelson's Feasts and Festivals*; *Lock of Human Understanding*; *Turner's Surgery*; *Jacob's Continuation of Hale's Pleas of the Crown*; and *Chambers' and Littleton's Dictionary*. I do not think Carte's list represents an eccentrically antiquarian interest; this sort of book represents one of the staples of the trade. But such books are not discussed by Woodmansee and Rose, which is one reason why their account of authorship is skewed.

[16] Pope particularly insisted on the right to control publication of his letters. See Rose, 'The Author in Court: Pope v. Curll (1741)', and James McLaverty, 'The First Printing and Publication of Pope's Letters', *Library*, 6th ser. 2 (1980), 264–80.

[17] *Authors and Owners*, 13. The *Spectator*'s discussion of 'genius' coincides interestingly with the first Copyright Act.

seditious.[18] Sedition, heresy, and libel remained important issues in the early eighteenth century, with authors vulnerable to legal action for what they said. The third sort of responsibility, to criticism, need not be confined to the narrowly aesthetic. It should encompass a wide range of evaluative terms: true, false, accurate, careless, servile, arrogant, beautiful, ugly, sentimental, cold, conventional, original, persuasive, dull, and so on. Rose, arguing on the narrower base, of aesthetic responsibility, claims that prosecutions for heresy, sedition, or libel, where writers take responsibility for what they *do*, precede authors' aesthetic responsibility for what they *make*, own, and sell. Although I have reservations about the broader argument, Rose's observation seems to meet Pope's case exactly.[19] Except in the case of the theatre, early aesthetic evaluations tended to be face to face, private not public. At the time of the Copyright Act these evaluations were moving into the public sphere. The author was beginning to accept the same sort of responsibility for his work but before a larger jury, taking the same sort of public responsibility for his literary worth as Hooker did for his theological insight and orthodoxy.

At the time of Pope's publishing his *Works*, therefore, there are several different currents working together to move authorship into new areas of public responsibility: copyright, with the possibility of financial rewards and recognition through the market; Grub Street, not yet an issue for Pope; a growing interest in the lives of authors and the relation of their lives to their work (with Pope's friend Rowe being the first to include a life of Shakespeare in his edition of 1709); an interest in authentication of authorship (with Pope's enemy Bentley dramatically showing that the *Epistles of Phalaris* were not authentic); legal responsibility for one's work and the possibility of arrest for sedition or of an action for libel (Swift, was, of course, in this sort of danger at various points in his career, and Pope had played with these dangers in *The Key to the Lock*); religious and ethical responsibility for one's work (later to be a central matter for Pope, with Crousaz's critique of *An Essay on Man* and Warburton's response); and finally, the extension of aesthetic responsibility for one's work from circles of friendship and patronage ('Benè', 'Bellè', 'Pulchrè', writes Cromwell in the margin of 'Sapho to Phaon') to the measured public critique of the *Spectator* or the fury of John Dennis. The arena into which Pope's work emerged was probably broader than that encountered by works of medicine or even theology; it has been characterized by Habermas as the bourgeois public sphere, which he identifies as being present in exemplary fashion in early eighteenth-century England. The development of the journals, newspapers, and coffee houses had led to a new publicity for literature and a new awareness of public judgement. Habermas's classic account emphasizes the importance of judgement on public matters by private citizens. The functions of legitimization and criticism are carried out by a large group of enfranchised citizens, who have an acknowledged stake in their politics and culture.

---

[18] Defoe's early argument for copyright comes straight out of this situation: if you can suffer for being an author, you ought to be able to profit from it.

[19] Labour as well as objects can be sold, and aesthetic responses to literature may take it as an action. Rose says, 'The distinguishing characteristic of the modern author... is proprietorship' (*Authors and Owners*, 1), but proprietorship can be transferred whereas authorship cannot.

I suspect it is no coincidence that this is the era in which politics also becomes individualized through the person of the Prime Minister and rival political personalities. An interest in individuality, personality, and even 'stars' is generated by the absence of face-to-face contact in a situation where judgements are being made. The question 'Who is Pope?' becomes important when his readers cease to have a direct social link to him. This development of the public sphere, to which the development of copyright is entirely complementary, makes this the beginning of the period in which the author is publicly individualized through the projection of public personality. It is to this moment that Pope addresses his *Works*.

The premature publication of a volume of *Works*, then, was Pope's way of embracing and gaining some control over the new public responsibility of the author. He was part of the general development in literary publishing pioneered by Tonson and Lintot, and he was exploiting the provisions of the Copyright Act, but the crucial development that disturbed and excited him was the new era of publicity that was opening up. His response in the *Works* was a characteristic decision to construct his authorship from the outside. His response to difficulties and challenges throughout his career is to turn things inside out. He tries persistently to be on the other side of himself, outside, looking in: if possible, he writes his own notes; paints his own portraits; drafts his own biography; writes his own criticism; publishes his own libels. He attempts to shape every Pope or *Pope* the public might encounter. David Piper is quite right in saying that like Napoleon he takes the crown out of the Pope's hands and crowns himself.[20] *The Works* is less an attempt to resist and regulate the new currents than to do something exciting and risky with them himself. The volume gave Pope the opportunity to fashion a large book that was to represent the author himself. It is a mode of self-expression, but it goes beyond that to present a reception of that self-expression. In this one volume Pope was able to define a canon, publish an image of himself as man and writer, shape his relations with his reader, and guide the interpretation of individual poems through illustration and annotation. The major, if not pervasive, theme of the volume is fame. This is a classical theme, but for Pope it had a strong contemporary resonance: authorship was entering a new sphere of publicity and, from this point on, his concern was to exploit its potentialities while avoiding its dangers. The *Works* marked Pope's acceptance of a new public sort of responsibility for his writing.

THE PREFACE

Pope's clearest attempt to explain his conception of the *Works* and define the nature of his authorial responsibility is through the preface, where the pressure of the new publicity shows most clearly through those sections that he rejected in draft but were later printed by Warburton in 1751 and by Mack in a transcription of the manuscript.[21] The preface negotiates edgily between public and private. Fear of public

---

[20] *The Image of the Poet: British Poets and their Portraits* (Oxford: Clarendon Press, 1982), 57.
[21] For a valuable discussion see also Maynard Mack, 'Pope's 1717 Preface with a Transcription of the Manuscript Text', in *Collected in Himself*, 159–78.

judgement, often treated lightly or humorously, pervades it. Again and again the preface submits the volume to public judgement, protesting its unwillingness and inability to constrain it, and again and again it attempts to colour that judgement. Although Pope begins his preface by talking simply of writers and readers, he soon moves on to recognize the very public nature of the sort of writing he has in mind: 'I am inclined to think that both the writers of books, and the readers of them, are generally not a little unreasonable in their expectations. The first seem to fancy that the world must approve whatever they produce, and the latter to imagine that authors are obliged to please them at any rate' (*Twickenham*, 1, lines 1–5). The tone is informal, witty in its antitheses, wryly observant, trying to place itself above the fray. Consequently 'The world' enters the preface with a touch of mockery—only the conceited author could conceptualize his audience so grandly—but it remains to be taken seriously as the main way of referring to the public that literature serves. In the course of the essay Pope refers to 'the world' many times: 'the world has no title to demand, that the whole care and time of any particular person should be sacrificed to its entertainment' (8–10); bad writers may persist in writing because of poor advice from their friends 'and the rest of the world in general is too well bred to shock them with a truth, which generally their Booksellers are the first that inform them of' (46–9); a man's reputation depends upon 'the first steps he makes in the world' (54); the author who is led to hope 'he may please the world' is in for a tough time (61–2); 'the present spirit of the learned world is such, that to attempt to serve it (any way) one must have the constancy of a martyr' (92–4); it is to Pope's credit that 'the world has never been prepared for these Trifles [his poems]' by the means he actually does employ in the *Works* (101); 'the world is under some obligation' to him for the poems he has burned (164–5); if judgement goes against him, 'I declare I shall think the world in the right' (192–3). Sometimes Pope uses another word for the same or an allied concept. In the draft he follows the section in which booksellers have to take the responsibility of disillusioning their authors with 'This is not till the town has laughd at them' (64–6), but 'town', which would have served as Restoration shorthand for the sophisticated consumers of culture, does not find its way into the final version; it is too narrowly socially coloured. Pope does, however, use the word 'publick' at key points in the conclusion of his published essay. Moving to his close, he writes, 'The only plea I shall use for the favour of the publick, is, that I have as great a respect for it, as most authors have for themselves; and that I have sacrificed much of my own self-love for its sake, in preventing not only many mean things from seeing the light, but many which I thought tolerable' (151–6). And in his concluding sentence, which I think strikes an awkwardly defiant note, he says that the failure of the collection would teach contemporaries 'that when real merit is wanting, it avails nothing to have been encourag'd by the great, commended by the eminent, and favour'd by the publick in general' (211–15). In his use of 'the world' and 'the publick' Pope is helping to develop a concept of a wide-based critical community. His use of 'the world' seems mainly to follow *OED*'s sense 15 'The body of living persons in general; society at large, "people", the public; often with reference to its judgement or opinion', with some colouring from sense 18 'High or fashionable society', which has supporting quotations from Addison, Swift, and Lady Mary Wortley Montagu in the early 1700s. The discussion of the expectations of 'the world' may carry with it some implication that these people are pampered and unreasonable, but 'the world' must be larger than the 'learned world', not just some alternative group. It must mean something very like 'publick' in *OED*'s sense B1b 'The community as an aggregate, but not in its organized capacity; hence, the members of the community'. The first illustration of this sense comes from Robert Boyle in 1665 and the second from Steele in the *Spectator*, 258, in 1712. The *OED*'s idea of aggregation without organization supplements the implied contrast with private: what is being discussed is not family, friends, or allies, nor is it a particular political or social group; it is that heterogeneous collection of readers and critics that make up a poet's reputation.

Though the preface places great emphasis on public judgement, Pope acknowledges that there is also an inescapably private side to literature that is in conflict with this emphasis. Pope devotes an independent paragraph to the thought:

I am afraid this extreme zeal on both sides is ill-plac'd; Poetry and Criticism being by no means the universal concern of the world, but only the affair of idle men who write in their closets, and of idle men who read there. (22–5)

This is a powerful insight, compressing two thoughts. The first, and less interesting, is that the whole of the public is not interested in poetry; the 'world' consists of overlapping sub-worlds. Pope is right—the public sphere is not just one thing—but the thought remains undeveloped. The second thought, a much more vital one, is that there is something irreducibly private about modern literature as an activity of individual writing and reading. The private experience is not, of course, incompatible with a public verdict.[22] Public judgements are made by private citizens; the mingling of the two is necessary for the creation of the public sphere. But a consequence is that Pope is always conscious of the intimacy with the reader available to an author. This intimacy may be exploited in subtle ways by shaping the physical book that helps determine the reader's experience, or it may be engaged more directly by confessional writing. Pope recognized the growing interest in the private author writing in his closet, and one of the initial aims of the preface was to allow that writer to engage directly with his reader.

In the draft of the preface Pope furthered his intimacy with the reader by toying with a quasi-confessional mode. He saw this mode as a possible consequence of the decision to publish the collection, something he 'must' do. The preface falls into two parts: the first deals with the situation of writers and readers in general; the second deals with Pope's particular practices and expectations. The draft made this division much clearer, with the second half of the essay, beginning 'I believe, if any one, early in his life should contemplate the dangerous fate of authors, he would

---

[22] Nor is it necessarily 'idle'. As his preface makes clear, poetry was a strenuous and dedicated activity for Pope, and his allusions to the classics carry with them a high value for literature. There is a regrettable tendency in the preface to carry over self-deprecation into deprecation of poetry itself.

scarce be of their number on any consideration', moving rapidly on to the poet's own experience:

For my part I confess, had I seen things in this view at first, the publick had never been troubled either with my writings, or with this Apology for them.
    I am sensible how difficult it is to speak of one's self with decency: but when a man must speak of himself, the best way is to speak truth of himself, for all manner of Tricks will be discoverd. I'll therfore make this Preface a general Confession of all my Thoughts of my own poetry, resolving with the same freedome to expose myself, as it is in the power of any other to expose that.[23]

The publication of the *Works* constitutes a situation in which 'a man must speak of himself', a phrase echoed in the draft of *To Arbuthnot*, fifteen years later. There is a fear of exposure and the recognition that free self-exposure is the best way that publicity (in its modern sense) can be controlled. Shortly after this passage Pope sketches an intimate account of his development in the margin of the draft before cancelling it. This account is little enough known to merit extensive quotation, but since the passage is complex in Mack's transcription, I give Warburton's version of it.

In the first place I thank God and Nature, that I was born with a love to poetry; for nothing more conduces to fill up all the intervals of our time, or, if rightly used, to make the whole course of life entertaining: *Cantates licet usque (minus via laedet.)* 'Tis a vast happiness to possess the pleasures of the head, the only pleasures in which a man is sufficient to himself, and the only part of him which, to his satisfaction, he can employ all day long. The Muses are *amicae omnium horarum*; and, like our gay acquaintance, the best company in the world as long as one expects no real service from them. I confess there was a time when I was in love with myself, and my first productions were the children of self-love upon innocence. I had made an Epic Poem, and Panegyrics on all the Princes in Europe, and thought myself the greatest genius that ever was. I can't but regret those delightful visions of my childhood, which, like the fine colours we see when our eyes are shut, are vanished for ever. Many trials, and sad experiences have so undeceived me by degrees, that I am utterly at a loss at what rate to value myself.

This is a touching and persuasive self-examination, with an evocation of childhood vision qualified by amusement at the failure of childhood judgement. The emphasis on the pleasure of poetry informs the motto of the *Works* and finds its way into the finished preface, but the concern with 'the pleasures of the head' (as opposed to the pleasures of the body) is rarely mentioned elsewhere, either by Pope or his critics, even though it informs many of the poems of this collection. Like the confession of self-love, it offers some potential ammunition to Pope's opponents, and I suspect Pope came to feel that it offered too much intimacy and made him too vulnerable to attack. He consequently modified his plans. The object of the preface and the

---

[23] Mack, 'Pope's 1717 Preface', lines 283–91. The version I have given comes at the end, a rewriting of 119–31. In this quotation and others from Mack's transcription, I have simplified the text and removed Mack's apparatus. Readers need to consult the full apparatus for a proper understanding of the draft.

volume as a whole was still to project the person of the poet, to control the public recognition of the personality, but this did not have to involve the poet speaking of himself or self-exposure. The volume as it was completed provided an exposition of the poet through picture, annotation, and tone, without resorting to confessional mode.

At various points in the preface Pope toys with the idea that the judgement of 'the world' does not have any great value, but the underlying assumption is that it is vital because fame depends on it. He devotes a separate paragraph to a wry confession that the *Works* is a bid for fame: 'In this office of collecting my pieces, I am altogether uncertain, whether to look upon my self as a man building a monument, or burying the dead?' (175–7). Either his works will perish to be heard of no more, or they will provide him with a continuing representation; he will have achieved fame. Fame is inescapably bound up with the verdict of the world; it comes to genius and is the only reliable endorsement of genius. 'What we call a Genius, is hard to be distinguished by a man himself' and therefore a writer is dependent on the verdict of others (33–7). Under the influence of Addison in using these key terms, 'genius' and 'fame', Pope is anxious to disclaim any vanity or excessive regard for fame, even though at a deeper level it seems to be what publication is about.[24] In a curious passage he tells us, 'I could wish people would believe ... that I have been less concern'd about Fame than I durst declare till this occasion', since his works have now been published and 'had their fate already' (94–9). The declaration would be more telling if the 'less' had something interesting to attach itself to. The extent of Pope's ambition is best expressed in the passages on the Ancients, where recognition of their achievement does not prevent him from suggesting comparison of their work with his own. In a telling passage Pope asks the reader to reflect:

that the Ancients (to say the least of them) had as much Genius as we; and that to take more pains, and employ more time, cannot fail to produce more complete pieces. They constantly apply'd themselves not only to that art, but to that single branch of an art, to which their talent was most powerfully bent; and it was the business of their lives to correct and finish their works for posterity. If we can pretend to have used the same industry, let us expect the same immortality ... (116–23)

It is pretty clear that Pope could pretend to such industry and consequently could expect immortality. That is why the preface closes with a complete submission to the judgement of the public. The *Works* is a bid for immortality by building a monument to represent a poet of Genius and mark his fame, but it was to be a sprightly, individual monument, combining the public and the personal.

---

[24] Addison's essay on Pope's *Essay on Criticism* begins with a discussion of fame and fame is his topic also for the following three papers, Nos. 253, 255–7; 20, 22, 24, 25 Dec. 1711 (2. 481–99). His most important essay on Genius is in No. 160, 3 Sept. 1711 (2. 126–9), but both are recurrent topics. Addison is emphatic that the man who is eager for fame is unlikely to achieve it ('It is *Sallust*'s Remark upon *Cato*, that the less he coveted Glory the more he acquired it'), and he departs from classical precedent by regarding desire of fame as an imperfection in character.

PICTURING THE AUTHOR

The monument Pope built reveals the hand of its designer at every point. Martha Woodmansee uses an interesting quotation to illustrate her argument that in the 1750s in Germany the author was 'just one of the numerous craftsmen involved in the production of a book—not superior to, but on a par with other craftsmen'. She supports her argument with a quotation from the definition of 'book' in *Allgemeines oeconomisches Lexicon* (1753).

either numerous sheets of white paper that have been stitched together in such a way that they can be filled with writing; or, a highly useful and convenient instrument constructed of printed sheets variously bound in cardboard, paper, vellum, leather, etc. for presenting the truth to another in such a way that it can be conveniently read and recognized. Many people work on this ware before it is complete and becomes an actual book in this sense. The scholar and the writer, the papermaker, the type founder, the typesetter and the printer, the proofreader, the publisher, the book binder, sometimes even the gilder and the brass-worker, etc. Thus many mouths are fed by this branch of manufacture. (425)

The quotation is hardly indicative of the status of the author in Germany in 1753, as Woodmansee takes it to be; it looks at authorship from a particular economic angle. A modern account of the book as a manufacture or commodity would take the same line. Yet, though I see no unfavourable consequences for the author's status in a definition like this, it is almost as though Pope had read the passage and responded quite differently. He guards against losing status by controlling production as much as possible. Hence the precision in the contract about the nature of the paper to be used in the edition, and his notes to Bowyer giving specific instructions about typography which I quoted in the Introduction:

I desire, for fear of mistakes, that you will cause the space for the initial letter to the Dedication to the Rape of the Lock to be made of the size of those in Trapp's Prælectiones. Only a small ornament at the top of that leaf, not so large as four lines breadth. The rest as I told you before.

I hope they will not neglect to add at the bottom of the page in the Essay on Criticism, where are the lines 'Such was the Muse whose rules,' &c., a note thus: 'Essay on Poetry, by the present Duke of Buckingham,' and to print the line 'Nature's chief masterpiece' in italic. Be pleased also to let the second verse of the Rape of the Lock be thus,

> What mighty contests rise from trivial things. (*Correspondence*, 1. 394)

There are still many people engaged on the book, many mouths fed, but one person is more important than the others; their activity is being regulated by the author. It is as though, anticipating the cinema, Pope had invented the *auteur* approach to book production. In this passage in his correspondence his control of the book extends beyond the poem, which he is revising, to the notes, the founts, the illustrations, and the layout. Pope embraces the role of public author and controls as many aspects of projection and reception as possible.

The *Works* is an exceptionally well-integrated volume, with the personality of the author at its centre. The various elements combine to emphasize the gentility of the

FIG. 3. Frontispiece to *Works*, 1717 (Bodl. Vet. A4 d. 140; 375 × 275 mm).

author, the classical range and dignity of his work, and—in spite of his youth—his entitlement to fame. The ordering of the items is traditional. A central role in the projection of the author is played by the frontispiece (Fig. 3). The reader opening the volume and unfolding the large engraving (measuring 37.5 × 27.5 cm) finds, as Vincent Carretta and others have argued, the portrait of a young gentleman: elegant, bewigged, but at ease, young, serious, good-looking, the long curling wig falling down over his shoulders and hiding his hump, if Pope had a hump to hide. The frontispiece is a refutation of attacks on Pope's person by his critics. How could John Dennis have suggested he looked like a monkey? It is also a subtle mingling of the personal and the monumental. The original portrait by Jervas, now in the Bodleian Library, conveys greater intimacy: its colours, especially the pink of the cheeks and lips, create a delicate, unsevere, even vulnerable Pope. Not so Vertue's engraving, which firms up the lower part of the face, making the lower lip protrude slightly and the chin a little more aggressive. The nose is given a slight hook, and the whole expression becomes a touch haughty.[25] The engraving is surrounded by an oval frame representing a wreath of oak leaves, the award given to the victor of the Capitoline poetry contest.[26] It also follows the conventions of frontispieces by providing a supporting tablet, as though the picture belonged to some sort of monument. When Pope designed Buckingham's *Works* six years later, he used his funerary monument as a frontispiece. What is most striking about the frontispiece, however, is its size—so big that it has to be folded twice in order to fit into the quarto. It is as though the representation of the author is grander than the book itself. There is some truth in that idea, for the engraving was originally independent of the book. Wimsatt points out that it was sold independently and existed before the collection. It was advertised in the *Daily Courant*, on Saturday, 20 August 1715: 'On Tuesday next will be Published, A Print of Mr. Alexander Pope, done from the Original painting of Mr Jervasi, by Mr. Vertue.'[27] It was a companion to the Homer translation, and it shows Pope already in the limelight. The author/translator, far from being anonymous, is a figure of interest in his own right; he has become a celebrity. Of course, the visual celebration of authors is not new. David Piper sketches a long history, with evidence of statues by as early as 500 BC. But this continuity disguises an important difference between the classical and the modern. Statues and paintings in the ancient world were not required to resemble their authors: the representation of an author might amount to nothing more than a kind of personification of the text. By the time of Virgil, depiction of authors was common, but we have no idea what Virgil looked like. The appearance of the man was not important, though the representation of the poet was. But the engraving in the *Works* resembles the Mr Pope you might meet in the coffee house or at the play. The frontispiece stands on the border of public and private. Although the painting is by a friend, it is a public not an intimate portrait, and although Pope's fame as a writer generated the engraving, it does not represent him as a writer. Pope wishes to create a connection between the man and the work (the man is honoured by the work), but not to allow the former to

---

[25] I am grateful to Kelsey Thornton for suggesting to me that some of these effects are simply the consequences of line-drawing rather than painting, but I suspect Vertue and Jervas could have avoided these consequences if they had wished.

[26] See the entry under oak in Michael Ferber (ed.), *A Dictionary of Literary Symbols* (Cambridge: Cambridge University Press, 1999). Laurel would be inappropriate because Pope was not, and could not be, the poet laureate.

[27] Wimsatt, *The Portraits of Alexander Pope*, 18. Wimsatt notes further engravings.

be subsumed in the latter. Habermas's commentary on a passage from Goethe's *Wilhelm Meister* in which Wilhelm rejects bourgeois life is relevant here. A nobleman can make his authority immediately present; he embodies it in his movement, voice, and manner. The burgher is unable to do so because his personal qualities have no appropriate public expression. 'The nobleman was what he represented; the bourgeois, what he produced.'[28] Pope's depiction in the frontispiece, in keeping with his stance in his preface, makes it clear that though his work gains its identity from him, he is not to be identified with what he has produced; it is part only of his public identity. Pope was not to be an anonymous craftsman, alienated from his work, nor was he to be its creation. In the course of his career this assertion of personal power over the work, while deriving fame and fortune from it, grew more difficult, with Pope drawing more and more on his intimate life as a way of defending the work and its author from criticism. But at this point, he may have felt he had solved the problems of modern authorship by presenting the poetry as one of the forms of expression of the public man.

The representation of authors is a motif picked up in the illustrative headpieces to the books, and particularly by the first page of *An Essay in Criticism* (Fig. 4). In the headpiece to this poem we find two busts on pillars, Homer and Virgil. Before them, directed by Fame with her trumpet, is a young man, kneeling and doing obeisance to them. The young man looks very much like a classicized version of the idealized Pope of the frontispiece. He has the long curled hair of Pope in his wig and a similar profile. Fame is instructing him to respect the ancients, even as Pope advises readers of his poem:

> Be *Homer*'s works your study, and delight,
> Read them by day, and meditate by night,
> Thence form your judgment, thence your notions bring,
> And trace the Muses upward to their spring.
> Still with itself compar'd, his text peruse;
> And let your comment be the *Mantuan* Muse. (124–9)[29]

The picture, Carretta points out, is modelled on one in an edition of Boileau's *Works* in which Apollo points the reader to the busts of Homer and Virgil. This young man (and the Pope of the frontispiece) does resemble Phoebus as he is described in Pope's poem in the contemporary *Poems on Several Occasions*:

> Great God of art, whose locks unshaven grow,
> In graceful curls, and o'er thy shoulders flow . . .[30]

---

[28] *The Structural Transformation of the Public Sphere: An Inquiry into a Category of Bourgeois Society*, trans. Thomas Burger, with Frederick Lawrence (Cambridge: Polity Press, 1989), 13. Wilhelm intends to abandon bourgeois activity for the stage, a hollowed-out form of public representation that imitates aristocratic qualities. The life of the writer could have similar qualities of representation without embodiment.

[29] Quotations are taken from the *Works* (Griffith 80), with line numbers from the Twickenham edition.

[30] For the poem and for arguments in support of the attribution to Pope, see Ault (ed.), *Pope's Own Miscellany*, pp. xlv–xlvi and 49–50.

AN
ESSAY
ON
CRITICISM.

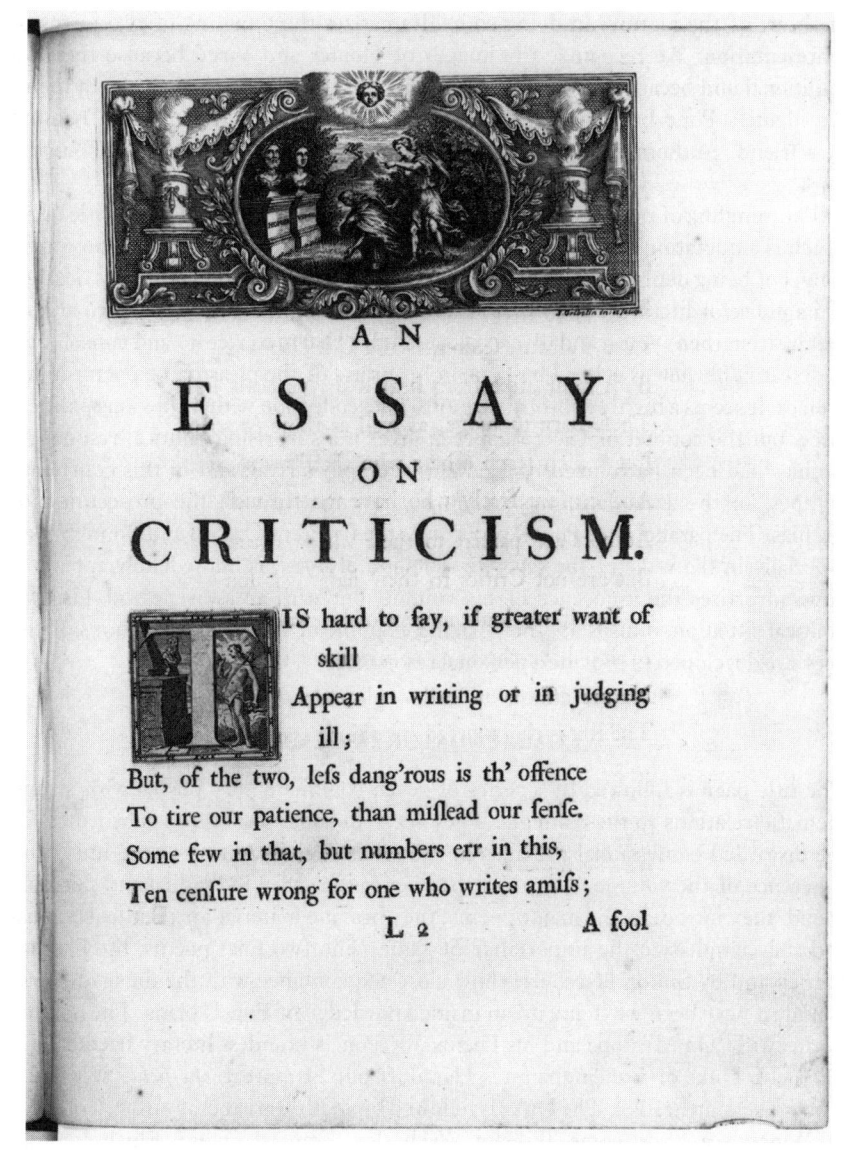

'TIS hard to say, if greater want of skill
 Appear in writing or in judging ill;
But, of the two, less dang'rous is th' offence
To tire our patience, than mislead our sense.
Some few in that, but numbers err in this,
Ten censure wrong for one who writes amiss;

L 2   A fool

FIG. 4. First page of *An Essay on Criticism*, *Works*, 1717 (Bodl. Vet. A4 d. 140; 290 × 218 mm).

But the figure can hardly be Phoebus/Apollo because he would have no need of instruction from Fame. I suspect the figure is Pope, and resembles Phoebus/Apollo in appearance as Pope resembles him in art. The image appropriate to the poem might show Pope pointing the reader to the busts, but he chooses instead a humbler image of himself that nevertheless places his attitudes and activity at the heart of the picture. In doing so he dramatizes the mingling of two forms of representation. We recognize the images of Homer and Virgil because they are traditional and because the busts have 'Homer' and 'Virgil' written beneath them. We identify Pope because we recognize a link with a portrait of him painted by a friend. Authorship is being reinterpreted to include the man as well as the work.

The mingling of personal and impersonal continues in the motto on the title page, which is a quotation from Cicero's defence of the Greek poet Archias, who was in danger of being deprived of his Roman citizenship. The quotation is famous because of its praise of literature even when literature lacks political utility; Cicero argues that it strengthens youth and diverts old age, adds charm to victory and consolation to defeat. The view is echoed by Pope in his praise of the pleasure of poetry in his preface. It seems a highly appropriate motto for a collection with no obvious political force, but the context of Cicero's speech gives it a surprising political resonance. Archias had been patronized by Lucullus, Pompey's rival, and in this court case Pompey, in the ascendant, was taking his revenge through the prosecution of Archias. The parallel with Pope's position as the friend of Oxford and Bolingbroke, especially in the wake of the Jacobite rebellion of 1715, is immediately apparent. Pope advertises the innocence of his volume, but with an awareness of his own political situation. This is another striking example of how the conventions of the book are developed to provide a personal resonance.

COMMENDATORY POEMS

The title page is followed by a series of seven commendatory poems with strong thematic relations to the contents. They serve in some respects as substitutes for the discarded confessional passages in the preface, contributing to the unity and coherence of the volume. They link the poet to the social individual and personal friend; they introduce the major poems; they provide material for the illustrations; and they emphasize the importance of fame. The two final poems, by Thomas Parnell and by Simon Harcourt, show close acquaintance with the finished *Works* and must have been written with an inside knowledge of Pope's plans. The volume begins with 'On Mr Pope and his Poems', by Pope's grandest literary friend, John Sheffield, Duke of Buckingham (as Harcourt puts it, 'Great *Sheffield*'s Muse the long procession heads'). The Duke is followed by another aristocrat, Anne, Countess of Winchilsea; then come William Wycherley, Francis Knapp, Elijah Fenton, Parnell, and Harcourt. The only figure from outside Pope's circle is Francis Knapp, of Killala in County Mayo, and it is notable that his poem, written in praise of *Windsor-Forest*, is less thematically significant than the rest, though it follows on neatly from Wycherley's praise of the *Pastorals*. The poems commend Pope's moral or personal qualities as well as his artistic ones; this volume is the collected poems of a gentleman. Buckingham praises him as 'A good Companion, and as firm a Friend'; Wycherley 'Pays what to friendship and desert is due'; Parnell sings to 'Shew my own love'; while Harcourt says that Pope's place in the temple of Fame is with the '*Good* and *Just*'. Only Winchilsea, more reserved than the others (which is probably why she was dropped in 1736), says she is not good at panegyric and gives as advice what others give as praise, telling him to 'gain applauses by desert'. Pope had originally intended to seal this emphasis on his personal and social status by closing the volume with a poem which responded to Buckingham's, binding the volume in a circle of friendship:

> MUSE, 'tis enough: at length thy labour ends,
> And thou shalt live; for *Buckingham* commends.
> Let crowds of criticks now my verse assail,
> Let *D----s* write, and nameless numbers rail:
> This more than pays whole years of thankless pain;
> Time, health, and fortune, are not lost in vain.
> *Sheffield* approves, consenting *Phoebus* bends,
> And I and Malice from this hour are friends.

Unfortunately the late arrival of *Eloisa to Abelard* made that poem the last item in the collection and thereby spoilt the plan.

Winchilsea's more abstract and analytic poem identifies purposes in the commendatory poems other than friendship. She shrewdly focuses on the relations of public and private, and says that the poet's aim is always in some way public:

> THE Muse, of ev'ry heav'nly gift allow'd
> To be the chief, is publick, tho' not proud.
> Widely extensive is the Poet's aim,
> And, in each verse, he draws a bill on fame.
> For none have writ (whatever they pretend)
> Singly to raise a Patron or a Friend;
> But whatsoe'er the theme or object be,
> Some commendations to themselves foresee.

These lines are sharply opposed to Pope's on Buckingham, which take one man's praise as an adequate reward for years of endeavour, but they chime with the preface's emphasis on the poet's appeal to the public. Winchilsea seems to address Pope as though he is still a novice, but the others are entirely confident of Pope's stature as a poet approved by Apollo and worthy of a place among the Ancients; his youth merely makes his achievement more remarkable. He is worthy, therefore, of an edition such as this. Buckingham calls him a genius and says that Apollo commands other poets to praise him; Wycherley compares him to Virgil for success in pastoral, predicting that Pope's Muse 'shall, like his, soon take a higher flight'; Knapp begins, 'Hail, sacred Bard' and goes on to compare him with Titian for painting. Fenton devotes his whole poem, 'In Imitation of a *Greek* Epigram on Homer', to an equivalence between Phoebus, Homer, and Pope. Phoebus tells 'the nine harmonious maids' that a wandering, blind Greek poet heard him singing the song of Troy and repeated it so that it was taken for his own. But now Pope's translation will be taken for the work of Phoebus himself:

> Fame, I foresee, will make reprizals there [in the West],
> And the Translator's Palm to me transfer.
> With less regret my claim I now decline,
> The World will think his *English Iliad* mine.

Even Fenton's praise seems tame in the light of the poem by Parnell that follows and compares Pope with several of his classical predecessors. Horace would acclaim Pope as critic; Ovid would envy the Lodona episode from *Windsor-Forest*; Callimachus' praise of Berenice's hair is overshadowed by Pope's praise of Belinda's; Virgil, finding similarity between his own measures and Pope's, would place him next to him in the temple of Fame; while

> In *English* lays, and all sublimely great,
> Thy *Homer* warms with all his ancient heat . . .

Harcourt echoes Parnell's emphasis on the range of Pope's achievement, before placing his friend with the good and just.

> Say, wondrous youth, what Column wilt thou chuse,
> What laurell'd Arch for thy triumphant Muse?
> Tho' each great Ancient court thee to his shrine,
> Tho' ev'ry Laurel thro the dome be thine;
> (From the proud Epic, down to those that shade
> The gentler brow of the soft *Lesbian* maid) . . .

Harcourt's poem, which reviews the volume as a whole while giving centrality to the *Temple of Fame*, works with the other commendatory poems to give an overview of the volume, while touching on key elements. What Pope cannot say directly for himself in the preface, his friends can say for him: he is a youthful genius approved by the gods; he has achieved fame and is ready to rank with the Ancients; he is a good man and a valued friend. The ordering of the volume revealed by the table of contents that follows confirms some of these emphases.

THE CONTENTS

Pope says in a letter to Swift of 16 February 1733, anticipating the second volume of *Works*, that his 'works will in one respect be like the works of Nature, much more to be liked and understood when consider'd in the relation they bear with each other, than when ignorantly look'd upon one by one' (*Correspondence*, 3. 348). The point is a good one and Pope illustrates it by suggesting a different order for his poems of the 1730s, but the analogy with nature conceals a vital issue—selection. The *Works* of 1717 attempts to fix a canon. Although Pope's preface is emphatic in its submission to public judgement, in one respect it hopes to constrain it: 'For what I have publish'd, I can only hope to be pardon'd; but for what I have burn'd, I deserve to be prais'd. On this account the world is under some obligation to me, and owes me the justice in return, to look upon no verses as mine that are not inserted in this collection' (162–6). In the first draft he went on to give this as the main reason for the collection: 'That was chiefly my view in making this collection tht nothing might [*Here Pope leaves a space to be filled in*] wch is not contained in it. For in the prest [present] Liberty of the press, I am forc'd to appear as bad as I am not to be thought worse' (269-74). I take it the thought here is that Pope must submit all his poems, make a frank disclosure even if some are bad, because the alternative is to have worse poems attributed to him. This is an important claim to his moral rights as author: the true version of Pope has to be based on a true version of his writings. Unfortunately, the *Works* is not a true version. Norman Ault has argued very persuasively that, only six weeks after the *Works*, Pope published another volume called, significantly, *Poems on Several Occasions*, which contained thirty-seven unacknowledged poems. Many of these were love poems that he had written in his youth, some were mildly improper, one was 'To Mr Pope on his Translation of Homer'. Pope had not been able to burn them; the self-love he writes about in the preface was too strong; he had to give them their chance of finding their way in the world. The two volumes are like twins: one the respectable side of the tapestry of Pope's writing, the other the underside of apprentice work. At one level, the *Works* preface is simply untruthful in its claims to openness, modesty, and self-discipline, but the two volumes are easily seen as two different forms of publication. One, exemplified by *Poems on Several Occasions*, simply utters the poems in public; the other, exemplified by the *Works*, is an act of self-definition. It would be difficult to deny Pope his right to publish a version of himself, but unfortunately the *Works* presents itself as the only version.

In his letter of 1733 Pope recommends a departure from chronological order. The reordering of the poems in 1717 is not as radical as it was to be in 1735, but nevertheless their disposition highlights some emphases. There are seven sections, though the last two, evidently serving to collect a number of shorter poems together, are listed in larger type: 'Pastorals'; 'Windsor-Forest'; 'Essay on Criticism'; 'Rape of the Lock'; 'Temple of Fame'; 'Translations'; 'Miscellanies'. The first point that emerges is the sheer range of the young man's achievement. The necessary context of this volume is, of course, the *Iliad* translation, in which Pope showed his command of epic style and presented Greek literature to an English readership. The *Works*, as the congratulatory poems insist, show complementary success in other genres. Here Winchilsea's poem comes into its own with its advice to encompass many kinds of poetry: to inform the head, dissolve the heart, inflame the soldier, elate the young, warm the sage, allure women, and describe the forest. We see in the contents poems that cover this range. The order is roughly chronological, with Pope providing dating for each poem and insisting on the poet's youth at the time of writing. The collection begins, like Virgil's career, with *Pastorals* (1704, Pope says), including a reconciliation of classical and biblical in 'Messiah' (published 1712), and follows that with a poem that could be represented (following Pope's lead) as Georgic, *Windsor-Forest* (written 1704 and 1713, Pope). The *Pastorals* have their own poem of congratulations, by Wycherley, in which the comparison with Virgil is duly made, and so does *Windsor-Forest*, even if Knapp stops short of the appropriate classical parallels. There follows a Horatian essay in *An Essay on Criticism* (1709,

Pope) ('*Horace* himself wou'd own thou dost excell | In candid arts to play the Critic well', Parnell) and a mock-epic in *The Rape of the Lock* (1712) that was thought to excel the work of Boileau and Callimachus. The next poem in the table reflects on those that have gone before. The *Temple of Fame* (1711, Pope) appears in its chronological place of publication (1715), but it is a much less predictable item in the sequence than its predecessors. Like 'Messiah' it might have been placed in the section headed Translations, and it was in fact relegated to that section in 1736. Its role is thematic: it comes at the end of the major individual poems and marks, as Harcourt suggests, Pope's having achieved Fame and the right to a place in the temple he describes. The two major thematic preoccupations of the *Works* are fame and love, and of these fame is much the more important. The key position occupied by this poem is one indication; the role of fame in the commendatory poems is another, and the illustrations are yet another. The two concluding groups, Translations and Miscellanies, are hold-all sections, though in this respect Pope seems to have been of two minds, affording some poems more dignity in their place in the volume than he has in the table.

THE HEADPIECES

The *Works* is very carefully divided up. Each major poem has its own section title (in the style of a half-title), usually with a dating, and several poems have introductory notes of some sort on the verso of the section title. The first page of a major poem, following the section title, has an engraved ornamental headpiece and initial. Although some of these ornaments are borrowed from Joseph Trapp's *Praelectiones Poeticae* (his lectures as Oxford's Professor of Poetry) or the *Iliad*, their choice is always intelligent and closely related to the themes of the volume.[31] The pattern of illustration has obviously been thought through so that connections are established between poems, and it seems to have been planned in close relation to the commendatory poems; either Pope's friends were shown the volume in advance or Pope was influenced by what they said. I suspect a mixture of the two. The theme of Fame, and Fame in relation to classical tradition and predecessors, is persistent. The title page has a small ornament, taken from Trapp, but entirely appropriate. It has a lyre and laurel leaves, representing poetry, and crossed trumpets representing Fame. The lyre and trumpet also figure in subsequent illustrations, either in the central oval illustration or in the rectangular frame in which it is set.

The two most interesting uses of these motifs are in the headpiece to *An Essay on Criticism*, which has already been discussed as a possible portrait of Pope himself, and in the headpiece to the preface. In the latter the lyre and trumpet figure either side of the centre oval, which has a scene that seems to derive its inspiration from the poems by Buckingham and Fenton. It represents Apollo and the nine Muses, Apollo singing

---

[31] There are also tailpieces, some of them engraved, others printer's ornaments. The tailpieces are often large and sometimes vaguely thematic. They are, in my opinion, a blot on the book, and I have not taken space to discuss them here.

and the Muses engaged in playing instruments, reading, or writing. Buckingham says that he writes on the instruction of Apollo. To a genius like Pope's

> Poets are bound a loud applause to pay;
> *Apollo* bids it, and they must obey.

Apollo's approval is central to Fenton's poem too and the headpiece seems to reflect his opening:

> When *Phoebus*, and the nine harmonious maids,
> Of old assembled in the *Thespian* shades . . .

If this reading is correct, I think the implication is that Apollo and the Muses are already praising the writer of the preface. The general evaluation implicit in the book is made explicit by the headpiece, echoing the preceding poems. Pope may be diffident about his achievement; the book is not. The same headpiece is used, neutrally, at the beginning of the 'Ode for Musick'.

Apollo features only once more in the illustrations (in the 'Discourse on Pastoral' and *Essay on Criticism*, he appears in an initial, standing naked before a bust that I take to be of Theocritus), but Fame has a larger role to play. In the headpiece to *An Essay on Criticism* she introduces the poet to busts of Homer and Virgil, while Pegasus in the background suggests the importance of divine inspiration. The *Temple of Fame* shows her at the centre of her temple, on a pedestal under a baldachin. She is blowing her trumpet, and so are minor deities around her. In the surround of this representation are instruments of battle on one side, and the lyre and instruments of peace on the other. The engraved initial on this page represents her again with her trumpet, but with an inappropriate *Iliad* ship in the background.

Several poems have illustrations that I take to be quite straightforwardly illustrative. 'Spring' in the *Pastorals* has the three shepherds, Damon, Daphnis, and Strephon, and the bowl and the lamb for which they compete. 'Messiah' has an initial with a boy leading what must be a tiger, but looks like a leopard, by a flowery chain. *Windsor-Forest* has Father Thames and three nymphs, with Windsor Castle in the background. 'January and May' has a scene in which May climbs on January's back in order to join her lover in the tree. 'Sapho to Phaon' has an illustration of the deserted Sapho with putti, one of whom holds the letter (that is the poem) to the lover sailing away in the background. Two of the headpieces link up to lines on the *Rape of the Lock* in Parnell's commendatory poem:

> But know, ye fair, a point conceal'd with art,
> The Sylphs and Gnomes are but a woman's heart.
> The Graces stand in sight; a Satyr-train,
> Peeps o'er their head, and laughs behind the scene.

The comment is interesting for its stress on satiric purpose and on woman's nature being the satiric object. It is difficult not to see Pope implicated in Parnell's image

THE
RAPE of the LOCK.

CANTO I.

WHAT dire Offence from am'rous causes springs,
What mighty contests rise from trivial things,
I sing——This verse to *C---*, Muse! is due:
This, ev'n *Belinda* may vouchsafe to view:
Slight is the subject, but not so the praise,
If She inspire, and He approve my lays.

R   Say

FIG. 5. First page of *The Rape of the Lock*, *Works*, 1717 (Bodl. Vet. A4 d. 140; 290 × 218 mm).

—as one of the satyr train laughing behind the scene—and, therefore, as implicitly represented by satyrs in two of the headpieces. The headpiece to the *Rape of the Lock* (Fig. 5) shows the sylphs disporting outside Hampton Court, while either side of the central oval half-naked satyrs look on holding masks (comedy and tragedy) in either hand. At the top of the oval is Mercury's winged helmet and at the bottom is a face that is like a cross between a dog and a human. The headpiece is on one level a formal declaration that this is a satire, but it inevitably also suggests that the relation of the poet and his readers to the characters is one of derision, and even derision with some force of sexual energy. The headpiece to the 'Wife of Bath's Prologue' displays some of the same motifs and hints at prurient ridicule. In the centre Venus looks at herself in a mirror held by Cupid, as in the famous paintings by Titian and Velázquez (a motif found in both the 1717 headpiece and the 1714 frontispiece); a quiver lies on the ground beside them, suggesting the dangerous power of love. But to the right of a screen of trees four satyrs creep up on the scene: they seem to be the train of satyrs in Parnell's poem who peep and laugh. Did Parnell misremember what he had seen, or did Pope and Gribelin transfer the image to a different poem? The Wife's prologue, of course, contains no such female vulnerability within its account; the only vulnerability is to the ridicule of the poem.

Pope chooses to dignify some poems with section titles and illustrations, even though the organization in the table suggests that they do not merit it. I suspect the illustrations are a better indication of Pope's fondness for these poems than the summary judgement conveyed in the table is. The Chaucer translations, 'January and May' and 'The Wife of Bath her Prologue', both have section titles and engraved headpieces, just like the major poems, and so does 'Sapho to Phaon'. Chaucer gains his place in the collection as an author worthy of imitation partly on the recommendation of Dryden. Pope says in his advertisement to the 1736 edition of his early pieces that 'Mr. *Dryden's Fables* came out about that time, which occasion'd the Translations from *Chaucer*'. In the preface to his *Fables*, Dryden says that as Chaucer 'is the Father of *English* Poetry, so I hold him in the same Degree of Veneration as the *Grecians* held *Homer*, or the *Romans Virgil*'.[32] Chaucer takes his place in the *Works*, therefore, as another great writer against whom Pope can be measured. But in the light of Dryden's further comments, the choice of poems for imitation looks like a daring (even naughty) bid for popularity:

> I have confined my choice to such Tales of *Chaucer* as savour nothing of immodesty. If I desire'd more to please than to instruct, the *Reve*, the *Miller*, the *Shipman*, the *Merchant*, the *Sumner*, and above all, the *Wife of Bathe*, in the Prologue to her Tale, would have procur'd me as many Friends and Readers, as there are Beaux and ladies of Pleasure in the Town. (Spurgeon, *Chaucer Criticism*, 279)

The choice of two of these popular and immodest poems for translation (the 'Wife of Bath's Prologue' and the 'Merchant's Tale') is highlighted by the illustrations, which convey by narrative depiction or symbol the sexual interest of the pieces. Three other poems are dignified by section titles, even though they do not have their own headpieces: 'The First Book of Statius', 'Part of the Thirteenth Book of Homer's Odysses', and *Eloisa to Abelard*. *Eloisa to Abelard* would doubtless have had

---

[32] Caroline F. E. Spurgeon, *Five Hundred Years of Chaucer Criticism and Allusion, 1357–1900*, 3 vols. (Cambridge: Cambridge University Press, 1925), 276. This remains an absorbing collection.

a headpiece if it had not arrived so late, and its prefatory note does have a large illustration, an engraving of Helen and Paris from the *Iliad*. Once again the illustration gives emphasis to transgressive sexuality.

THE TEMPLE OF FAME

Two poems are marked out for special treatment in these *Works*: the first is the *Temple of Fame* because of its position in the collection, closing the section of major poems, and its thematic significance; the second is 'Messiah', which is the only poem given the complex apparatus an elaborate *Works* edition makes possible, anticipating developments later in Pope's career. The *Temple of Fame* allows Pope to meditate on human greatness, and, in a characteristic manœuvre placing the poet at the centre of the poem, to dramatize his own conception of fame. Fame is subject to erosion by Time, and the original decisions of the goddess are not just. However, writers are at the heart of her temple:

> But in the centre of the hallow'd quire,
> Six pompous columns o'er the rest aspire;
> Around the Shrine it self of Fame they stand,
> Hold the chief honours, and the fane command. (178–81)

The pillars are dedicated to Homer, Virgil, Pindar, Horace, Aristotle, and Tully. Pope is not invited to join these poets as he is in the poems of Parnell and Harcourt; nor is he in the band of the Good and Just who ask Fame for compensation for their lives. But at the end of the poem he is approached by a figure who asks, 'Art thou, fond Youth, a Candidate for Praise?' Unlike Chaucer's dreamer, Pope's narrator is allowed a reply, and it is a persuasively honest one:

> 'Tis true, said I, not void of hopes I came,
> For who so fond as youthful bards of fame? (501–2)

The resulting speech anticipates a theme of Pope's later poetry by expressing the desire for an honest fame or none, but more significant is the representation of fame in terms of social success:

> How vain that second life in others breath,
> Th' estate which wits inherit after death!
> Ease, health, and life, for this we must resign,
> (Unsure the tenour [tenure], but how vast the fine!)
> The great man's curse without the gains endure,
> Be envy'd, wretched, and be flatter'd, poor;
> All luckless wits our enemies profest,
> And all successful, jealous Friends are [at] best. (505–12)

For the first time here, I think, Pope compares property in literature to property in land, making a sharp contrast between material gains, with the passing on of estates, and the breath of praise that will remain to poets after death. Success in other spheres is depicted as substantial, success in wit as fragile, and endangered by civil strife. And yet Pope is willing to associate himself with these Wits, not only through the stance of the 'I' in the poem, but also through the use of the first person plural, 'for this we must resign', 'all luckless wits our enemies'. When he revised the poem in 1736, he changed to the third person, 'for this they must resign', 'all luckless wits their enemies'. Perhaps it was because the forecast of 1717 had proved so accurate that he felt it necessary to distance the lines from the specificity of his own experience. What the change does emphasize is Pope's acute consciousness of the measure of self-dramatization in this central poem.

'MESSIAH'

'Messiah' is a striking example of the syncretic cast of Pope's mind and the use of typography to give it expression. The poem is a reconciliation of Virgil's Pollio (Eclogue 4) with Isaiah. Pope uses the language of the Bible as the means of creating an English imitation. Virgil and Isaiah are, as it were, fused into Pope's poem, but it is vital to his conception that the reader remains conscious of the independence of the originals. The *Pastorals* section of the *Works* begins with an analogous, if much less dramatic, attempt to reconcile diverse traditions. In the 'Discourse on Pastoral Poetry' Fontenelle and Rapin are reconciled, and what seems on the surface to be a critique driven by Fontenelle turns out to be an expression of Rapin's ideas. In this case Pope uses a few footnotes to point the reader to an understanding of the harmonization that is taking place, but the format of the *Works* gave him the opportunity to experiment with the potentialities of annotation to a much greater extent in 'Messiah', where he creates a work that typographically insists on its relation to two other texts. In this respect the printing of the poem lays the ground for later declarations of intertextuality—the parallel passages from Chaucer, the imitations of Horace, and the *Dunciad Variorum*—but in other respects it is unique. According to Pope, the whole point of 'Messiah' is its lack of originality. Other imitations manifest their author's wit in devising or discovering parallel circumstances and persons; the past is applied creatively to the present day. 'Messiah' is put forward as the authorial discovery of pre-existing parallel texts; the poet's skill consists in making the parallelism acutely present to us. Pope's poetry is merely the medium in which a fusion can be seen taking place. All he has to offer is an enabling idiom; the more conventional his own contribution the better. For this reason, much of the criticism of the poem for its diction, especially its epithets, is beside the mark. The success of Pope's poem depends less on its quality as an aesthetic whole than on its skill in realizing its idea of powerful congruencies. His presentation of the poem always recognized the need to supplement the reader's own information with references, but the *Works* format gave him the opportunity to develop this to a new level. Other texts require further textual knowledge for their full appreciation —translations are an example—but the meaning of 'Messiah' is its relation to the other texts.

'Messiah' was first published without the full apparatus that accompanies it in Pope's *Works*. It appeared in the *Spectator*, 378, 14 May 1712, and Steele's introduction to it touches on some of those notes—retired independence, integrity, and wit—that were later to form an important part of Pope's self-presentation: 'I will make no Apology for entertaining the Reader with the following Poem, which is written by a great Genius, a Friend of mine, in the Country; who is not ashamed to employ his Wit in Praise of his Maker.'[33] This introduction, though, seems to point to a poem simpler than the one we find. Its heading is '*MESSIAH.* | A sacred Eclogue, compos'd of several Passages | of *Isaiah* the Prophet. | *Written in Imitation of* Virgil's POLLIO.' The awkwardness of the conception is vividly conveyed in the juxtaposition of 'compos'd' and 'Written'. In one sense Pope is responsible for the writing, in another he is not, being an arranger of pre-existing materials. The nature of the 'composition' is made quite clear even in this first printing by the use of frequent references to chapter and verse of Isaiah, references that are so important to the poem's meaning that they are placed at the end of the relevant lines of verse, rather than as footnotes or endnotes.

The *Works* five years later allowed Pope to be much more explicit about both the conception of the poem and its relation to its originals. The advertisement is sufficiently intricate to merit quotation in full:

In reading several passages of the Prophet *Isaiah*, which foretell the coming of Christ and the felicities attending it, I could not but observe a remarkable parity between many of the thoughts, and those in the *Pollio* of *Virgil*. This will not seem surprizing when we reflect, that the Eclogue was taken from a *Sybilline* prophecy on the same subject. One may judge that *Virgil* did not copy it line by line, but selected such Ideas as best agreed with the nature of pastoral poetry, and disposed them in that manner which serv'd most to beautify his piece. I have endeavour'd the same in this imitation of him, tho' without admitting any thing of my own; since it was written with this particular view, that the reader by comparing the several thoughts might see how far the images and descriptions of the Prophet are superior to those of the Poet. But as I fear I have prejudiced them by my management, I shall subjoin the passages of *Isaiah*, and those of *Virgil*, under the same disadvantage of a literal translation.

Although this introduction preserves something of the air of the amateur hinted at by Steele, the complexity of the information is at odds with any simple purpose and suggests a background more complex than the original introduction suggested. It turns out that in several respects 'Messiah' reflects the interests of Pope's circle, and particularly of the *Spectator*'s editors, at this time.

'Messiah' is one of a number of poems on religious subjects published by the *Spectator* in this period. Addison had said in No. 160, 3 September 1711, that 'in the Old Testament we find several Passages more elevated and sublime than any in Homer' (2. 127), and Pope's imitation was followed by a series of biblical paraphrases of passages from the Old Testament.[34] Addison himself published his most famous poem on a sacred subject, 'The spacious firmament on high', only three months after 'Messiah', on 23 August 1712. 'Messiah', therefore, seems to share in a religious

---

[33] *Spectator*, ed. Bond, 3. 419. Pope had revised the poem in the light of Steele's comments.

[34] Of The Song of Solomon, Proverbs, and of particular psalms; see *Spectator*, ed. Bond, 3. 455, 514, 533; 4. 50, 127, 144.

culture promoted by the *Spectator*. Pope's correspondence shows that for a short while he even shared Addison and Steele's enthusiasm for William Whiston, the distinguished mathematician, astronomer, and theologian who had been expelled from his chair at Cambridge for his heterodox theological views, 'This minute, perhaps, I am above the stars, with a thousand systems round about me, looking forward into the vast abyss of eternity, and losing my whole comprehension in the boundless spaces of the extended Creation, in dialogues with W[histon] and the astronomers' (Correspondence, 1. 185). Whiston's astronomical lectures had a powerful influence on Pope's imagination and also helped to shape Addison's hymn, and it is likely that Pope was influenced too by the debates generated by Whiston's Boyle lectures of 1707 on *The Accomplishment of Scripture Prophecy*, with their interest in a key passage in Isaiah: 'Behold, a Virgine shall conceive and beare a Sonne, and shalt call his name Immanuel' (7: 14). Whiston's insistence on the literal fulfilment of prophecy provoked an attack from Anthony Collins (1724) and a response to that from Edward Chandler in his *Defence of Christianity* (1725).[35] Chandler's *Defence* could not, of course, have influenced Pope, but it reveals a great deal about the state of biblical and classical scholarship at the time and the context to which Pope's poem is addressed. A major section of Chandler's book is devoted to the relations between Isaiah's prophecy and Virgil's fourth eclogue, his aim being to show that the coming of Christ fulfilled Jewish expectations of the coming of the Messiah, those expectations having been filtered down to us through Greek and Roman writing. A long history of Christian interpretation, going back to the fourth century, had regarded Virgil's eclogue as an inspired and independent Christian vision. Chandler was attempting to construct a line of transmission compatible with a new rigour in scholarship.[36]

From the perspective of these contemporary debates about prophecy, Pope's advertisement to his poem, with all its apparent simplicity, emerges as a masterpiece of tactful reticence. The advertisement does not recognize any Christian tradition in interpreting the fourth eclogue. 'Messiah', it seems to claim, emerges out of isolated personal experience. 'In reading several passages . . . I could not but observe a remarkable parity . . .'; the insight seems not to be sought or even to have been prepared for by earlier reading. If Pope is making a contribution to current debate, it is as an observant reader, not as a theologian. His position skilfully confines itself to a new textual awareness without settling distinctly for either the older view (that

---

[35] There is a good account of this aspect of Whiston's thought in James E. Force, *William Whiston: Honest Newtonian* (Cambridge: Cambridge University Press, 1985). In quoting from the Bible, I have used A. W. Pollard's facsimile of the 1611 Authorized Version (1911). Pope's own citations are from a later, unidentified version, and his writing was influenced by the Douai translation (see *Twickenham*, 1. 100–2).

[36] Wendell Clausen gives an interesting history of interpretation in *A Commentary on Virgil's Eclogues* (Oxford: Clarendon Press, 1994), 127. As late as 1767 Heyne rejected the Christian interpretation while saying most of the learned believed in it. Edward Coleiro, in *An Introduction to Vergil's Bucolics with a Critical Edition of the Text* (Amsterdam: B. R. Grüner, 1979), 232, offers an interpretation that is in many ways similar to Chandler's, which Sir Leslie Stephen disparages rather too freely in his *History of English Thought in the Eighteenth Century*, 3rd edn., 2 vols. (London: Smith, Elder, 1902), 1. 218.

Virgil was divinely inspired) or a view like Chandler's (that Virgil's poem emerged indirectly from biblical prophecy). Notably, the issue which is debated by Whiston and Collins is taken for granted by Pope. The passages in Isaiah 'foretell the coming of Christ'; on this issue his position is straightforward and orthodox. However, the parity between the thoughts of Isaiah and the thoughts of Virgil that Pope has observed does seem to require some explanation, and Pope hints at one without specifying what it is: the parity 'will not seem surprizing when we reflect, that the Eclogue was taken from a *Sybilline* prophecy on the same subject'. This is compatible both with the modern view, represented by Chandler, that the Sibylline prophecies originated from the Hebrew scriptures, or with the older view, that the ancient world had it own sources of inspiration. The effect, in either case, is to honour, through Christianity, both the Hebrew and Graeco-Roman worlds. Isaiah may be more sublime than Virgil, but both have access to the same spiritual truth. Isaiah's picture of the world awaiting Messianic redemption is confirmed, and Virgil is dignified with the status of a minor prophet of Christianity.

Pope's own role is to display the congruence of ideas and simultaneously display Isaiah's superiority. He says that he has imitated Virgil selectively, but that he has done so solely through the language of Isaiah 'without admitting any thing of my own'. This is a surprising view of the extent to which his own art has contributed to a poem of 108 lines. Even if he had merely disposed Isaiah in a different order he would have introduced a large measure of his own judgement, but he had done much more than that, modifying and developing Isaiah's language. Pope is anxious to guard against charges of spiritual pride—after all, he might be claiming to have a vision that encompassed both that of Virgil and that of Isaiah—and the annotation partly functions as a badge of modesty: 'But as I fear I have prejudiced them [the Prophet's images and descriptions] by my management, I shall subjoin the passages of *Isaiah*, and those of *Virgil*.' But the annotation is also the display of an implicit thesis—that the writers share the same vision—and the poem, rather than any technical argument, in uniting the two, constitutes the demonstration of that unstated thesis.

Pope selects the material that best serves his purpose. The new apparatus gives the evidence for his thesis, and also displays the structure of the poem, which, after the invocation, is not otherwise divided into sections. The annotation that in the *Spectator* drew attention to the parallels to Isaiah is now scattered at the foot of the page, but Pope has added endnotes giving both the passages from Virgil that are imitated and the parallel passages from Isaiah. The reader is invited to an analytical reading, moving backwards and forwards between the poem and its annotation, recreating the process by which one text is integrated with the other.

The opening invocation of 'Messiah' is representative of Pope's method: it shows him pushing his poem to the boundary of the classical tradition he inherits from Virgil and then breaking through a little beyond it.

> Ye Nymphs of *Solyma!* begin the song:
> To heav'nly themes sublimer strains belong.[37]

---

[37] In quotations the text is taken from the *Works* and the line numbers are from *Twickenham*.

Virgil appealed to the Sicilian Muses to sing a loftier strain in tribute to Pollio, but Pope's poem requires different muses altogether, those of Jerusalem. In invoking nymphs at all, of course, he remains within the imaginative world of paganism, and a feature of the poem is that the conception of agency is not generally restrictedly Christian. In line 18, for example, 'Returning Justice' lifts her scale and in line 20 'white-rob'd Innocence' descends from heaven. These personifications were not acceptable to Johnson when he translated the poem into Latin as an academic exercise,[38] but Pope, like Milton before him, regarded such figurative writing as compatible with Christian belief, and the employment of them is part of the work of reconciliation that the poem is undertaking. At the end of his invocation, he is still able to move into prayer to the Holy Spirit, without any sign of discomfort:

> The dreams of *Pindus* and th' *Aonian* maids,
> Delight no more—O thou my voice inspire
> Who touch'd *Isaiah*'s hallow'd lips with fire! (4–6)

The transition is unproblematic because the analogy with Isaiah is structurally pervasive.

Pope's selection of passages for imitation is by no means a simple matter. The six sections of the poem indicated by the apparatus focus on religious ideas that are developed with pastoral elaboration in images drawn from Isaiah. The first section is concerned with the return of the Virgin, the birth of a new generation, freedom from guilt, and the government of the earth in peace. Pope retains these ideas in his imitation, but the first two are developed through passages from Isaiah on a system of association: Virgin bear a son, Jesse's root, sacred flower, Dove descending, kindly shower, healing plant. The second section, concerned with Nature's flowering in response to the child, is dealt with very briefly, the poem representing the equivalents in Isaiah while adding characteristic personification ('See lofty *Lebanon* his Head advance'), but the following, third section, involves considerable elaboration, five lines in Virgil become thirty-eight in 'Messiah'. Pope chooses to imitate three lines from Eclogue 5 (62–4) because they resemble the famous verses in which a voice crying in the wilderness proclaims the exaltation of every valley (Isaiah 40: 3–4).[39] Pope's imitation, while deploying Isaiah, keeps close to Virgil:

> Hark! a glad voice the lonely desart chears;
> Prepare the way! a God, a God appears;
> A God, a God! the vocal hills reply,
> The rocks proclaim th'approaching Deity. (29–32)

---

[38] See Maurice J. O'Sullivan, jun., 'Ex Alieno Ingenio Poeta: Johnson's Translation of Pope's *Messiah*', *Philological Quarterly*, 54 (1975), 579–91.

[39] These lines also represent Virgil's 50–1,

> aspice convexo nutantem pondere mundum
> terrasque tractusque maris caelumque profundum

(Behold the world bowing with its massive dome—earth and expanse of sea and heaven's depth!), though these are not annotated.

Once again, and rather startlingly, Pope allows the poem to be flavoured with polytheism, reflecting Virgil, but within a few lines that tendency has been counteracted, for he writes 'The Saviour comes! by ancient Bards foretold' (37), Virgil having already figured as 'the Bard' in line 7. The poem could have been thoroughly Christianized, but Pope prefers to preserve pagan–Christian tensions within it before resolving them finally in the Christian direction. Pope concludes this section with the adaptation of less famous verses from Isaiah:

> And they shall builde houses, and inhabite them, and they shall plant vineyards, and eate the fruit of them.
> They shal not build, and another inhabit: they shall not plant, and another eat: for as the daies of a tree, are the dayes of my people, and mine elect shal long enjoy the worke of their hands. (Isaiah 65: 21–2)

> the joyful Son
> Shall finish what his short-liv'd Sire begun;
> Their vines a shadow to their race shall yield,
> And the same hand that sow'd shall reap the field. (63–6)

Pope, who was fully aware of the sequestration of Catholic property after 1688 and was later to write, 'My lands are sold, my Father's house is gone', is here using Isaiah as a critic of the present age. But the criticism is muted: there is no reference to 'my people' in the poem, and the most potentially biting line, 'They shal not build, and another inhabit', is reflected only indirectly. Maynard Mack has acutely pointed to the relationship between 'Messiah' and the apocalyptic elements of *Windsor-Forest*; in this section 'Messiah' is equally alive to the political implications of pastoral.

Section 4 develops the topos of fertility already touched on in section 2, but section 5, corresponding to Virgil's lines 21, 22, 24, and 25, develops Virgil's image of nature at peace into a more radically transformed world corresponding to Isaiah 11: 68.

> The lambs with wolves shall graze the verdant mead,
> And boys in flow'ry bands the Tyger lead;
> The steer and lion at one crib shall meet,
> And harmless serpents lick the pilgrim's feet.
> The smiling infant in his hand shall take
> The crested Basilisk and speckled snake;
> Pleas'd the green lustre of the scales survey,
> And with their forky tongue and pointless sting shall play. (77–84)

This passage is a good example of Pope's regular supply of epithets—verdant, flow'ry, harmless, smiling, crested, speckled, forky—which has been the feature of his poem most criticized.[40] In this passage, however, it is more than a tic of Augustan style; there is a characteristic attention to visualization, especially an emphasis on the visual beauty of nature. This passage is a notable example of such visualization and stimulates the illustration of the book, which begins the poem with an initial

---

[40] See *Twickenham*, 1. 102–7 for a valuable discussion of these criticisms.

showing a boy leading a 'tiger' with a flowery chain. Maurice O'Sullivan, in his article on Johnson's youthful translation of the poem makes a theological issue of this representation of nature: 'When Pope must distinguish the transcendent state from the condition of perfection in the world, however, he wavers between Virgil's view of the millennium and that of Christianity.'[41] But even if so sharp a divorce is theologically correct (and I am far from convinced it is), Pope's writing is close enough to Isaiah and distant enough from practical and moral life to disqualify this objection as a challenge to his orthodoxy.

If admiration of nature is taken too far in section 5, Pope compensates with a depiction of the new Jerusalem in the final section that owes little to either Virgil or the attractions of nature. Pope refers to four lines in Virgil's poem (5, 9, 12, and 52) but says they bear no comparison with the elevation in the thought of Isaiah. The concluding section draws heavily on chapter 60 of Isaiah, and the passage on light firmly expresses a Christian orthodoxy:

> No more the rising Sun shall gild the morn,
> Nor ev'ning *Cynthia* fill her silver horn,
> But lost, dissolv'd in thy superior rays,
> One Tyde of glory, one unclouded blaze
> O'erflow thy courts: The Light himself shall shine
> Reveal'd, and God's eternal day be thine! (99–104)

The presence of Cynthia as our final goddess marks the rejection of the classical conception in the apocalyptic vision. As in other sections of the poem, we move from a step of comparison to a step of transcendence.

The annotation of 'Messiah', which honours Pope's poem with the kind of attention Virgil's own work might merit, also admits the reader to a special intimacy with its creator. We are allowed behind the scenes and given access to the raw material of the creative process. In this respect 'Messiah' is representative of the *Works*' interest in the relation between private and public, and in patrolling the boundary between them. It is an example of Pope's use of the resources of the book to expand the possibilities of his art, engaging with ages and voices other than his own. The notes to 'Messiah' open up possibilities that were to be developed by the annotation of later poems.

---

[41] 'Ex Alieno Ingenio Poeta', 586. However, I am persuaded O'Sullivan represents Johnson's view of these matters.

4. The Dunciad Variorum: *The Limits of Dialogue*

THE *VARIORUM*, DIALOGUE, AND THE NOVEL

The Dunciad Variorum, first published in early April 1729, has self-evident claims to be a great polyphonic text. The title itself proclaims a plurality: 'The Dunciad, Variorum. With the Prolegomena of Scriblerus.' Variorum abbreviates 'editio cum notis variorum', an edition with the notes of various commentators, and the name of the chief commentator, Martinus Scriblerus, appears on the title page, whereas that of the author does not. The book is so complex and various that it contains not one list of contents but two. An idea of its scope, but not of its appearance, with each page accompanied by a great crypt of footnotes (see Fig. 6), is conveyed by reprinting the contents lists.

PIECES contained in this BOOK.
THE Publisher's Advertisement.
A Letter to the Publisher, occasioned by the present Edition of the Dunciad.
The Prolegomena of Martinus Scriblerus.
Testimonies of Authors concerning our Poet and his Works.
A Dissertation of the Poem.
Dunciados Periocha: Or, Arguments to the Books.
The DUNCIAD, in Three Books.
Notes Variorum: Being the *Scholia* of the learned M. Scriblerus and Others, with the *Adversaria* of John Dennis, Lewis Theobald, Edmund Curl, the Journalists, &c.
Index of Persons celebrated in this Poem.
Index of Things (including Authors) to be found in the Notes.
Appendix.

PIECES contained in the APPENDIX.
PREFACE of the Publisher, prefixed to the five imperfect Editions of the *Dunciad*, printed at *Dublin* and *London*.

A List of Books, Papers, *&c.* in which our Author was abused: with the Names of the (hitherto conceal'd) Writers.

William Caxton his Proeme to *Æneidos*.

Virgil Restored: Or a Specimen of the Errors in all the Editions of the *Æneid*, by M. Scriblerus.

A Continuation of the Guardian (N° 40) on Pastoral Poetry.

A Parallel of the Characters of Mr. Dryden and Mr. Pope, as drawn by certain of their Cotemporary Authors.

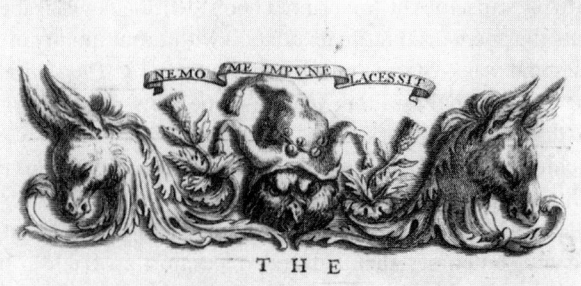

FIG. 6. First page of *The Dunciad Variorum*, 1729 (Bodl. CC 76 (1) Art; 249 × 188 mm).

A List of all our Authors Genuine Works hitherto published.
INDEX of Memorable things in this Book.[1]

---

[1] Quotations from the *Variorum* are from the facsimile, *Alexander Pope: The Dunciad Variorum 1729* (Menston: Scolar Press, 1968). References will be given by page number, followed where appropriate by book and line numbers. A text of this version of the poem and of the four-book *Dunciad* can be found in James Sutherland's edition, *Twickenham*, 5. Valerie Rumbold's edition of *The Dunciad in Four Books* (Harlow: Longman, 1999) provides scrupulous annotation of this already-annotated text.

84 The Dunciad Variorum

The variety of these materials makes the *Variorum* a model of the organizational and schematizing capacities of the printed book identified by Elizabeth Eisenstein,[2] but it also sets the poem in an intimate relation with a multiplicity of voices. A difficult, and in some ways vital, issue for criticism of *The Dunciad* is the nature of Pope's engagement with these other voices—opponents, predecessors, rivals, scholars, critics—drawn into his work.

The coincidence of the *Variorum* with the rise of the novel opens up the possibility that in this complex re-utterance of his poem Pope had discovered a dialogism complementary to the novel's, a means by which poetry could recognize and engage with diverse and sometimes alien cultural voices, just as, Bakhtin has argued, the novel had done.[3] The history of the planning and publication of *The Dunciad*, from its conception in 1719–20, through its first appearance in octavo in 1728, to the quarto in four books of 1743 fits in neatly with the development of the canonical English novel: *Robinson Crusoe* was published in 1719, *Pamela* in 1740, and *Joseph Andrews* in 1742. Pope's poem shares with the developing genre an attention to social detail, particularly the symbolism attaching to London's topography; it draws into its orbit not only the contemporary literary scene, but eventually education, science, medicine, and the Grand Tour; and there are particular links between the poem and novels and novelists. Defoe and Haywood are ridiculed in *The Dunciad*, while Pope's translation of Homer is warmly praised in *Joseph Andrews*, which also attacks the man who was shortly to become the hero of the four-book *Dunciad*, Colley Cibber. But the most important links between *The Dunciad Variorum* and the novel concern its relation to epic and to what Bakhtin calls heteroglossia, the variety of dialects or social languages within a national language.[4] For Bakhtin, it is the novel's positive relation to these languages that distinguishes it from the epic. The distinctions developed through Bakhtin's mapping of the novel's relation to heteroglossia help to define both Pope's achievements in dialogue with other voices and the restrictions he failed to escape. They also engage in new ways with questions of the *Variorum*'s relations to its celebratees or victims, questions which are raised most acutely by the title of Pope's second index, 'Index of Things (including Authors)'.

Bakhtin has written various accounts of the history of the novel, not all of them easy to reconcile, but for the purposes of this investigation of Pope the details of his historical analysis and his relation to Marxian critics can be put aside. What is needed is a grid on which Pope's work can be plotted. The long, almost short-book-length,

---

[2] Elizabeth L. Eisenstein, *The Printing Press as an Agent of Social Change*, 2 vols. (Cambridge: Cambridge University Press, 1979).

[3] Although I have Ian Watt's classic account in *The Rise of the Novel* (London: Chatto, 1957) in mind, I am conscious of the alternative accounts, especially Rosalind Ballaster's *Seductive Forms*, Catherine Gallagher's *Nobody's Story*, William B. Warner's *Licensing Entertainment*, and Michael McKeon's *The Origins of the English Novel, 1600–1740* (Baltimore: Johns Hopkins University Press, 1987). McKeon makes good use of Eisenstein (43–5).

[4] A particularly clear account of heteroglossia is to be found in Simon Dentith (ed.), *Bakhtinian Thought: An Introductory Reader* (London: Routledge, 1995), esp. 35–8, and there is another admirable introduction in the chapter on 'Novelness as Dialogue' in Michael Holquist, *Dialogism: Bakhtin and his World* (London: Routledge, 1990), 67–106.

essay 'Discourse in the Novel' provides such a grid. Its scheme is compatible with the more abstract one presented in *Problems of Dostoevsky's Poetics* in the chapter on 'Discourse in Dostoevsky', and I have preferred it here because of its historical emphasis and its explicit interest in the epic.[5] Bakhtin distinguishes four categories of discourse through their relation to heteroglossia: poetry; novels of the first line of stylistic development; novels of the second line of stylistic development; and rhetoric. By heteroglossia Bakhtin means the stratification of language into various dialects or languages. These dialects are social, attached to groups or generations; they express points of view, ideological positions; they compete to describe the world.[6] Any utterance takes its meaning from its relation to both prior and alternative utterances; one speaks always in response to an already spoken question. A literary genre can recognize and build positively on this dialogue or try to suppress it. Any utterance, great or small, is subject to centripetal forces trying to unify language (or gain hegemony for one dialect) and centrifugal forces recognizing social diversity and development.[7]

Bakhtin's division of genres is sharp, but it is not designed to reinforce existing taxonomies (*Eugene Onegin*, for example, is classed as a novel) and it leaves freedom in the location of the *Variorum*. Poetry is not dialogic. Some poets (Bakhtin mentions Pushkin and Byron) write dialogically, but poetry, schematically conceived, is monologic. This is difficult for poetry, because as the poetic word makes its way to its object it is 'continually encountering someone else's word', but the record of this is cleared away by the creative process, which fuses the word with its object. In poetry the artist seems to be expressed directly, without mediation or distance. Poetry is unaware of other languages. Bakhtin provides no examples of such poetry—all his examples have fallen into dialogue—but his purpose is not to describe poems but to set up a polar opposite to the achievement of the novel.

Novels of the first line of stylistic development (from the 'Sophistic novels' to Richardson and Radcliffe) are distinguished from poetry by their clear recognition of external heteroglossia. Sometimes they incorporate semi-literary genres from everyday life (for example, conversations, letters, diaries), but as they do so they 'ennoble' them, purifying their language to conform to a literary language.[8] Heteroglossia is the recognized context in which these novels operate, but it is not directly incorporated. Their language, unlike that of poetry, never takes itself to be the only language.

---

[5] *The Dialogic Imagination: Four Essays by M. M. Bakhtin*, ed. Michael Holquist, trans. Caryl Emerson and Michael Holquist (Austin: University of Texas Press, 1981), 259–422; *Problems of Dostoevsky's Poetics*, ed. and trans. Caryl Emerson (Manchester: Manchester University Press, 1984), esp. 181–200. An important consideration of Pope in relation to Bakhtin is Peter Stallybrass and Allon White, *The Politics and Poetics of Transgression* (London: Methuen, 1986), 109–18.

[6] In this essay I use 'ideology' as Bakhtin uses it, to mean the manner of thinking characteristic of a person or group, and not in a strict Marxian sense.

[7] For an attempted assessment of the philosophical, Neo-Kantian, underpinning to Bakhtin's view of language and literature, see Michael Holquist, *Dialogism: Bakhtin and his World*.

[8] 'Discourse in the Novel', 410. This represents Bakhtin's summary position. His discussion of the sentimental novel and its relation to conversation and to the small space, one's own room, is actually more subtle than this suggests (396–7).

Whereas novels of the first line approach heteroglossia from above, novels of the second stylistic line (from *Parzival* to *Tom Jones*) approach it from below. This line originates in parody, which presents the language of others as something opaque rather than transparent; it double-voices the displayed language, presenting it as the speech of both mocker and mocked. In this tradition we are concerned with images of languages rather than with individuals—artistically shaped and orchestrated images of languages. In this line of development, the novel becomes a 'microcosm of heteroglossia'.[9]

Rhetoric is less directly the object of Bakhtin's discussion than poetry and the novel, but it occupies an important position on the grid. In some respects it lies closer to novels of the second line than it does to poetry or novels of the first line; it can be dialogic but only in strictly limited ways. Under rhetoric and rhetorical genres Bakhtin groups biographies, autobiographies, confessions, and juridical and political speeches. The basic distinction is between the novelist with 'multi-languagedness surrounding and nourishing his own consciousness' and the rhetorician's 'superficial, isolated polemics with another person'. Bakhtin's example is the double-voicing of Lensky's poetic symbols in *Eugene Onegin* (representing Pushkin's viewpoint as well as Lensky's), which he contrasts with 'superficial, rhetorical parody or irony'.[10] The rhetorician is characteristically limited by too narrow and specific a concern with another's words, the concern remains asocial. The rhetorical genres embody a limited conception of human beings, a conception which excludes development: 'Thus the hero can be evaluated only as *exclusively* positive or *exclusively* negative. Rhetorical-judicial categories predominate in the conception of human beings.'[11]

The Dunciad Variorum in some respects looks like a work attempting to break the bounds Bakhtin sets for poetry by coming to terms with heteroglossia. Its prolegomena, appendices, and footnotes show it to be a truly many-voiced work. And yet its very bibliographical complexity suggests a reservation that must enter the analysis and may ultimately determine its outcome. The *Variorum* is more aware of other books, and of itself as a book, than it is of other voices. In many ways this adds a welcome refinement to Bakhtin's account, which tends to remove authors and their voices from the muddle of publication, with its economics, censorships, and corruptions. The utterances Bakhtin accounts for are idealized, the products of some sphere of equality and freedom, whereas those Pope is concerned with are commodities, 'He had once a mind to translate the *Odyssey*, the first Book whereof was printed in 1717 by B. Lintott, and probably may yet be seen at his Shop' (11; 1. 106). Theobald's aspirations are transformed into something you could buy over a counter. Pope's are perhaps the only footnotes to a major poem to go into such commercial detail, sometimes even telling you the price of their sources. The problem for the authors embedded in the notes and apparatus, and for the dialogic power of the work, is that even when they do not remain imprisoned within their own

---

[9] 'Discourse in the Novel', 411. See also 400–1 and 416–17.
[10] Ibid. 327–9. See also 406–7.   [11] Ibid. 407. See also 354.

titles and formats, with the status of mere objects, they are trapped within the structure of the *Variorum* itself. This enables the poem to place and control them; they speak but their speech is confined within quotation marks and the quotations are further typographically compartmentalized. The dunces are in the *Variorum* but they cannot get at the poem. The organization and typography of the book afford Pope a triumph, but at the cost of a fuller engagement with his opponents.

THE SCHOLARLY EDITION

The Dunciad Variorum exercises its control over other authors as a scholarly edition, or as a parody of one. Bakhtin writes of some utterances as double-voiced, as being observably inflected, for example, by both the narrator and character. The essential quality of *The Dunciad Variorum* might be conveyed by saying it is double-printed. It is simultaneously an annotated edition of the poem, supplying interpretation and information about context and allusion, and a parody of the scholarly editions of the classics promoted by figures like Bentley. The key to an understanding of its mode of operation lies in Pope's sense of his relation to Boileau and, in particular, to contemporary editions of Boileau.[12] Boileau emerges as an alter ego at the end of the 'Letter to the Publisher' in the *Variorum*. He is the first of two figures (Dryden is the other) that the work presents in parallel with Pope; criticism and abuse are seen not as personal but as tired repetition of the opposition to any important writer. Under Pope's reifying gaze, attacks are drained of their personal energy and become mere formulaic husks. The 'Letter to the Publisher' draws in Juvenal and Boileau as precedents for Pope, and then continues in what is a prolonged (and, therefore, powerfully motivated) digression:

Having mention'd BOILEAU, the greatest Poet and most judicious Critic of his age and country, admirable for his talents, and yet perhaps more admirable for his judgment in the proper application of them; I cannot help remarking the resemblance betwixt Him and our Author in Qualities, Fame, and Fortune . . . But the resemblance holds in nothing more, than in their being equally abus'd by the ignorant pretenders to Poetry of their times; of which not the least memory will remain but in their own writings, and in the notes made upon them. What BOILEAU has done in almost all his Poems, our Author has only in this: I dare answer for him he will do it in no more . . . However, as the parity is so remarkable, I hope it will continue to the last; and if ever he shall give us an edition of this Poem himself, I may see some of 'em treated as gently (on their repentance or better merit) as Perault and Quinault were at last by BOILEAU. (13–14)

The honorific capitals reinforce the comparison between the authors, but the emphasis on notation and on 'COMMENTARY' has run through the 'Letter', as it runs, necessarily, through the 'variorum' edition. Boileau had been edited in a spectacularly

---

[12] I first discussed this topic in 'The Mode of Existence of Literary Works of Art: The Case of the *Dunciad Variorum*', *Studies in Bibliography*, 37 (1984), 82–105, and hope to compensate for deficiencies in that discussion here. For a fascinating study of scholarly editions in this and other periods, see E. J. Kenney, *The Classical Text: Aspects of Editing in the Age of the Printed Book* (Berkeley and Los Angeles: University of California Press, 1974).

detailed fashion by Claude Brossette in a two-volume edition whose appearance coincided with the publication of the first volume of Pope's works in 1717: *Oeuvres de M^R. Boileau Despréaux. Avec des Éclaircissemens Historiques, Donnez par Lui-même* (Geneva: Fabri & Barrillot, 1716). Pope had been given a copy, now at Mapledurham House, by his friend James Craggs,[13] and this example of what could be done in editing a contemporary author affected his thinking about his works for the rest of his career. A letter to Tonson about Theobald's Shakespeare and Bentley's Milton showed his awareness of the *Dunciad*'s special position in what was to become an editorial tradition:

> I think I should congratulate your Cosen on the new Trade he is commencing, of publishing English classicks with huge Commentaries. Tibbalds will be the Follower of Bentley, & Bentley of Scriblerus. What a Glory will it be to the Dunciad, that it was the First Modern Work publish'd in this manner?[14]

The letter provides an index of Pope's subtlety in defining his relation to textual scholarship. An editor himself and the supervisor of elaborate editions of his own texts, he nevertheless established himself as the major critic of 'verbal criticism' and the scholarship of Bentley and Theobald. If the wit and subtlety of the *Variorum* format are of characteristic Popeian kind—a type of irony that might be called 'have your cake and eat it'—they nonetheless create a fascinating, doubly inflected book, in which information and parody jostle one another and sometimes blend. Pope advertises his practice in the second page of the preliminary advertisement to the *Variorum*:

> The *Imitations* of the Ancients are added, to gratify those who either never read, or may have forgotten them; together with some of the Parodies, and Allusions to the most excellent of the Moderns. If any man from the frequency of the former, may think the Poem too much a *Cento*; our Poet will but appear to have done the same thing in jest, which *Boileau* did in earnest; and upon which *Vida*, *Fracastorius*, and many of the most eminent Latin Poets professedly valued themselves. (4; italics reversed)

A cento is a work composed of quotations from other authors. The form provides a challenge to a theory such as Bakhtin's, but the implications need not be worked out here, because Pope's quotation, like the rest of his material, is separated out and subjected to special regulation by the form of the book.

The details of the apparatus derive from Brossette, who draws attention to the division of his notes into three types: *Changemens*, *Remarques*, and *Imitations*. The division was sufficiently impressive for the Amsterdam edition of Boileau (1718) to apologize for departing from it, and Pope intended to adopt it for his own *Works*, making a start with Remarks and Imitations in the *Variorum*. The appeal of the

---

[13] The volumes, which are now in the library of Mapledurham House, are listed by Maynard Mack, 'Appendix A: A Finding List of Books Surviving from Pope's Library with a Few That May Not Have Survived', No. 26, in *Collected in Himself*, 399.

[14] *Correspondence*, 3. 243–4. Pope must have been pleased to have anticipated the nation's leading publisher in this enterprise.

divisions lay both in their thoroughness and their compartmentalization. Pope seems to have aimed at a fullness of information, and a recuperation of the processes of composition, from the time of circulating the manuscript of the *Essay on Criticism* with variants, through the *Key to the Lock*, to Warburton's extensive annotation of his works. An edition that separated the mediation of the text to the reader (in the Remarks) from the task of relating it to the poetic tradition (Imitations) and the developing talent of the poet (Variations) was generally desirable—the pattern was followed in Warburton's 1751 edition of the *Works*—and was especially needed in a case like the *Variorum*'s.[15] The parallels with Brossette's edition of Boileau may be briefly summarized:

1. The notes are at the foot of the page. I suspect *The Dunciad* was the first English poem to receive extensive annotation of this kind.
2. The notes are sectionalized by headings in italic capitals.
3. Two of these headings, 'Remarques'/'Remarks' and 'Imitations', are the same. Pope has no *Changemens*; this was a task allotted to Jonathan Richardson for future works.[16]
4. The two columns of notes are divided by a vertical rule (the Boileau divides only *Remarques*). (Pope must have insisted on this. I doubt whether eighteenth-century printers were any keener on vertical rules than twentieth-century ones are; they must have been difficult to keep in place.)
5. The lemma and subsequent quotation from the text are printed in italic.
6. The lemma begins with Vers or Verse in full.
7. There are no catchwords; this is the first Pope text to drop them.

The first inflection of this printed form is as a repetition or parallel. The Geneva Boileau has as its frontispiece a splendid portrait of the author with the inscription:

> Boileau sut remplacer Horace,
> Seul il sut remplacer et Perse et Juvenal;
> Mais de cet auteur sans egal
> Qui remplira jamais la place?

In the *Variorum* Pope takes that place. He is supposedly able to do so through the support of friends. Boileau had one friend and annotator, Brossette; Pope has many. Although in truth the voices of the *Variorum* are largely those of Pope and his enemies, the theory of the volume is different. The various commentators are presented as Pope's friends. The 'Letter to the Publisher' says,

Such Notes as have occurr'd to me I herewith send you; you will oblige me by inserting them amongst those which are, or will be, transmitted to you by others: since not only the Author's

---

[15] Warburton uses the threefold division for *The Dunciad* in his edition; for other poems, Remarks simply follow the lemma, without a heading.

[16] For the notes to Pope's *Works*, see Chapter 8. Many of the surviving Pope manuscripts come down to us from Jonathan Richardson, jun., given him as a reward for collation. See *Richardsoniana* (London, 1776), 264. For an account of Pope's manuscripts, see Mack, *Collected in Himself*, 322–47, and *The Last and Greatest Art*.

friends, but even strangers, appear ingag'd by humanity, to some care of an orphan of so much genius and spirit . . . (5–6)

The advertisement reassures the reader of the value of such a commentary, saying that although it was sent from several hands and is consequently 'unequally written', it will 'have one advantage over most commentaries, that it is not made upon conjectures, or a remote distance of time' (3). The friends and well-wishers supply the place of Brossette, who presents himself as the author's friend, anxious to keep his work fresh for posterity. His information is from the author himself: 'Ce n'est donc pas ici un tissu de conjectures, hazardées par un Commentateur qui devine: c'est le simple récit d'un Historien qui raconte fidellement, & souvent dans les mêmes termes, ce qu'il a apris de la bouche de l'Auteur original' (p. vi). These words are faintly echoed in the passage from the advertisement just quoted, with the characteristic Popeian irony that these words actually are from the mouth of the author himself.

The second inflection of the *Variorum* is as spoof or parody of the scholarly edition. It ridicules the attempt to supplement and explain the work, even as it supplements and explains it. The complexity of the *Variorum* apparatus and notes turns out not to derive from Pope's attempt to disguise his own interpreting voice in those of many friends, but from the attempt to incorporate the voices of his opponents and make them also work towards his purposes. The various commentators turn out to be enemies rather than friends. Some enemies—Dennis, Curll, Theobald—are real; one, Martinus Scriblerus, is an established invention put to new use. In this respect the *Variorum* is powerfully dialogic. It recognizes that it takes its meaning from prior and alternative speech (Boileau's Œuvres, Bentley's Horace, the attacks of Gildon, Concanen, and Moore); it speaks in reply to questions that are already posed; and it goes beyond mere recognition of surrounding heteroglossia—it attempts to absorb and revoice surrounding attitudes and positions. The *Variorum* has much in common with novels of the first line of development in that it recognizes semi-literary genres and incorporates them, but it does not ennoble them. Letters, testimonies, a dissertation, a preface, and a bibliography all appear, but without modification to harmonize them with the poem. The major techniques of the *Variorum* are display and parody. In so far as the display is tightly controlled and objectivizing, the work belongs with Bakhtin's rhetorical genres: Pope is polemically engaged with figures who have no opportunity of development. But in so far as the parody takes us beyond individuals to underlying positions and attitudes, it moves towards novels of the second line of development and presents 'images of languages'. A journey from the poem, through the footnotes, to the apparatus turns out to be a movement from the less dialogic to the more so; one that passes through varying levels of ambition in incorporation.

THE POEM AND DIALOGUE

The poem itself is both threatened and guarded by its apparatus. The first page (Fig. 6) shows something of the nature of the threat. The signature, 'F', shows that we are already some way into the book (over twenty-nine pages, because there are preliminary gatherings) before the poem can make its start, and yet it is still threatened with swamping. Only two lines of verse sit on top of the two pillars of commentary. The nature of the first note reinforces the sense of danger. Theobald proposes a change of the very title to 'Dunceiad'. The note here skilfully drifts from an excessively fusspotting tone into direct quotation from *Shakespeare Restored*, suggesting there is no space between Theobald the critic and Pope's ludicrous image of him (I have placed the quotation in italic):

Nor is the neglect of a Single Letter so trivial as to some it may appear; *the alteration whereof in a learned language is an Atchivement that brings honour to the Critick who advances it; and Dr. B. will be remembered to posterity for his performances of this sort, as long as the world shall have any Esteem for the Remains of Menander and Philemon.*[17]

Pope precedes the quotation with a neat echo of Theobald's repeated concern with emendation of one or two letters (for example, *Shakespeare Restored*, pp. vi, 5, 59, 165). The title is saved from this 'correction' by Scriblerus' intervention in the following note—his rule is always to follow the manuscript—but, of course, the title is already protected by the fixity of print. This is a work that has already made its way through the snares of textual criticism to the printed page. The poem and its title proceed irresistibly on their way: the title in the headline of every page, guarded by two rules, and the poem gaining an increased ascendancy over its commentary. Pope could have made things otherwise. His close relationship with the printer of the *Variorum*, John Wright, whose shop seems to have been pretty much dedicated to printing for Pope, made any form of experimentation possible: the poem could have been broken with asterisks like *A Tale of a Tub*, or the commentary could have taken over for a number of pages, but, even when the poem is hardest pressed, as on the appearance of Tibbald shortly after Dennis on page 10, continuity is carefully preserved.

The poem that survives the threat of its apparatus has a mock-epic character that suggests it may belong to a Bakhtinian transitional group, in which parodic stylizations of canonized genres provide an important stage in the development of the novel. In the periods of preparation for the novel's ascendancy, we find parodies and travesties of the high genres, of the genres themselves, not just of individual authors or schools. In Bakhtin's view such parodies are the precursors, 'companions' to the novel.[18] The relation of mock-epics like *The Dunciad* to parody is, of course, a complex one. Boileau made a central distinction in his preface to *Le Lutrin* in 1674: 'C'est un burlesque nouveau, dont je me suis avisé en notre langue: car, au lieu que dans l'autre burlesque, Didon et Enée parlaient comme des harengères et des crocheteurs, dans celui-ci une horlogère et un horloger parlent comme Didon et Enée.'[19] Boileau's

---

[17] Lewis Theobald, *Shakespeare Restored*, facsimile edition (London: Frank Cass, 1971), 193 (page numbers are those of the original).
[18] 'Epic and Novel', in *The Dialogic Imagination*, 6.
[19] Nicolas Boileau-Despréaux, *Œuvres*, 1: *Satires-Le Lutrin*, ed. Jérôme Vercruysse (Paris: Garnier-Flammarion, 1969), 182. Geoffrey Tillotson's observations on this topic are still valuable, see his

distinction is of central importance because it suggests that the disruptive parodic forces that might unsettle traditional styles and genres can be appropriated for their defence. But this distinction, valuable though it is, is not the same as the crucial one between those works that satirize the received literary language and style and those that do not. Both his kinds show a disruption of literary unity: a crack has appeared; there is a misfit between language and character. The first of his kinds may pose more of a threat to received literary culture because admired, heroic figures are degraded, but the satire in the second need not fall foursquare on the *horlogère* and *horloger*—epic language may also be exposed to ridicule. The first kind puts epic figures at risk; the second endangers epic language. Key questions for *The Dunciad* are how far it participates in this ridicule of the epic, showing its conception and language to be only one among many possibilities (and a flawed one at that), and how far it adapts the tactics of ridicule to a defence of the established literary order.

Although the poem is mock-epic, it occasionally touches on burlesque. An example is found in book 2, lines 79–88:

> A place there is, betwixt earth, air and seas,
> Where from Ambrosia, Jove retires for ease.
> There in his seat two spacious Vents appear,
> On this he sits, to that he leans his ear,
> And hears the various Vows of fond mankind,
> Some beg an eastern, some a western wind:
> All vain petitions, mounting to the sky,
> With reams abundant this abode supply;
> Amus'd he reads, and then returns the bills
> Sign'd with that Ichor which from Gods distills.

This passage bears one of the small-format, first edition's rare notes, 'See Lucian's Icaro-Menippus', to justify its vulgarity by reference to classical tradition.[20] There is undoubtedly an undermining of classical dignity here—a degrading of the gods to parallel the depiction of Saturn with a cold in Tassoni's *La secchia rapita*—and, beyond that, there is mockery of a particular epic style. Pope told Spence that there were passages in Hobbes's translation of Homer that might have been written to ridicule the poet. As James Sutherland notes, one of them is the passage which explains that the blood divine is

---

edition of *The Rape of the Lock*, Twickenham, 2. 107. Claude Rawson's 'Heroic Notes: Epic Idiom, Revision and the Mock-Footnote from the *Rape of the Lock* to the *Dunciad*', in Howard Erskine-Hill (ed.), *Alexander Pope: World and Word*, Proceedings of the British Academy 91 (Oxford: Oxford University Press for the British Academy, 1998), 69–110, has stimulating analyses of the poetry and some important reflections on 'chit-chat'. For an alternative view of mock-epic, one emphasizing the importance of particulars, see Gregory G. Colomb, *Designs on Truth: The Poetics of the Augustan Mock-Epic* (Philadelphia: Pennsylvania State University Press, 1992).

[20] The first Dunciad is now available in an excellent facsimile edition, *Pope's Dunciad of 1728: A History and Facsimile*, ed. David L. Vander Meulen. Vander Meulen provides the best account of the history of the composition of the poem. The octavo served as a large-paper version of the duodecimo; Pope was already thinking of a stratified readership.

> Not such as men have in their Veins, but *Ichor*.
> For Gods that neither eat Bread nor drink Wine
> Have in their Veins another kind of Liquor . . .[21]

The Dunciad lines have their origin in the materiality of this passage: the investigation of the digestive processes of the gods calls out for completion: eating and drinking lead to defecation as well as blood. But although Pope's lines depict a god in a position associated with human indignity and introduce an element of burlesque, nothing is allowed to debase their tone and style. The poem might have been invaded at this point by the matter-of-factness that makes the Hobbes passage potentially so amusing, or, as Warburton points out, it might have followed Dryden's 'MacFlecknoe' and spoken directly of 'Reliques of pies and martyrs of the bum'; but the dignity of the language is unvarying. Hobbes's language, which is essentially the object of ridicule, has to become an idea, by translation into an action, before it can enter the poem.[22] In this case Pope has created an episode in which excrement becomes writing, reversing the poem's customary transformation of self-expression and writing into excrement.

Pope's language, then, corresponds to Bakhtin's idealization of poetic language in one vital respect but not in another:

> And poetry also sees its own language surrounded by other languages, surrounded by literary and extra-literary heteroglossia. But poetry, striving for maximal purity, works in its own language *as if* that language were unitary, the only language, as if there were no heteroglossia outside it. ('Discourse in the Novel', 399)

Heroic language remains the only language, however inappropriate its subject. This is the point reinforced by the Imitations in the footnotes, which again and again draw attention to parallels in heroic writing, chiefly in Virgil; it is important that we identify the language as that of the heroic tradition. But the consequence of this one language, highlighted by the apparatus, is that this poetry cannot express the poet's intention in a pure and apparently unmediated way; the poet's language cannot be an Edenic language in which the word is 'co-extensive with its object' ('Discourse in the Novel', 331). Instead we are made aware of the great disparity between the language and its object, and in a shadowy way we become conscious of other languages that might be more appropriate. Some sort of dialogism is at the gates.

There are occasions when *The Dunciad* succeeds in grasping its subject directly and asocially. When Dulness first appears, for example, 'she beholds the Chaos dark and deep, | Where nameless somethings in their causes sleep' (1. 53–4). But for the most part its pure and heroic language is forced into contact with the socially specific and the socially repulsive. The claim in the publisher's preface that *'the Poem was not made for these Authors, but these Authors for the Poem'* (appendix, p. 90) cannot entirely

---

[21] *Twickenham*, 5. 108. Spence, *Anecdotes*, No. 451. Osborn's notes give details of the passage and Pope's response.

[22] This is one reason that many of Pope's jokes remain private, revealed only by conversation and commentary. Absurdities of style have to undergo intellectual translation before they can enter the poem, where, of course, they may cease to be recognizable.

protect it from contact with particular and known authors, even though they invade the poem as persons, not as languages. The poem's beginning relates the hero to a known place,

> Books and the Man I sing, the first who brings
> The Smithfield Muses to the Ear of Kings

but not to a known language. This is the style of Virgil, not that of Theobald, Gildon, or even Dennis. Characters in the poem speak, but not in their own voices. The leading example is Tibbald's prayer to Dulness in book 1. The passage has some variation in style. The beginning, 'Great Tamer of all human art!' (143) is heroic, followed rapidly by an allusion to Virgil, Theocritus, and Horace, 'With whom my Muse began, with whom shall end!', pointed up in the footnotes. The initial dignity, sustained by references to the 'Helvetian and Batavian land', is retained in images of struggle that follow, but fades as Tibbald moves to describe his own activities, though even here the eccentricities of word order, 'studious I', retain some epic gravity.

> Here studious I unlucky moderns save,
> Nor sleeps one error in its father's grave,
> Old puns restore, lost blunders nicely seek,
> And crucify poor Shakespear once a week.
> For thee I dim these eyes, and stuff this head,
> With all such reading as was never read;
> For thee supplying, in the worst of days,
> Notes to dull books, and prologues to dull plays;
> For thee explain a thing till all men doubt it,
> And write about it, Goddess, and about it . . . (1. 161–70)

With the punning of 'And crucify poor Shakespear once a week' (to afflict with pain and to afflict with a crux) and 'And write about it, Goddess, and about it' (to discuss and to circumlocute) the tone becomes relaxed and even downright ('dim these eyes, and stuff this head'). But in none of these styles do we hear Theobald's own voice, or even a vestige of his voice within a Popeian ridicule. These are Pope's opinions and Pope's puns. Tibbald is not an opposing voice but a puppet; this is the purest ventriloquism. Tibbald is made to pronounce Pope's own verdict of condemnation on himself. When Tibbald finally is quoted in the verse of *The Dunciad*, the words are put into the mouth of Settle at the end of the vision of the triumph of Dulness:

> For works like these let deathless Journals tell,
> 'None but Thy self can be thy parallel. (3. 271–2)[23]

The words, from Theobald's 'Shakespear's' *Double Falshood*, are not engaged with in any way; they are not even parodied; they are merely displayed, as their quotation marks suggest. The technique is that of the rhetorician, quoting a statement in a new context, holding it up to be judged. Significantly, the line does not appear in the first

---

[23] This shows Pope's preferred system of quotation marks: marks at the beginning of each line of quotation but no closing marks.

edition, where the satire is rarely personalized in quite this adversarial way; it takes its colour from the technique we come to associate with the *Variorum* footnotes.

Within the generally homogeneous tone of *The Dunciad* there are variations in the presentation of voices: sometimes a character rises to epigram, 'Dulness is sacred in a sound Divine' (2. 330); or there is mimicking of literary eccentricity, 'But who is he, in closet close y-pent' (3. 181), the equivalent of the 'yclep'd' and 'yclep't' of *Shakespeare Restored* (31); or there is a descent into absurdity, 'And am I now threescore? | Ah why, ye Gods! should two and two make four?' (2. 275-6). But the voice of a character never achieves a higher status than display as a thing; most characters have no individual voice at all. *The Dunciad* as poem operates as the rhetorician does, not as the novelist. This problem of speakers who are not allowed to speak in the poem, and are, therefore, in some way inadequately discovered, is one that the notes go some way to solving.

THE NOTES AND DIALOGUE

The notes to the *Variorum* are something like a talking bibliography. The book contains the first bibliography of attacks on the *Dunciad*, a list of some fifty-six books, and its bibliographical authority both powers and restricts its achievements. The preoccupation of the commentary with journals and books goes beyond the mere noting of sources and extends to conveying the business and background noise of contemporary publication. We are flooded with details such as: Dennis has written in the *Daily Journal* of 11 June 1728 (6; 1. 61); the Duchess of Newcastle produced eight folios, usually adorned with gilded covers and her coat of arms (12-13; 1. 122); Blome's books were remarkable for their cuts (13; 1. 126); great numbers of Ward's works were sold into the plantations (17; 1. 200); the *Cave of Poverty* was printed in octavo in 1715 (19; 1. 226); Haywood's works were printed in four volumes duodecimo, with her picture in the frontispiece (36; 2. 152); Concanen attacked Pope in the preface to *A Collection of Verses, Essays, Letters, &c.*, printed for A. Moore, page 6 (74; 3. 299). And these are merely strong examples of a pervasive interest in the goings on of the book trade. The wealth of detail in the commentary suggests that Pope had something of the train-spotter's enthusiasm for making records not only of journeys that involved him but of the general traffic of his time.

The Remarks in the commentary fall into three main types: notes that are drawn from Giles Jacob's *Lives of the English Poets* and other dictionaries; notes that quote Pope's opponents and their attacks on him (Theobald figures largely in these); and the notes of the official editor, Martinus Scriblerus. Jacob's biographical dictionary was a major stimulus to the creation of the original *Dunciad* in 1719-20, through its attempt to sum up and evaluate contemporary literature and its adoption of a 'Whig' view of literary history.[24] Jacob's most important service to Pope was to provide a vision of literary progress that generalized individual misjudgements. The preface

---

[24] I have discussed the relationship in more detail in 'Pope and Giles Jacob's *Lives of the Poets: The Dunciad* as Alternative Literary History', *Modern Philology*, 83 (1985-6), 22-32.

to Jacob's second volume confidently claims that the reader will have 'the Satisfaction of observing the Progress and Improvement of our English Poetry' (2, p. xiv) and the dedication of the first volume to George Granville, Lord Lansdowne, goes still further:

ÆNEAS in VIRGIL is made to look with Joy on the Heroes of his Family, who were to pass into the World, and do Honour to his Name; Your Lordship, from this backward View of Your Predecessors in Poetry, may receive a Pleasure of another kind: The Line of *Rome* began with Him, the Line of *Dramatick Poets* is crown'd and compleated in *You*. (1, pp. iii–iv)

This passage forecasts the action of book 3 of *The Dunciad* in which, like Aeneas, Tibbald is taken to the underworld, but, like Granville, is invited to review the line of writers that ends in him:

> And are these wonders, Son, to thee unknown?
> Unknown to thee? These wonders are thy own. . . .
> These, Fate reserv'd to grace thy reign divine,
> Foreseen by me, but ah! with-held from mine. (3. 269–74)

Jacob's claim in his dedication is appropriated and parodied in the general structure of *The Dunciad* rather than in particular episodes and styles, but other of Jacob's remarks are directly revoiced.[25] Jacob himself, in a letter to John Dennis, said that Pope had endeavoured to render him 'the Author of his Scandal',[26] and Pope's 'Index of the Author's of the Notes' shows Jacob to have been given a major role in the commentary, telling us about Dennis, Theobald, Cibber, Ward, Budgell, Ozell, Oldmixon, Centlivre, and Jacob himself. Jacob is often useful to Pope in supplying information and informing the reader, but frequently the information is accompanied by some technique of satiric display. Sometimes Jacob is accompanied by commentary, sometimes the entry is extended, sometimes it is inserted in a new context. The best example is in the treatment of John Dennis. Jacob had generally got his basic biographical information from the authors themselves:

As to the Accounts of the *Living* AUTHORS, most of them came from their own Hands, excepting such Parts as relate to the Fame of their Writings, where I thought my self at liberty to give such Characters of Praise or Disprais, as the best Judges before me had pass'd upon their Performances. (1, p. xi)

Pope had collaborated himself by correcting the proof of his own entry. Unfortunately for Dennis, Jacob let slip the method of compilation in discussing Dennis's entry: 'In the Account this Gentleman sent, he omitted, but for what Reason is unknown to us, a Play wrote by him' (1. 286). Pope saw his opportunity to hold Dennis responsible for the whole account:

---

[25] The parody of Jacob's conception is related to the question of hero, which I discuss later in the chapter. Granville was Pope's friend, and so Pope could not have written a pastiche *Aeneid* with Granville as hero. I suspect Pope may at an early stage have intended a patrician, and perhaps the King.

> This, this is He, foretold by ancient rhymes
> Th'Augustus born to bring Saturnian times . . . (3. 317–18)

[26] *The Critical Works of John Dennis*, 2. 372.

For his character as a writer, it is given us as follows. 'Mr. *Dennis* is *excellent* at pindarick writings, *perfectly regular* in all his performances, and a person of *sound Learning*. That he is master of a great deal of *Penetration* and *Judgment*, his criticisms (particularly on Prince *Arthur*) do sufficiently demonstrate.' From the same account it also appears, that he writ Plays 'more to get *Reputation* than *Money*.' DENNIS *of himself*. See *Jacob*'s Lives of Dram. Poets, page 68. 69. *compared with* page 286. (11; 1. 104)

The way this note puts words in Dennis's mouth is different from the ventriloquism afflicting Tibbald in the poem itself, but there are similarities in the technique. Jacob merely revealed inadvertently that Dennis had reviewed his own entry and provided information for the *Lives*, but Pope presents the entry, including the evaluations, as Dennis's own speech. Dennis is presented as speaking for himself, though the words are another's. The consequence is that the reader's attention is drawn to the extravagance of the praise: what in Dennis's mouth would be wildly boastful cannot be acceptable in Jacob's either because it must show a failure of discrimination. A similar tactic is employed in the account of Cibber, where Pope ridicules the assessment by adding an absurd one of his own.

Mr. *Colly Cibber*, an Author and Actor; of a good share of wit, and *uncommon vivacity*, which are much improved by the *conversation* he enjoys, which is of the *best*. JACOB *Lives of* Dram. Poets. p. 38. Besides 2 Volumes of Plays in 4°, he has made up and translated several others. Mr. *Jacob* omitted to remark, that he is particularly admirable in Tragedy. (20; 1. 240)

Cibber regarded this ironic praise of his performance in tragedy as particularly unfair, but Pope's main aim in this section was to add an utterly absurd judgement, with the implication that having progressed so far in absurdity Jacob should accept no limit.[27] The Cibber entry is genuinely informative, providing characteristic bibliographical detail, but Pope uses typography to make its own comment. The evaluations are in italic, highlighting (it is emphatic) the effusive nature of the praise.

In Jacob's account of Cibber the evaluation is coloured by certain social assumptions that Pope is unwilling to share. Cibber's own voice emerges in the claim to enjoy the best society, but only for the whole entry to be deflated by Pope's sarcasm. The *Variorum* allows repeated glances at what Pope takes to be the dunces' social milieu. A telling entry in the 'Index of Things (including Authors)' is under *Alehouse*:

Alehouse, The Birth-place of many Poems, i. 202.
—— And of some Poets, ii. 130.
—— One kept by *Taylor* the Water-poet, ii. 325.
—— and by *Edward Ward*, i. 200.

Ale figures significantly in *The Dunciad*, from Tibbald's address to his children, who grew in the 'alehouse', to

> Flow Welsted, flow! like thine inspirer, Beer,
> Tho' stale, not ripe; tho' thin, yet never clear . . . (3. 163–4)

[27] For Cibber's complaint and detailed commentary, see Rumbold's edition, 1. 250 and nn.

The entry for Edward Ward makes clear Pope's disapproval of attempts to disguise these social realities. 'He has of late Years kept a publick house in the City (but in a genteel way) and with his wit, humour, and good liquor (Ale) afforded his guests a pleasurable entertainment, especially those of the High-Church party. JACOB *Lives of Poets* vol. 2. p. 225' (17; 1. 200). Pope has added one word here, '(Ale)', but in doing so he displays for criticism the fake gentility of Jacob's account. The technique is one of forced distancing from the cosy world of mutual congratulation opened up by Jacob. At one point the disagreement emerges in a plain sarcasm. Pope says of Oldmixon, 'This person wrote numbers of books which are not come to our knowledge. "Dramatick works, and a volume of Poetry, consisting of heroic Epistles, *& c.* some whereof are very well done," saith that great Judge Mr. JACOB' (39; 2. 201). Jacob was not a great judge, but Pope usually gives his own verdict indirectly, by exaggerating his weakness for praise and thus holding his excesses up to ridicule.

Pope's games with dunces' own words in the notes are surprisingly mild, as can be seen from the case of the poem's hero, Lewis Theobald. Theobald's *Shakespeare Restored* haunts the pages of the *Variorum*, yet its treatment is almost good natured. The satire of Theobald in the first note, where exaggeration slides into quotation, is not untypical. Early in the commentary, Theobald is called on to defend Pope against Curll's complaint, 'This is an infamous Burlesque on a Text in Scripture, which shews the Author's delight is Prophaness' (5; 1. 48). Pope proceeds to quote six examples of Shakespeare's using scripture, often jocosely. As he acknowledges, these are drawn from *Shakespeare Restored*, 144, where Theobald has listed some fourteen. The pleasure in this citation comes chiefly from employing Theobald as a shield against criticism, but it is linked to ridicule of his habit of 'luxuriating' in examples, something Pope has great fun with elsewhere in the *Variorum*. In book 2 at line 177 there is a note on 'His rapid waters in their passage burn', with an alternative reading, 'glow', being investigated by the commentary. The note claims that 'every reader of our Poet must have observ'd how frequently he uses this word *glow* in other parts of his works' and cites seven examples from Homer before concluding this section, 'I am afraid of growing too luxuriant in examples, or I could stretch this catalogue to a great extent, but these are enough to prove his fondness for this *beautiful word*, which therefore let all future Editions re-place here' (38). The passage refers directly to *Shakespeare Restored*, 11, where Theobald is discussing coining new verbs from substantives. After the eighteenth example, he says, 'I am afraid of growing too luxuriant in Examples of this Sort, or I could stretch out the Catalogue of them to a great Extent', but he then goes on to give seven examples of making verbs out of adjectives. Pope's mockery in no way exaggerates the manner of the original; it rather underplays it. The technique is once again that of presenting the material in a new context. The context of a known reading demonstrates that the frequent occurrence of a word is not good evidence that it is the right reading in a particular place. A more powerful critique comes with the attack on the play Theobald attributed to Shakespeare, the *Double Falshood*. Pope summarizes and trivializes Theobald's four reasons for accepting the attribution, and then emphasizes the final one 'that he has a *great mind* every thing that is good in our tongue *should be* Shakespeare's'. He then proceeds to invent errors and emendations for two columns, admirably capturing Theobald's vocabulary of certainty: 'Read it, without contradiction'; 'Correct it boldly'; 'Restore it, with the alteration but of two letters' (72; 3. 272). This shows a development from the technique of display to one of parody, but once again language is directly applied to a different object. There is no engagement with the editorial suppositions that so separated Theobald's textual criticism from Pope's.

MARTINUS SCRIBLERUS

The contribution of Scriblerus to the *Dunciad Variorum* is much more important than that of the other annotators; I suspect it merits a full-length study. Unintentionally, Pope created through Scriblerus the critical voice that came to dominate mid-twentieth-century criticism of his poetry. The most striking example of this is Aubrey Williams's brilliant development of 'Martinus Scriblerus, Of the Poem', with its declaration that 'the Action of the Dunciad is the Removal of the Imperial seat of Dulness from the City to the polite world'.[28] More recently Frederick M. Keener has argued for a view of Scriblerus that refines and develops Williams's positive evaluation and takes 'Of the Poem' to be largely Pope's own explanation of his poem, using concepts he had drawn, quite straightforwardly, from René Le Bossu's *Traité du poëme epique*.[29] This interpretation of Scriblerus seriously reduces the dialogic energy of this section of the apparatus by identifying a single, shared critical position, and it leads to a diminished role for the particular in the poem. I believe Pope's general attitude to Le Bossu is respectful, without inhibiting him from having some fun at Le Bossu's expense, but that Scriblerus highlights a major point of disagreement with Le Bossu's theory of epic. Both in Scriblerus' general approach to the poem, and in his particular comments, Pope's stance is parodic.

The *Traité*'s treatment of epic is moral in its emphasis. Le Bossuean epic aims at reform of manners and teaching truth. This truth is then represented in a general narrative (the fable), which is constructed to convey it. The final stages in writing an epic are the identification of figures in history or myth who can fulfil the function of the figures in the narrative, and the development of the action, with its episodes, which finally gives particular embodiment to the moral and fable. Keener notes that, in Le Bossu's view, epic derives from the historical condition of the author's country and addresses contemporary problems; the moral solves these problems. The poet has two audiences and two messages in mind, both of them in need of the poem's moral. 'Martinus Scriblerus, Of the Poem' certainly follows Le Bossu, as the notes added to the second edition of the *Variorum* make clear. The poem is declared to be a comic epic. The problem faced by the nation is the easy availability of paper and print

---

[28] *Dunciad Variorum*, 24. Williams, *Pope's 'Dunciad'*. Gregory Colomb has some good comments on this school of criticism, which he calls the 'higher humanism', in his *Designs on Truth*.
[29] 'Pope, *The Dunciad*, Virgil, and the New Historicism of Le Bossu', *Eighteenth Century Life*, 15 (1991), 35–57.

and the consequent deluge of authors covering the land. Dulness and Poverty are the causes of these authors and so an allegory has been constructed in which these goddesses join together:

The great power of these Goddesses acting in alliance (whereof as the one is the mother of industry, so is the other of plodding) was to be exemplify'd in some *one, great* and *remarkable action*. And none cou'd be more so than that which our poet hath chosen, the introduction of the lowest diversions of the rabble in *Smithfield* to be the entertainment of the court and town; or in other words, the Action of the Dunciad is the Removal of the Imperial seat of Dulness from the City to the polite world; as that of the Æneid is the Removal of the empire of *Troy* to *Latium*. (24)

Scriblerus does not follow Le Bossu precisely here because he sweeps the fable into the action. Le Bossu does sometimes refer to the fable as the action, but his fable is always general. Scriblerus' account causally links his goddesses to a particular topography. The next step, however, is nicely in line with Le Bossu. Scriblerus argues that a bad writer must be found to support the action, and that Pope finds in Tibbald a bad journalist, playwright, poet, and critic (24–5). He then goes through the machinery, episodes, characters, and so on in Le Bossuean fashion. Later on Scriblerus is acute in finding a Le Bossu-style 'disguised sentence' at 1. 109, but, according to Keener, he misses one important aspect of the poem. An epic addresses and represents the ruler, 'Any Bossuean epic written in a monarchy had to be about the particular monarch' (54). *The Dunciad*, therefore, is about George II and is 'antimonarchical or at least anti-Georgian' (55). But this oversight need not, Keener believes, diminish the respect we accord to Scriblerus: 'We might well consider it [*The Dunciad*] not so much a mock epic as a comic Le Bossu epic, with an Action, a Fable, and Morals distinctly its own, relatively fully explained by Scriblerus, and conceivably substantial when one attends to them' (49).

I agree that the characterization of Scriblerus as a commentator on the *Dunciad* needs to be interpreted in the light of Pope's treatment of Le Bossu elsewhere, but that treatment is far from consistently positive. Pope did abstract parts of the *Traité* for 'A General View of the Epic Poem and of the Iliad and Odyssey' prefixed to the first volume of the *Odyssey*, published in April 1725, but that account would be more impressive as an indicator of Pope's attitude to the *Traité* if it endorsed and incorporated Le Bossu's point of view within its own account. Instead it is merely 'Extracted from Le Bossu'. It was certainly an act of deference to the critic, but it was also an easy way of supplying a preface. As the *Twickenham* editors note, 'The English follows rather closely the translation by "W.J." (London, 1695)' (9. 3 n. 1). Pope also drew on Le Bossu's *Traité* in the comic 'Receipt to Make an Epic Poem', first published in the *Guardian*, 10 June 1713, and reprinted in *Miscellanies. The Last Volume* in March 1728. The 'Receipt' has some fun at Le Bossu's expense in its structure and general stance, and particularly in its treatment of the hero. The mood is one of holiday rather than satire; cynical and easy, rather than solemn and strenuous composition. Clearly the advice, 'For the Moral and the Allegory. These you may extract out of the Fable afterwards, at your leisure: Be sure you *strain* them sufficiently', is not Le Bossu's or even a valid development of Le Bossu's, but mockery is rarely fair.[30] Like the headings (Fable, Episode, and so on, as used by Scriblerus), the conception of epic is his, and I suspect the beginning of the 'Receipt' refers to him:

An Epic Poem, the Critics agree, is the greatest work human nature is capable of. They have already laid down many mechanical rules for compositions of this sort, but at the same time they cut off almost all undertakers from the possibility of ever performing them; for the first qualification they unanimously require in a Poet, is a *Genius.* (*Prose Works*, 2. 228)

Although Le Bossu's aim is understanding, and providing a foundation for explaining, the *Aeneid*, his method is to lay out a process of composition for the poet. So W.J., translator of *Monsieur Bossu's Treatise of the Epick Poem* (1695), in a dedication to Blackmore that Pope must have found risible, praises the author of *Prince Arthur* for confining himself 'to the Rules and Precepts which *Aristotle* and *Horace*, and even our *Bossu*, have prescrib'd to the *Epick* Poem'.[31] It is this sense of prescription that Pope exploits comically in the 'Receipt'.

The key to Pope's critique of Le Bossu in the *Dunciad Variorum* lies in the relation of the hero to the fable. In Le Bossu the fable comes first, the hero and the action belonging to a second order of representation. The oddity of this conception—Le Bossu's accounts of fables look suspiciously like summaries of the poems with the names left out—seems emphasized even in Pope's appropriation for the 'General View of Epic'. W.J.'s translation of the key passage about the *Iliad* reads, 'He has given the Name of *Achilles* to a valiant and angry Phantom; that of Agamemnon to his General, that of Hector to the Enemies Commander, and others to the rest' (19). In the 'General View' this becomes, 'To a Phantom of his brain, whom he would paint valiant and cholerick, he has given the name *Achilles*; that of Agamemnon to his General; that of *Hector* to the Enemies Commander, and so to the rest' (9. 7). The phantom of the fable is a problem to Pope. He cannot accept a cipher in a scheme as a basis for creativity, so he insists on the phantom's place in the imagination, but, even as he does so, the naming gets causally linked to characterization. This phantom is not already valiant and choleric; he is made so in the particularization. This feature of Le Bossu's theory, which caused problems even in translation, is picked out for ridicule in the 'Receipt': 'Then take a Hero, whom you may chuse by the sound of his name, and put him into the midst of these adventures' (*Prose Works*, 2. 228). This parodies Le Bossu's own view, 'We may disguise the Fiction, render the Action more singular, and make it a *Rational Fable* by the Names of Men invented at Pleasure' (16). Pope is not happy with a hero arbitrarily chosen, and the point is picked up again in the presentation of Scriblerus.[32]

---

[30] *The Prose Works of Alexander Pope*, 2, ed. Rosemary Cowler (Oxford: Shakespeare Head for Basil Blackwell, 1986), 229.

[31] *Treatise*, A3ʳ, italics reversed.

[32] Of course, the name of the hero changed in 1743, which gives Scriblerus' attitude a specious credibility. In 1743 Pope created a new commentator, Richardus Aristarchus, who condemns the influence of Le Bossu and the '*Phantom of a Hero*, only raised up to support the Fable' (*Twickenham*, 5. 254). The

Scriblerus' popularity with modern critics is surprising not merely because of his name and general status in Pope's writings. In the *Variorum* Pope surrounded his criticism with warnings. The most important clue to the Scriblerian misconception of the poem is in the index (p. cxxiv): 'TIBBALD, why he was made Hero of this Poem? according to *Scriblerus*. Prolegom. p. 25. The true reason. Book i. 102. and iii. 319.' Scriblerus' reason follows on from his account of the truth, allegory, and action of the poem, an account which the second edition of the *Variorum* linked to Le Bossu by footnotes:

> A *Person* must be fix'd upon to support this action, who (to agree with the said design) must be such an one as is capable of being all three [party-writer, dull poet, and wild critic]. This *phantom* in the poet's mind, must have a *name:* He seeks for one who hath been concerned in the *Journals*, written bad *Plays* or *Poems*, and published low *Criticisms*: He finds his name to be *Tibbald*, and he becomes of course the Hero of the poem. (24–5)

The reference to the phantom picks up the translation of the 'General Account'. On this interpretation, the personal satire of the poem is incidental to the general design; the poem is conceptual, and names which appear to be of persons refer at a deeper level to concepts. Against this interpretation Pope sets the personal and particular. His note on Tibbald deals in personal and representational matters:

> During the space of two years, while Mr. *Pope* was preparing his Edition of *Shakespear*, and published Advertisements, requesting all lovers of the Author to contribute to a more perfect one; this Restorer (who had then some correspondence with him, and was solliciting favours by Letters) did wholly conceal his design, 'till after its publication. Probably that proceeding elevated him to the Dignity he holds in this Poem, which he seems to deserve no other way better than his brethren; unless we impute it to the share he had in the Journals, cited among the *Testimonies of Authors* prefixed to this work. (11; 1. 104)

The index's 'true reason' is a matter of conduct, fixed precisely in time and mores. Theobald behaved badly enough to be chosen hero. The contrary sense, that he did not deserve the status, comes from his pettiness. Pope is anxious to deprive this evil of any grandeur. The note on 3. 319 furthers this debunking: 'Mr. *Eusden* was made *Laureate* for the same reason that Mr. *Tibbald* was made *Hero* of This Poem, because there was *no better to be had*.' Theobald is guilty of his crime but unworthy of his poem.

Scriblerus recognizes no particulars. The seeds of Pope's conception of Scriblerus may be found in the *Memoirs of Scriblerus*, though the characteristic patterns of thought Scriblerus later adopts belong in that book to his schoolfellow Crambe:

> Martin's understanding was so totally immers'd in *sensible objects*, that he demanded examples from Material things of the abstracted Ideas of Logick: As for Crambe, he contented himself with the Words, and when he could but form some conceit upon them, was fully satisfied.

commentator is thus set against the grain of the poem, just as I believe he is in the 1729 *Variorum*. There is some critical tendency to read 1729 in the light of 1743: Pope did not then know he would change his hero; what was at first a joke later became an opportunity.

Thus Crambe would tell his Instructor, that All men were not *singular*; that Individuality could hardly be praedicated of any man ... few men have that most valuable logical endowment, Individuality.[33]

It is easy to see how Le Bossu's thinking about the hero could be assimilated to this failure of integration. This foolish division between the material and the abstract —either side, we are to understand, is wrong—is maintained in Scriblerus' commentary to the *Variorum*, but now Crambe's opposition to materiality and individuality becomes Scriblerus'. *The Dunciad*, of course, contains its own 'phantom' episode. In the first of the games in book 2, Dulness places a poet's form before the dunces:

> Never was dash'd out, at one lucky hit,
> A Fool, so just a copy of a Wit;
> So like, that criticks said and courtiers swore,
> A wit it was, and call'd the phantom, More. (2. 43–6)

The 'phantom' is one in a sense quite different from Le Bossu's: a fake, a pretender to wit, Dulness's attempt at a Wit. The phantom is identified as the plagiarist James Moore-Smythe, but the identification is then denied on the grounds that Pope could not have been bothered to vindicate himself against Smythe. When Scriblerus enters the debate, it is in order to deny any possibility of identification: '*The Phantom, More.*] It appears from hence that this is not the name of a real person, but fictitious.' He gives an etymology and quotations from Erasmus to support his case. Pope, having already established the identification, mocks the reduction of his socially specific poem to an abstraction. The mockery is reinforced in the note to line 118:

Breval, Besaleel, Bond.] I foresee it will be objected from this line, that we were in an error in our assertion on verse 46. of this Book, that *More* was a fictitious name, since these persons are equally represented by the Poet as phantoms. So at first sight it may seem: but be not deceived, Reader! these also are not real persons. [The text then identifies two of them.] Thou may'st depend on it no such authors ever lived: All phantoms!

Le Bossu's understanding of Achilles has been picked up, analysed as an intellectual error, and then parodied in Scriblerus.

The critique is developed a little further in Scriblerus' footnote to 1. 109. Former commentators idly supposed 'supperless he sate' in the description of Tibbald to imply 'that the Hero of the Poem wanted a supper'. Scriblerus knows better, 'In truth a great absurdity! ... much more refin'd, I will venture to say, is the meaning of our author: It was to give us obliquely a curious precept, or what *Bossu* calls a *disguised sentence*, that "Temperance is the life of Study" ... to represent a Critic encompast with books, but without a supper, is a picture which lively expresseth how much the true Critic prefers the diet of the mind to that of the body' (11–2; 1. 109). I doubt even critics of a Scriblerian/Bossuean cast of mind can believe that interpretation.

---

[33] *Memoirs of the Extraordinary Life, Works, and Discoveries of Martinus Scriblerus*, ed. Charles Kerby-Miller (New York: Russell & Russell, 1966), 119.

This mode of interpretation, denying the social and material reality of Pope's references, runs through Scriblerus' commentary. According to Scriblerus, 'all the Mighty Mad in Dennis rage' is 'by no means to be understood literally, as if Mr. D. were really mad; Not that we are ignorant of the *Narrative* of Dr. *R. Norris*, but it deserveth no more regard than the *Pop upon P.* and the like idle Trash, written by *James Moor*, or other young and light Persons, who themselves better deserve to be blooded, scarified, or whipped, for such their ungracious merriment with their Elders' (9; 1. 104). In discussion of the important question of whether Curll's 'rapid waters' 'burn' or 'glow', Scriblerus is confident of his author's personality: 'For surely every lover of our author will conclude he had more humanity, than to insult a man on such a misfortune or calamity, which could never befal him purely by his *own fault*, but from an unhappy communication with another' (38; 2. 177). Similarly Eusden should not be taken as drunk, the poet is teaching us not to judge by appearances (52–3; 2. 395). Finally (for I am afraid of growing too luxuriant in examples) the vision of book 3 is no more than 'the Chimera of the Dreamer's brain, and not a real or intended satire on the Present Age' (54; 3. 5, 6). While the notes from Jacob connect the work with representations of social reality, and the references to Theobald connect it to alternative scholarship, the Scriblerus notes try, ludicrously, to free the poem from its contemporary setting. They are of one piece with 'Martinus Scriblerus, Of the Poem'.

The Scriblerian voice is the most successful aspect of the *Variorum*'s dialogism. There is an intellectual consistency combined with a special tone, elevated and antiquarian but also betraying personal impatience and irascibility. The urge to abstraction and the closing of the world of the poem go alongside an apparent sympathy with bad writers. But the concern with 'humanity' as a quality does not extend to young writers who deserve to be whipped, and the closed world of the work does not save other scholars from being criticized. It follows from this view of Scriblerus that I believe his interpretation of the poem is not Pope's and possesses only a small degree of truth. That does not mean that the allegory and action Scriblerus outlines are to be obliterated from the reader's consciousness. They remain, in the old structuralist sense, under erasure, a potential for moral exemplification within the poem. Le Bossu would have been happy for a satire to differ from his account of the epic. Of reference, for example, he says that sometimes 'the *Things* and the *Names* of the Persons are *singular* and *true*, and not feigned or invented by the Poet. The *Satyrists* make use of this sort' (36). If, as I believe, *The Dunciad* is mock-epic rather than comic epic, as Scriblerus maintains, its closeness to the *Aeneid*, especially books 5 and 6, is fully justified, and Pope alerts us to this quality of his poem by adding to Scriblerus' account of the action the simile, 'as that of the *Æneid* is the Removal of the empire of *Troy* to *Latium*'. This is a better indication of the structure of the poem than the misapplied Bossuean theory. As the poem is not comic epic, it is free to deal more directly with the social problems it confronts; even if it were required by Le Bossu that an epic should address and represent the ruler, this poem need not. But, as it happens, the ruler is not one of Le Bossu's two addressees for the *Iliad* anyway, and address and representation are separate matters. I should have been inclined to criticize Pope unfairly and argue that the presentation of Scriblerus' voice was unsubtle and his monomania too plain, had not critical response since 1945 shown me to be mistaken.

APPARATUS BEFORE AND AFTER THE POEM

The various material that surrounds *The Dunciad* and its commentary largely falls into the category of the rhetorical, rather than following the lines of the novel's development. The 'Testimonies of Authors' and the 'Parallel of the Characters of Mr. Dryden and Mr. Pope', for example, both present judgements in unusual contexts. The comments are abstracted and abbreviated and then placed in a pattern of juxtaposition on the page. The technique is essentially one of display and the voices are never able to engage with one another. There are two exceptions to this restricted dialogue in the apparatus. One, the superb *Guardian* essay on pastoral, succeeds in mastering the tone of its opponents and developing it by almost indiscernible moves into absurdity. Pope was obviously very fond of this essay or he would not have reprinted it in the *Variorum*, where it does not belong. The other is William Cleland's 'Letter to the Publisher'. Like the advertisement to the *Variorum*, the 'Letter' has a decidedly juridical flavour, but it also gives Pope, who doubtless wrote it, the opportunity for a delicate exercise in the mode of the gentlemanly amateur. The stress is on justice, more particularly on fair play. The poem is an orphan, deserving protection, and Cleland sees himself as championing the oppressed. He is not a professional annotator; he acts from friendship and a sense of honour. He has esteemed Pope as a good man, and his reputation is bound up with that of his friend. Throughout the piece Pope captures perfectly the reference of ethical questions to an inner circle of chums and the tone of lofty disdain ('These people') towards outsiders. At one point (in dealing with those of rank and fortune who have been attacked in the poem) the code emerges even more openly: 'these I was sorry to see in such company. But if without any provocation, two or three gentlemen will fall upon one, in an affair wherein his interest and reputation are equally embark'd; they cannot certainly, after they had been content to print themselves his enemies, complain of being put into the number of them?' (11). They have let the side down and must take the consequences. With the weapon of this ideology Pope mounts knock-down arguments against his opponents. The dunces are the aggressors, they have 'insulted the Fallen, the Friendless, the Exil'd, and the Dead' (8); they may be obscure, but they cannot be too low for satire, for that would mean leaving the rabble and domestic servants unchastised; they may be poor, but they are poor because they are bad writers, and bad writing, not poverty, is the object of the satire. They may be dull, but they are not ridiculed for being dull but for being dullards pretending to be wits. This letter is double-voiced, for Pope is much more knowing, much more caught up in this jostling world of letters than Cleland. His aim is partly to invoke a class position in favour of his own, but Cleland's outlook is more than a mere class one. Pope is taking the rare step of invoking the disinterested criticism of the citizen of the public sphere; Cleland has a stake in society and speaks honestly and clearly. The essay is a *tour de force* and it has been remarkably influential in Pope criticism; but I suspect that, like the interpretations of Scriblerus, it was close to being a mistake. Cleland is totally humourless and wholly respectable, whereas *The Dunciad* is humourful and far from respectable by Cleland's criteria. From the point of view of critical comfort and success the moral of this investigation may be that Pope should have made the *Variorum* not more dialogic but less.

The apparatus to *The Dunciad Variorum* has generally been undervalued, partly because it has not been easy to find it in readable form. Like *The Rape of the Lock*, the *Variorum* is an invaginated work. Outside material, often the words of others, becomes as much Pope's as the inside of the poem. And in some ways the outside circumstances—Theobald's treachery, Moore-Smythe's plagiarism—lie within the poem itself. Scriblerus draws our attention to this facet of the work through his denials of their reality. The notes are a genuine guide to the poem's intertextuality, its network of allusions (particularly to the *Aeneid*), and the context of printed controversy. Throughout, Pope is conscious of the interrelation of interpretation and evaluation. The apparatus and footnotes suggest modes of interpretation that engage issues of evaluation. The poem is a good poem, Scriblerus implies, because it has no contemporary reference; or it is a good poem, Cleland argues, because it is a justified response to provocation. Neither approach is wholly adequate, and neither voice is Pope's. Pope speaks through others, arranging real voices, Jacob's or Theobald's, to suit his design, or inventing true and false friends in Cleland and Scriblerus, each of whom defends the poem with some validity but misses the playfulness with which social reality is engaged. The resulting air of moral ambiguity was doubtless part of Pope's design. Its richness, or evasiveness, is a function of his command of typographical resources.

5. An Essay on Man *and Harte's* An Essay on Reason: *Title Pages and Implied Authorship*

Although Pope enjoyed marking his books with their author's personality, using typography as a form of self-presentation, he also relished anonymous and pseudonymous publication. This was particularly the case with *An Essay on Man*, culminating in a curious period in the mid-1730s when Pope seems to have thought of Walter Harte's *An Essay on Reason* as his own poem, or at least as a companion to his *Essay* and a poem that might as well have been his own. Anonymity had already ensured an unbiased reception for *An Essay on Man* between February 1733 and January 1734, but in February 1735 Pope was able to use his skills as a publisher to present Harte's poem so that it might be read as his own and reflect favourably on the piety and orthodoxy of *An Essay on Man*. The two poems differ in many ways—not least in their conception of reason itself—but they do breathe a common intellectual atmosphere. A comparison between them has much to say about *An Essay on Man*: about its status as an essay and its relation to philosophical argument; about its compatibility with Christianity and revelation; about its conception of reason; and about the unusual nature of its understanding of order. Although the impetus for this chapter comes from typography—Harte's title page makes *An Essay on Reason* look like a Pope publication—the subsequent discussion ranges very freely into a full comparison between the two poems, drawing intermittently on such matters as headnote précis, quotation marks, and capitalization. Finally a short epilogue considers Harte's recantation of the views expressed in his sermon on *The Union and Harmony of Reason, Morality, and Revealed Religion* (1737), a companion piece to *An Essay on Reason*, and an index of what accommodations a pious and orthodox Christian might make with deism in this period.[1]

THE BLURRING OF AUTHORSHIP

In his edition of the *Works of Alexander Pope* (1797), Warton relays an anecdote from Harte about Pope's response to requests for a Christian clarification of *An Essay on Man*:

---

[1] In his seminal study *Pope and the Context of Controversy: The Manipulation of Ideas in 'An Essay on Man'* (Chicago: University of Chicago Press, 1970), 11 n., Douglas H. White cites R. S. Crane's definition of deism, 'the essence of deism . . . lay in its radical assertion, against Christianity, of the principle that any religion necessary for salvation must be one that has always and everywhere been known to men', and adds, 'I define a deist as one who espouses the *sufficiency* of natural religion. An Anglican, therefore, who defended the *excellence* of natural religion but did not hold it to be sufficient for man's moral needs would not be a deist.' For a presentation of the relevant material, see also G. Douglas Atkins, 'Pope and Deism: A New Analysis', *Huntington Library Quarterly*, 35 (1972), 257–78.

Mr. Harte more than once assured me, that he had seen the pressing letter Dr. Young wrote to Pope, urging him to write something on the side of revelation; to which he alluded in the first Night-thought:

> O had he press'd his theme, pursu'd the track
> Which opens out of darkness into day!
> O had he mounted on his wing of fire,
> Soar'd when I sink, and sung immortal man!

And when Harte frequently made the same request, he used to answer, 'No, no! you have already done it;' alluding to Harte's *Essay on Reason*, which Harte thought a lame apology, and hardly serious.[2]

The anecdote had appeared in a slightly different form in Warton's *An Essay on the Genius and Writings of Pope*, without the 'No, no! you have already done it', but that omission was compensated at the close of another anecdote, this time about authorship:

Our author told Mr. HARTE, that in order to disguise his being the author of the second epistle of the Essay on Man, he made, in the first edition, the following bad rhyme:

> A cheat! a whore! who starts not at the *name*,
> In all the inns of court, or Drury-*Lane*?

And HARTE remembered to have heard it urged, in enquiries about the author, whilst he was unknown, that it was impossible it could be POPE's, on account of this very passage. POPE inserted many good lines in HARTE's Essay on *Reason*.[3]

What emerges from these anecdotes and from Pope's contemporary correspondence is that Pope was pursuing a policy of creating confusion about authorship of the *Essay on Man* right up to the publication of the *Works* of 1735, and that he intended Harte's poem to add a final twist to the plot by being mistaken for his own.

The anonymous publication of the *Essay on Man* is the outstanding success in the history of Pope's experiments in camouflage publication. The publication of the *Epistle to Burlington* in December 1731 had generated a frightening degree of controversy. As a result Pope held back the publication of *To Bathurst*, until 15 January 1733, and then followed that up rapidly with the *First Satire of the Second Book of Horace* on 15 February, and the first epistle of *An Essay on Man* on 20 February. *To Bathurst* was advertised on the title page as 'By Mr. Pope', while the *First Satire of the Second Book* was 'Imitated in a Dialogue between Alexander Pope of Twickenham ... and his Learned Council'; but *An Essay on Man. Address'd to a Friend. Part I* was anonymous. As Maynard Mack has shown, the anonymous publication of the philosophical poem (a genre Pope had not attempted before) was a complete success, earning praise for its 'greatness of Thoughts' (*Life*, 522). The second, third, and fourth epistles of the *Essay* also appeared anonymously, on 29 March 1733, 8 May 1733, and 24 January 1734, but when the complete four-book

---

[2] *Works of Alexander Pope*, ed. Joseph Warton, 9 vols. (London, 1797), 3. 10–11.
[3] *An Essay on the Genius and Writings of Pope*, 5th edn., 2 vols. (London, 1806), 2. 149 n. The anecdote appears in the first edition of vol. 2 (1782), 154 n.

poem was published on 20 April the appearance of 'To Henry St. John, L. Bolingbroke' on the title page constituted something like an open acknowledgement of authorship.

By this point in his career Pope knew quite well that anonymous publication involved more than the absence of his name from the title page; it also meant secret dealings with booksellers and printers, and a title and imprint that would throw readers off the scent. Pope was now in a good position to make such arrangements. From the publication of the *Dunciad Variorum* in 1729 onwards he had dealt mainly with a printer and a bookseller he had set up in business, John Wright and Lawton Gilliver. Together they constituted a sort of 'House of Pope', with Gilliver trading at the appropriately named Homer's Head in Fleet Street. They published Pope's own poems and the work of his friends and admirers. On 1 December 1732 Gilliver had been signed up to publish the *Essay on Man* and its companion epistles, paying £50 for the first year's rights to each epistle and promising to enter them all in the Stationers' Register. When *To Bathurst* appeared on 15 January 1733, it declared the author, printer, and bookseller openly on the title page: 'By Mr. Pope . . . London: Printed by J. Wright, for Lawton Gilliver at Homer's Head.' The imprint 'Printed by X for Y' is not usual in books of the period. Major productions, perhaps the work of a consortium, tend to bear it, as do works where the printing is particularly elaborate, but it is perhaps most often found when the author has dealt directly with the printer, paying the bill, so that the bookseller's role in the publication is diminished. I suspect the appearance of 'Printed by J[ohn] Wright for Lawton Gilliver' on so many of the title pages of Pope and his friends reflects this enhanced status of the author.[4] Even if Pope had not paid the printer, there was an arrangement that put Pope in charge. The title page of *To Bathurst*, therefore, declared itself pretty formally as a Pope poem from the 'House of Pope', and that opened up possibilities of deception in the two title pages of the following month.

The *First Satire of the Second Book* undoubtedly distracted attention from the *Essay on Man* by providing a controversial poem in a quite different genre, but it did so through a tease. Its title page is a masterpiece of ambiguity, leaving open a doubt about whether or not the poem is by Pope, and certainly implying that it is an unauthorized publication:

THE | FIRST SATIRE | OF THE | *SECOND BOOK* | OF | HORACE, | Imitated in a DIALOGUE between | *ALEXANDER POPE*, of *Twickenham in Com.* | *Midd.* Esq; on the one Part, and his LEARNED | COUNCIL on the other. . . . | *LONDON:* | Printed by *L. G.* and sold by A. DODD, near *Temple-Bar*; E. NUTT, | at the *Royal Exchange*; and by the Booksellers of *London* and | *Westminster*. M.DCC.XXXIII.

Although Pope's name is on the title page, he is not specified as the author; he is a participant in a dialogue. A similar puzzle is created in James Miller's *Are These Things So?* (1740), which has the subtitle 'The Previous Question, from an Englishman in

---

[4] See McLaverty, *Pope's Printer, John Wright*, and 'Lawton Gilliver: Pope's Bookseller', *Studies in Bibliography*, 32 (1979), 101–24.

his Grotto to a Great Man at Court', and which also implies an impersonation of Pope. In the first poem Pope represents himself; in the second Miller represents him. If the question of authorship is solved by a reading of the poem, the difficulty of authorization is not. Gilliver had paid his £50 for the *First Satire of the Second Book* on 4 March 1733, and he had already entered the poem in the Stationers' Register on 14 February, so he was fully entitled to have his name on the imprint, 'Printed for Lawton Gilliver'. The form the imprint takes, 'Printed by L. G.' (Gilliver did not print), is obviously misleading, an evasion. It suggests a piracy, with the name of a familiar book-trade figure being misspelled or abbreviated, as here. The presence of the names of two mercuries on the imprint further emphasized the likely unofficial status of the poem. I do not believe that Pope was actually unwilling to own, and accept responsibility for, *The First Satire of the Second Book*, but he was pretending to be. The hostile critics would have been congratulating themselves on finding Pope out, while the following poem, neutrally presented, slipped past unnoticed: 'An Essay On Man. Address'd to a Friend. Part I. London: Printed for J. Wilford, at the Three Flower-de-luces, behind the Chapter-house, St. Paul's. [Price One Shilling.]'. Gilliver paid Pope the £50 on 23 March but the poem had been entered in the Stationers' Register on 10 March by the John Wilford of the imprint. Wilford was the distributor of another Pope–Gilliver publication, the *Grub-street Journal*, and John Huggonson, the printer of the *Journal*, also printed the first epistle of *An Essay on Man*. Pope's involvement in the *Journal* was secret, and he could not have been identified through the bookseller and printer; the whole transaction was probably managed discreetly by Gilliver, without Pope's direct involvement. With both official and 'unofficial' poems by Pope in the public arena at the same time, the chance that the *Essay on Man* would be recognized as Pope's was small.

When Pope's authorship of *An Essay on Man* was admitted through the reference to Bolingbroke on the title page of the four-epistle edition on 20 April 1734, the imprint also acknowledged a 'House of Pope' publication:

AN | ESSAY on MAN, | Being the FIRST BOOK of | ETHIC EPISTLES. | TO | HENRY St. JOHN, L. BOLINGBROKE. | [engraving] | *LONDON:* | Printed by JOHN WRIGHT, for LAWTON GILLIVER, | MDCCXXXIV.

The splendid engraving of Apollo's head emerging in glory, surrounded by the Greek motto 'Know thyself', from a mass of clouds, cobwebs, and scholastic tags, stressed the dignity of the publication, while copies of this edition also originally bore at the end an 'Index to the Ethic Epistles' which laid out plans for Pope's *opus magnum*.[5] The first book of the *opus magnum* was to consist of the four epistles of *An*

---

[5] The leaf was cancelled. It survives in one copy, formerly in the possession of H. B. Forster, now in Cambridge University Library (Forster b. 129). The Scolar Press facsimile of this edition (Menston, 1969) reprints the index, as does Foxon, *Pope and the Book Trade*, 125. The leaf is transcribed from Bethel's copy in Spence's *Anecdotes*, No. 300, where Spence says that it was annexed to 'about a dozen copies' sent to particular friends. 'Most of these were afterwards called in again, but that which was sent to Mr. Bethel was not.' For an account of the project, see Miriam Leranbaum, *Alexander Pope's 'Opus Magnum'* (Oxford: Oxford University Press, 1977).

Essay on Man; the second book, 'Of the Use of Things', of nine poems, the first of which was called, 'Of the Limits of Human Reason'. This may have been what made Pope first consider, probably in May or June 1734, the possibility of appropriating Harte's poem in order to enhance his own reputation.

In 1734 Harte, of St Mary's Hall, Oxford, was still only 25. I do not know how or when he and Pope first met.[6] Spence records an anecdote from Harte about Pope in 1726 (*Anecdotes*, 556), the year in which he was 17. He is first noticed in Pope's *Correspondence* the following year, in a letter from Fenton to Broome of 28 March 1727, which announces mistakenly that 'little Harte the poet is dead in the country' (2. 429). Pope had certainly become involved in correcting Harte's *Poems on Several Occasions*, published by Bernard Lintot in 1727 (*Correspondence*, 2. 430 n.), and a verse in that collection, complimenting Pope, was later included in the Testimonies of Authors prefixed to the *Dunciad Variorum*:

> O! ever worthy, ever crown'd with praise!
> Blest in thy *life*, and blest in all thy *lays*.

And more in the same vein. He also wrote an *Essay on Satire* in defence of the *Dunciad*, published in 1730. Pope gives a touching short account of him in a letter to Burlington, trying to gain him a living:

My Zeal for him was not only moved by his Ingenuity & Morals, but by his great Piety to his Parents both whom he has maintain'd in their old age upon the *whole* of his small Income by pupils in the university: & left himself nothing but his Clothes & Commons many years. (*Correspondence*, 3. 332)

In 1734 he did obtain the living of Gosfield, Essex, probably through the good offices of Pope's friend Mrs Knight, James Craggs's sister Ann. Her husband John Knight, another of Pope's correspondents, who died 2 October 1733, is praised in *An Essay on Reason*. Pope was still discussing Harte's prospects of obtaining a further living in a letter of 25 November 1735. It is clear that, by the time of his appointment to Gosfield, Pope and Harte had become firm friends.

Pope's takeover of Harte's *Essay on Reason* is marked by a letter from Pope to Mallet which Sherburn dates May or June 1734, a month or two after the publication of the four-epistle *Essay on Man*. Pope has evidently taken charge of publication:

Pray tell Mr Harte I have given Gilliver his Poem to print, but whether he would chuse to publish it now, or next winter, let himself judge. I undertook to correct the press, but find myself so bad a Reviser by what I see has escaped me in my last thing, that I believe he had best have it sent him to Oxford, & besides that may be but an amusement to his or Your Eyes which indeed is a Pain to mine, since the frequency of my last Headakes. You will order Gilliver accordingly, & upon the whole let Mr Harte give him directions. I fancy the Title of an *Essay on Reason* is the best, & am half of opinion, if no Name be set to it, the public will think it mine especially since in the Index, (annext to the large paper Edition of the Essay on Man) the

---

[6] My guess would be that Harte sent Pope his poems and that Pope then recommended them to Lintot. Lintot paid Harte £30 for the poems on 18 Nov. 1726, a generous sum. It is unlikely that after the quarrel over the *Odyssey* Pope would have reviewed the poems at Lintot's request.

Subject of the next Epistle is mentioned to be *of Human Reason* &c. But whether this may be an Inducement, or the Contrary, to Mr Harte, I know not: I like his poem so well (especially since his last alterations) that it would no way displease me. (*Correspondence*, 3. 408–9)

Pope's relation to the poem is that of a modern editor or even publisher; Gilliver's position more that of a manager than the head of a business. Only Pope's headaches prevent him from standing in for the author as proofreader or reviser, and at the end of the letter he begins to toy with the idea of public confusion about authorship. Everything springs from the title page. The title, which Pope obviously has a role in choosing, affects the poem's perceived relations to other poems. The choice of *An Essay on Reason* will suggest the poem is a companion to *An Essay on Man*. If no author is declared on the title page, the inference will be that this is Pope's next poem, advertised in the Index to the *Essay on Man*. It may be that the 'last alterations' Pope approves of were among the 'many good lines' Warton claims Pope inserted in the poem. Sherburn records an autograph letter dated 9 Feb. 1727 from Harte to an unidentified correspondent, saying: 'My miscellany is now entirely printed off, except two sheets of Divine Poems. I thought to have published much sooner, but as Mr. Pope was pleased to correct every page with his own hand, I could not hurry him' (*Correspondence*, 2. 429–30 n. 4). Pope subscribed for four copies of that book; perhaps he was generous enough to correct every page of *An Essay on Reason* as well.

In his letter to Mallet Pope implies that he would not be displeased by a confusion about authorship of Harte's poem because of the quality of the writing, but that seems unlikely to be his real motive. Harte's poem is often harmonious and pithy in its sentiments, but 'competent' would not be an ungenerous overall verdict. He was more probably influenced by an awareness that *An Essay on Reason* had a colouring of Christian piety absent from *An Essay on Man*. Hence the response when asked to write on the side of revelation, 'No, no! you have already done it.' Identification of Harte's poem as Pope's, even temporarily, would therefore have established in the public mind the consonance of *An Essay on Man* with Christian teaching. Both *An Essay on Man* and its context suggest that Pope really believed in such consonance, but he was unwilling to make a positive commitment to a Christian understanding or development of the poem. His clearest statement on the issue was made to Spence, 21–5 February 1743, 'Some wonder why I did not take in the fall of man in my *Essay*, and others how the immortality of the soul came to be omitted. The reason is plain. They both lay out of my subject, which was only to consider man as he is, in his present state, not in his past or future' (Spence, *Anecdotes*, 306). But in an important sense, to include Christian doctrine would have been to narrow the scope of the poem rather than to broaden it. *An Essay on Man* is a very ambitious poem; it is for all men, and not just Christians. That Christians should find sustenance in the poem was desirable, and the encouragement Pope gave to Harte was therefore entirely appropriate. Later his semi-incorporation of Warburton's reading of the poem into his text through the notes carried this Christianization much further; but, as Warton was keen to emphasize, Pope never himself narrowed the poem to a Christian interpretation. A mistake over the authorship of *An Essay on Reason* would encourage one legitimate reading of *An Essay on Man* without disabling others. Less worthily, Pope must have been aware that a further mistake about authorship in relation to *An Essay on Man* would seriously discredit his critics.

Harte co-operated in disguising his authorship, accepting the title recommended by Pope, omitting his name from the title page, and allowing the 'House of Pope' imprint. Perhaps, like Broome and Fenton in the case of the *Odyssey* translation, he felt that a mistake in attribution would enhance his stature as a poet. The *Essay on Reason* appeared as part of a campaign of publication for the second volume of Pope's *Works*. The fullest picture emerges in an advertisement leaf that appeared with *To a Lady*, published 7 February 1735 (Fig. 7). A careful reading of the leaf suggests that the *Essay on Reason* is not part of Pope's *Works*, because it is not listed in any of the sections, but the very naming of the poem on this page suggests that somehow it does belong with these other poems. And, of course, the fact that parts may be had singly leaves open the possibility that, although this poem has not been allocated to any section of the new collection, collectors who want all Pope's poems should buy it and include it.

The *Essay on Reason* was entered in the Stationers' Register by Gilliver on the same day as *To a Lady*, 7 February 1735, and both poems were received by Pope's friend, Edward Harley, Earl of Oxford, the day before; they seem to have been twin publications. Harte's title page (Fig. 8) is typical of the 'House of Pope'. The ornament is the same that appears on the title page and throughout the *Dunciad Variorum*; it also appears on the title pages of *To Burlington* and *To Arbuthnot*, published just a month before, on 2 January 1735 (Fig. 8). The imprint is the same as that on *To Bathurst*, *To a Lady*, and *To Arbuthnot* (though in this last case the address is missing). The *Essay on Reason* looks like a Pope poem and has every appearance of being a part of the series. The day after publication Pope sent a letter to Caryll building on the implications of the title page:

> I send you constantly whatever is mine. The ludicrous (or if you please) the obscene thing you desired me to send, I did not approve of, and therefore did not care to propagate by sending into the country at all. Whoever likes it so well as to think it mine, compliments me at my own expense. But there is another piece, which I may venture to send you, in a post or two, *An Essay on Reason* of a serious kind, and the intention and doctrines of which I think you will not disapprove. (*Correspondence*, 3. 450)

The equivocation is masterly. The ludicrous or obscene thing that Pope does not approve of is *Sober Advice from Horace*, published on 28 December 1735, 'Imitated in the Manner of Mr. Pope', and in fact written by him. It does not look like a Pope poem, bearing the imprint of T. Boreman, to whom, with some others, Curll says Pope sold it 'by Agency' (Griffith 347). Caryll is left to work out for himself whether *Sober Advice* was not sent because it was not Pope's or because it was obscene. I suspect he would infer the latter. But I also suspect he would guess that the *Essay on Reason* was Pope's and that that was why Pope would send it. Pope also reveals why he was happy to have the *Essay on Reason* thought his: the intention and doctrine are

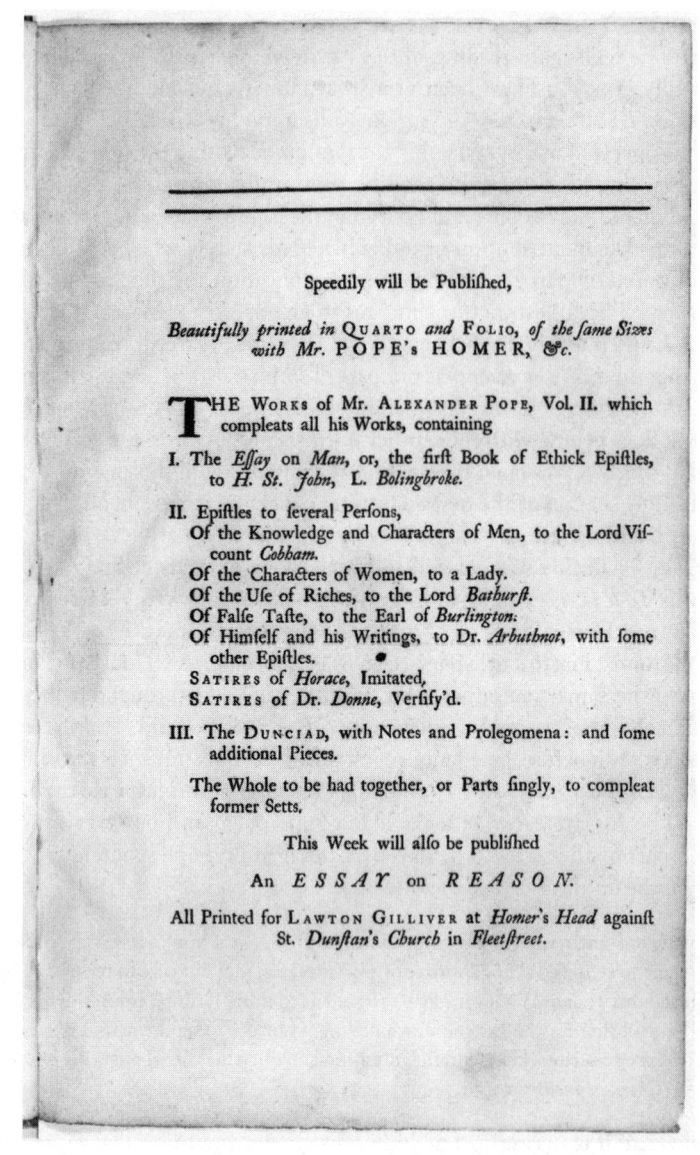

Fig. 7. Final-page advertisement from *An Epistle to a Lady*, 1735 (Bodl. Fol. Δ 760; 315 × 194 mm).

ones Caryll would approve. Long before Crousaz made the *Essay on Man* a subject of controversy, Pope is pleased to link it to more directly orthodox teaching. The *Essay on Reason* is the poem he might have written next in the sequence. Write something on the side of revelation? 'No, no! you have already done it.'

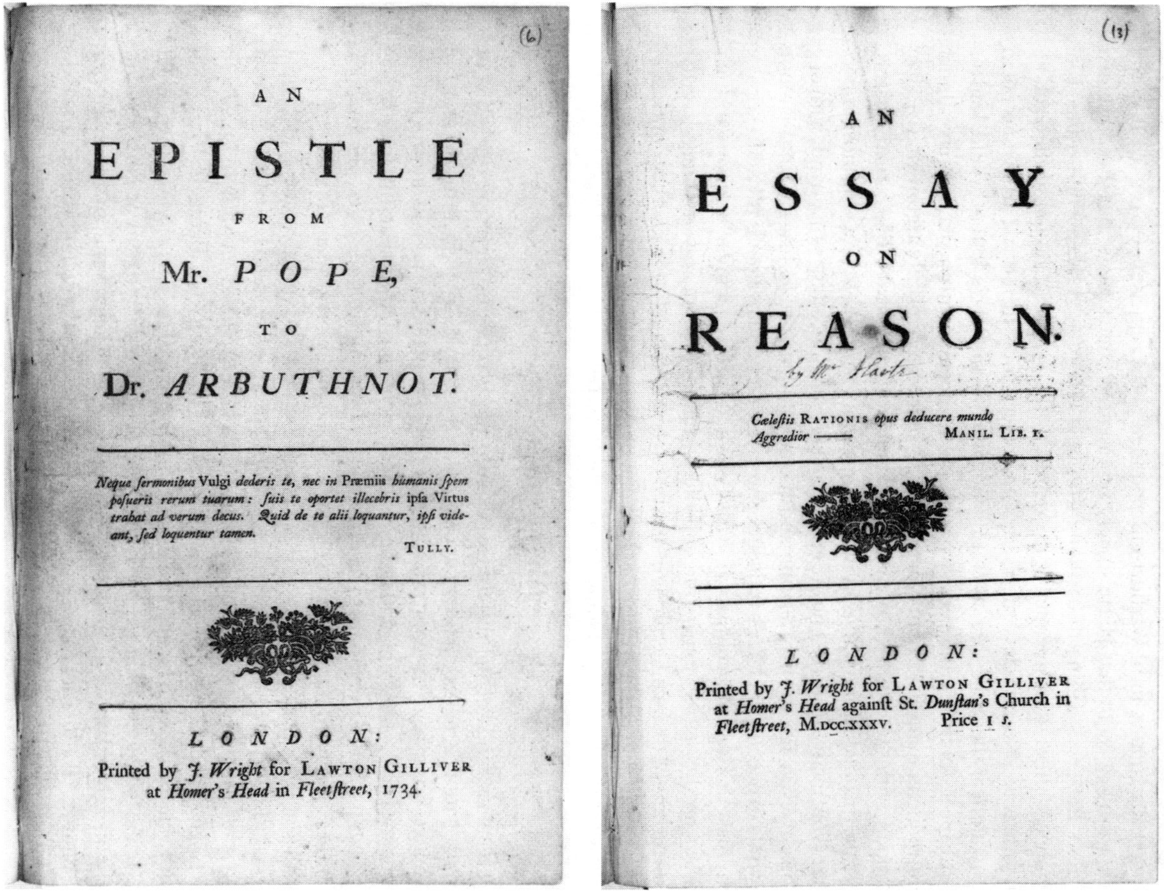

FIG. 8. Title pages of *An Epistle to Dr Arbuthnot* and *An Essay on Reason*, 1735 (Bodl. Vet. A4 c. 289 (6 and 13); 342 × 210mm).

Maynard Mack has suggested the *Essay on Man* is like the first series in a collection of eighteenth-century sermons. It presents the arguments for religion from nature and reason; the second series would show the compatibility of those arguments with revelation.[7] The *Essay on Reason* was as near as Pope got to completing that structure until Warburton responded to the criticism of Crousaz. It would be wrong to claim that Harte's poem represented Pope's views, but it represented views Pope was willing to be thought to hold. He revised the poem, arranged its publication, and commended it to friends; at the very least he found its teaching inoffensive. Nevertheless there are some discomforts in moving from one poem to another; they sit more awkwardly together than Pope's letters would lead us to expect. Both are pitched as philosophical poems, with a deliberately similar introductory apparatus, but Harte's essay is much more plainly argumentative than Pope's, with the author's authority fully behind the arguments. The etiolated nature of the theological and philosophical argument of *An Essay on Man* is its most important contrastive feature. The differences over revelation, on the other hand, with one positively in favour of revelation and the other silent, are less significant than one would expect, with Harte's reticence and tact (and perhaps Pope's revision) serving to make his account compatible with Pope's. There is a similar harmony in their ideas on the status of mankind and the powers of human reason, though, rather disturbingly, their conceptions of reason itself differ widely. Finally they contrast in their treatment of the 'reason of things', with Harte's orthodox account showing up Pope's insistence on a dynamic order in the universe.

THE POEMS AS ESSAYS

The most fundamental difference between the two poems—putting questions of imagination aside—is in their conception of the essay as a poetic form. *An Essay on Reason* brings out by contrast some of the informal, satiric, and improvisatory qualities of Pope's poem, and points to the suspension of argumentative force that is one of its distinctive modes. The difference is apparent from the titles, once we look at them carefully, but it then becomes disguised in the preliminary material. Pope's titles of this period maintain the tension between intimate epistle and formal essay. The first edition of epistle 1 was called 'An Essay on Man. Address'd to a Friend', the first editions of epistles 2, 3, and 4 were 'An Essay on Man. In Epistles to a Friend', and the first collected edition was 'An Essay on Man, Being the First Book of Ethic Epistles. To Henry St. John, L. Bolingbroke.' This poem is, therefore, much like the four poems that were written of the projected second book of ethic epistles: its title gives a topic (for example, 'Of the Use of Riches') and an addressee (for example, 'To the Right Honorable Allen Lord Bathurst'). Bolingbroke, as a person

---

[7] Mack says, 'Pope has telescoped the two and submerged the second' (*Life*, 741), but though I am a very warm admirer of Mack's treatment of the *Essay* in the *Life* (there are many valuable reflections on *Twickenham*, 3 (1)), there is no defence of revelation in the *Essay*.

of special importance to Pope, gets four epistles rather than one. Harte's poem is simply 'An Essay on Reason', and although at one point the dead husband of his patron Mrs Knight is addressed, the poem has no addressee and no tinge of intimacy; it is a transparently public performance.

Pope's essay comes closer to the style of Harte's in the preliminary material, which suggests that they are exercises in the same genre. After the parodic excesses of the preliminaries of *The Dunciad*—'Advertisement', 'Letter to the Publisher', 'Testimonies of Authors', 'Arguments to the Books'—Pope might have been expected to plunge the reader directly into the first epistle of *An Essay on Man*. But in the 1734 collection he provides first a general introduction, 'The DESIGN' (in the context, it must call to mind the idea of the artist as an all-wise creator), and then 'The CONTENTS', a section-by-section summary of the argument of the poem. The 'Design' begins modestly, with the present, abstract work characterized as a necessary precursor to the more socially directed ones to come:

HAVING proposed to write some Pieces on Human Life and Manners, such as (to use my Lord *Bacon*'s expression) *come home to Men's Business and Bosoms*, I thought it more satisfactory to begin with considering *Man* in the Abstract, his *Nature* and his *State*: since to prove any moral Duty, to inforce any moral Precept, or to examine the Perfection or Imperfection of any Creature whatsoever, it is necessary first to know what *Condition* and *Relation* it is placed in, and what is the proper *End* and *Purpose* of its *Being*.[8]

But already, by the end of this first paragraph, an introduction intended primarily as an explanation of the role of the poem is drifting into claims for its logical rigour and ratiocinative force: 'prove', 'inforce', 'examine', 'it is necessary'. Such a view of the *Essay* comes to a head at the end of the second paragraph of the 'Design', after an analogy with anatomy: 'If I could flatter my self that this Essay has any Merit, it is in steering betwixt Doctrines seemingly opposite, in passing over Terms utterly un-intelligible, and in forming out of all a *temperate* yet not *inconsistent*, and a *short* yet not *imperfect* System of Ethics.' The claim to take the path of moderation between doctrines or to find accommodations between them (for they are only 'seemingly opposite') points to the subtlety of the *Essay* itself, but the claim to system, though it is carefully balanced and hedged about with qualification, points instead in the direction of Warburton's later commentary and annotation, and makes Pope's work more like Harte's.

Warburton's approach to the *Essay* is anticipated in the contents that follow. This is surely a case where a decision about paratext has had a major influence on the critical history of a literary work. Warburton's *Commentary*, much criticized as an imposition, is an elaboration of an apparent systematization of the *Essay* in prose summary that Pope first attempted in the publication of the four-epistle poem on

---

[8] Quotations are from the four-epistle edition of *An Essay on Man* of 1734 (Foxon P853), in the Scolar Press facsimile, though in quoting from verse I shall give the line numbers of the *Twickenham* edition. References to Warburton's commentary and notes are to the edition of 1743/4 (Mack's 1743b). I have reversed the italics in quoting from the 'Design'.

20 April 1734.⁹ The model for the contents is almost certainly Michael Maittaire's edition of Lucretius, *De Rerum Natura* (1713), a copy of which is known to have been in Pope's library (Mack, *Collected in Himself*, Library, 110). The poem itself is one of the major influences on *An Essay on Man*.[10] Maittaire offers an initial summary of all six books, 'Brevis Rerum in Sex Lucretii libris comprehensarum Synopsis'. The style is that adopted by the *Essay on Man*, brief statements of the topics covered in the poem, followed by the lines involved:

Poeta, Venere invocatâ (v. 1.), suíque operis Memmio nuncupati Argumento breviter exposito (27.49), Epicuróque Relligionis hoste laudate (63.67), suae totius Philosophiæ fundamentum ponit, Nihil ex nihilo gigni aut in nihil reverti (151.217). Hinc affirmat omnia è Corpusculis adeò minutis, ut oculos fugiant, nata (269) . . .

And so on, with around twenty lines of summary for each book. Pope's contents, modelled on Maittaire's, is essentially paraphrase. It consists largely of headings and assertions, while sometimes manipulating its topics to create a pattern and parallelism not discernible in the poem itself:

The *Impiety* of putting himself in the place of *God*, and judging of the fitness or unfitness, perfection or imperfection, justice or injustice of his dispensations, 109 *to* 120. The *Absurdity* of conceiting himself the *final cause* of the creation, or expecting that Perfection in the *moral* world which is not in the *natural*, 127 *to* 164. The *Unreasonableness* of his complaints against Providence, while on the one hand he demands the perfections of the Angels, on the other the bodily qualifications of the Brutes, 165.

Less happily, the beginnings of the accounts of the first and fourth epistles summarize the poem as an argument, with Pope becoming his own first misreader:

Of *Man* in the *Abstract*. We can judge only with regard to our *own System*, being ignorant of the *Relations* of Systems and Things, VER. 17, *&c.* to 68. Man is not therefore to be deem'd *Imperfect*, but a Being suited to his *Place* and *Rank* in the Creation, agreeable to the *General Order* of Things, and conformable to *Ends* and *Relations* to him unknown. 69, *&c*. It is partly upon this *Ignorance* of future Events, and partly upon the *Hope* of a Future State, that all his Happiness in the present depends, 73, *&c.*

The initial impression given by this passage is that the poem will be very tightly articulated. The second section of summary is linked to the first with 'therefore', while the third picks up the initial claim and explores it from another angle. But this argument presented at the start of the contents is confusing. Does 'therefore' simply mean 'on that account' and exclude an adverse conclusion on the basis of man's ignorance? Or does it mean 'consequently' and claim to have established the conclusion in the second part of the sentence? From our being ignorant of other systems, we cannot conclude that Man is in his right place in the general

---

⁹ I discuss the suggestion that Jonathan Richardson may be responsible for the contents in Chapter 8. The speculation is implausible.

[10] The point was first made fully by Leranbaum, *Pope's 'Opus Magnum'*, 38–63.

order; we have no basis on which to conclude anything. Pope recognized the problem and omitted the 'therefore' when he revised the contents in 1735. Similarly in the summary of epistle 4 Pope says, 'God governs by *general*, not *particular* laws: intends Happiness to be *equal*; and to be so it must be *social*, since all particular happiness depends on general, 35.' The summary presents the positions as though they were interdependent, but general laws could result in an uneven distribution of happiness (many think they do), and, according to Pope, happiness must be social in order to be happiness; the issue of equality is not engaged by that requirement. This aspect of the contents provokes doubt as to how far Pope was fully in command of the structures of argument alluded to in his poem, but inadequacies of the contents are not inadequacies of the poem itself. The more subtle the poem, the greater the demands it makes on the reader, the more difficult the task of summary. The problem for Pope is that the contents may have fostered a misreading of the poem itself.

Harte has no 'Design' but he does have a contents which creates a significant typographical parallel with Pope's poem and was doubtless intended to do so, though Harte has each section beginning a new line, with a hanging indent, whereas Pope's sections are run on. Although the general style of Harte's contents is drawn from Pope, he is more distant from his material, paraphrases less, and avoids seeming to present the substance of his argument. What he provides is more like an academic summary:

Reason coeval with God, &c. Verse 1 to 6.
Transfused into *Adam*; Its particular Nature and Operations in the Paradisaical State, V. 7 to 52.
How impaired by the Fall, V. 53 to 62.
The Objection (V. 63 to 70) *Why* Adam *in a purely rational State, shou'd be judged by a positive Law*, casually obviated; and the Expediency of such a Law demonstrably proved, V. 71 to 115.

Sometimes the contents reads like lecture notes:

Author's State of the Case. I. Reason as in God, V. 309 &c.
  II. As in Things, Actions, Morality, Arts, &c. V. 315 to 346.
  [Farther State and Reply to the Objection above, V. 295, &c. casually taken in here, V. 347 to 366.]
  III. Reason as in Man. Its Perfection in the *abstracted* Sense, and its Modes and Degrees of Perfection and Imperfection in the *exerted* Sense, V. 367, &c.

This somewhat relaxed attitude to repetition and deviation in the contents is to be expected from a poet who can begin a verse paragraph with '[This by the way:]'. Harte's poem is, even on the evidence of the contents, more of a scholarly essay than Pope's. He knows what his headings are; he knows his opponents; and he knows which arguments he wants to rebut. The impression of Pope's poem from its contents is that it is driven forward by the impulse of its author's interests. The laxity of its articulation of some arguments suggests that associated ideas are presented with more power than can be captured in prose summary.

When we turn to the essay form of the poetry itself, similar contrasts emerge. The beginning of *An Essay on Man* shifts the emphasis to the epistle:

> AWAKE! my ST. JOHN! leave all meaner things
> To low Ambition and the Pride of Kings. (1. 1–2)

The opening address has a certain grandeur; but the tone soon becomes more intimate, and allusion draws us towards the gentlemanly pursuits of walking and hunting. Pope deploys a range of tone in the *Essay*, constantly responding to an imagined interlocutor, sometimes Bolingbroke, sometimes man in the person of the reader, sometimes merely those fools who think differently from the author. In this respect, the *Essay* is not an essay in the way that Harte's is; it is a satiric epistle like those that were designed to complement it.[11] Its concern with satiric targets can be disguised by what looks at first glance like a highly theological conclusion to the opening paragraph:

> Together let us beat this ample field,
> Try what the open, what the covert yield;
> The latent tracts or giddy heights explore,
> Of all who blindly creep, or sightless soar;
> Eye Nature's walks, shoot Folly as it flies,
> And catch the manners living as they rise;
> Laugh where we *must*, be candid where we *can*,
> But vindicate the Ways of GOD to Man. (1. 9–16)

The keyword in this section is 'vindicate', and the problem in interpreting it has been the powerful pull of the allusion to the end of the first paragraph of *Paradise Lost*, where Milton declares his intention to 'justify the ways of God to Men'. This allusion is supported by Pope's reference in line 8 to a 'Garden, tempting with forbidden fruit'.[12] But an assumption that 'vindicate' is an elegant variation on 'justify' produces a reading that fits oddly with the metaphors that introduce it (making the poem jolt from lively to severe), and it also opens up all those familiar issues of adequacy and orthodoxy that arise from reading of the poem as straight theodicy. Pope's metaphors are of the cheerful and vigorous hunting of game, and the game that is being pursued is human folly. Mack quotes an excellent parallel from Pope's letter to Gay of 2 October 1732: 'I advise you to make man your game, hunt

---

[11] The model is not exactly the same. For an excellent discussion of an important aspect of the poem's satire, see Douglas White and Thomas P. Tierney, '*An Essay on Man* and the Tradition of Satires on Mankind', *Modern Philology*, 85 (1987), 27–41.

[12] Harry M. Solomon, *The Rape of the Text: Reading and Misreading Pope's 'Essay on Man'* (Tuscaloosa: University of Alabama Press, 1993), 42–3, 180, points to Edward Young's *A Vindication of Providence: or A True Estimate of Human Life* (1728) as a further possible influence on the poem. Unfortunately Young never got beyond Discourse 1, outlining human frustration and disappointment, to develop his vindication. In his dedication to the Queen, Young uses vindicate to mean 'champion' when he suggests that her majesty, 'One of his most shining Representatives on Earth, patronize, and vindicate a *Vindication of His Providence*'.

and beat about here for coxcombs, and truss up Rogues in Satire' (*Correspondence*, 3. 318). Pope's quarry may include coxcombs ('catch the manners living as they rise'), but he is also in pursuit of intellectual folly, the ignorance that follows from too little ambition ('blindly creep') or too much ('sightless soar'). As he told Caryll in a letter of 6 February 1732, his design was 'rather to ridicule ill men than to preach to 'em' (*Correspondence*, 3. 173). It is this open and generous satiric attack, and not some theological argument, that must constitute the poem's vindication of God. 'Vindicate', then, has a meaning close to the Latin 'vindicare': to assert, to claim, to serve as champion, to deliver. Because Pope's use of the word is sometimes close to Latin meanings unusual in English, *OED* allows him to lead off a section of his own, 5b, 'To claim for oneself or as one's rightful property', with a quotation from the *Odyssey*, 'Affianc'd in your friendly power alone | The youth wou'd vindicate the vacant throne' (4. 224), and another from *An Essay on Man*, 'Is thine alone the Seed that strows the plain? | The Birds of heav'n shall vindicate their grain' (3. 37–8). These examples, even though they are special cases, are better guides to Pope's usage than are modern understandings of the word. None of the thirteen examples of Pope's use of 'vindicate' in his poetry carries the primary sense of 'justify'. By 'vindicate' Pope means roughly 'assert the rightness of some claim or allegiance by vigorous action': for example, 'To hell's abhorr'd abodes | Dispatch yon *Greek*, and vindicate the Gods' (*Iliad*, 5. 556), or, in what is perhaps a source for the passage in *An Essay on Man*:

> Why on those Shores are we with Joy survey'd,
> Admir'd as Heroes, and as Gods obey'd?
> Unless great Acts superior Merit prove,
> And Vindicate the bounteous Pow'rs above . . .

These lines appear in one of Pope's first published poems, 'The Episode of Sarpedon' (33–6), and later in the *Iliad*. Sarpedon's famous speech inciting his friend Glaucus to action against the Greeks reverberates engagingly with Pope's invitation to his friend Bolingbroke to join in the ridicule of folly. The conclusion of the first paragraph does not diminish the satiric impetus of the earlier lines. Ridicule is the weapon by which God's ordering of the universe is to be asserted. Although Pope divides his poem successfully into four sections—on the universe, on the individual, on society, and on happiness—he does not do so in order to develop and refine his argument. Each section proclaims God's providence and elaborates on it in relation to the new theme. The dominant mode of the essay, however, is assertion not argument, vindication not justification.[13]

An Essay on Reason by contrast switches between narrative and argumentative mode. The beginning provides a mythological start to the story, with Reason, essence

---

[13] Many of the speech acts are cast as imperatives, with instances of 'Say', 'Go', 'See', 'Cease', 'Trace', 'Look', 'Remember', and 'Know'. But these are not instructions that entertain the response, 'No I will not see that'; they lead the reader to accept the poem's many assertions, the chief of which are summarized in the poem's concluding lines.

of God, present at the creation and viewing past, present, and future. The poem then moves to Eden in order to examine the reason of unfallen man, but breaks off (as the contents points out) to consider objections to God's positive decree that the fruit of the tree should not be eaten.[14] After that the poem abandons any tracking of the biblical narrative in order to consider the nature of reason as found in man's fallen state and to defend the role of revelation. Then, in the middle of the poem, Harte allows his imagined opponents to raise an objection to the reality of truth, knowledge, and human freedom. This challenge to the rationality of the universe leads Harte into a discussion of the reason of things and of the distribution of reason among mankind, before moving on to consider the power of reason as it is manifested historically. The poem ends with a prayer. Although Harte allows himself some licence, he attempts a comprehensive treatment of his topic. His structure is essentially tripartite: he begins with an account of what revelation tells us about the fullness of reason, proceeds to consider its strengths and limits as it is found in fallen man, and then moves to consider the relations between reason and the world it seeks to interpret. Unlike Pope's poem, and befitting an exercise that has something in common with a dissertation, the *Essay on Reason* shows an awareness of surrounding controversy. It sees itself as challenged by wits and challenging deists; it seeks to avoid superstition and bigotry, while defending revelation.

Harte and Pope differ in the way they present intellectual arguments. Harte presents each argument as an opponent's or his own; Pope often gives an argument an equivocal status. One typographical feature highlights the difference between the two essays in this respect, and that is the use of quotation marks. Harte uses quotation marks in a simple and modern way, to set up a formal dialogue. They mark the speech of others, that is, those who have objections to the views he is putting forward. The objectors are given their say and then the poet responds directly with an argument or a comment (64–70, 347–60). He also places his concluding prayer in quotation marks. The prayer is in a different register, being addressed to God rather than to an opponent or the reader, but I think the quotation marks also function as they do when they highlight the beauties of a writer. Pope uses quotation marks in this way in his edition of Shakespeare and they were commonly used to mark sententious elements in speech. Pope's usage in *An Essay on Man* also draws on this practice.

In *An Essay on Man* Pope uses quotation marks for various speakers, including Nature, and to display sentiments, but he also uses them to bracket certain statements from the main discourse, giving them an equivocal status. Such speeches are not the responsibility of any particular person, including the author; they are something like sayings, the intellectual equivalent of folk wisdom. Pope does not wish to disown them, but he does not intend to argue for them either. The first speech in the *Essay* is by Pride, who claims that the heavens and earth exist for man's benefit

---

[14] For the debate about moral and positive duties, see White, *Pope and the Context of Controversy*, 114–25.

(1. 131–4).[15] Pope responds with a matter-of-fact-objection, which then leads to one of those equivocal quotations:

> But errs not Nature from this gracious end,
> From burning Suns when livid deaths descend,
> When Earthquakes swallow, or when Tempests sweep
> Towns to one grave, and Nations to the Deep?
> No ('tis reply'd) the first Almighty Cause
> 'Acts not by partial, but by gen'ral Laws;
> 'Th' Exceptions few; some Change since all began;
> 'And what created perfect?'—Why then Man? (1. 137–44)[16]

The reply is less a response from Pride as interlocutor than a satiric representation of a pat intellectual position on the problem of ontic evil. Pride would surely be entitled to object to the operation of God's providence in the natural world and claim that there should be no floods or earthquakes, or none that affect men. If we take this as a dialogue, Pride is unfairly made to provide a major plank in the defence of God's providence. But Pope blurs the responsibility for the quoted matter, writing simply, ''tis reply'd'. Our response to the short reply is ambivalent. The semicolons separating the four elements (three of them very short) suggest positions too stale to merit proper representation through grammatical expansion, and staccato speech is often a sign of Popeian ridicule.[17] Yet the first element ('gen'ral Laws') comes to play an important role in the essay, first being used to explain moral evil,

> If Plagues or Earthquakes break not heav'ns design,
> Why then a Borgia or a Catiline? (1. 155–6)

and then later to explain the distribution of happiness (4. 35–6). The second element, 'Th' Exceptions few', draws attention to the problem of unqualified endorsement of general laws. 'Not by partial, but by gen'ral Laws' might be taken to exclude miracles, and that would constitute a challenge to an important orthodox argument for revelation. This was an issue on which Pope wanted to avoid expressing a clear opinion, and the rapid quotation of a position and its potential qualifications was a way of achieving his end. The operation of these lines worried Warburton, who went to great lengths in his commentary and notes to make them orthodox: explaining that Pope's aim all along was to show that moral evil promoted good; claiming that this passage supplemented a demonstration that moral evil came

---

[15] A similar satiric vein is opened in epistle 3 when man's speech is followed by that of a goose: 'While Man exclaims, "see all things for my use!["] | "See Man for mine!" replies a pamper'd Goose' (3. 45–6). In the editions before the 1735 *Works* four more comic lines were devoted to the goose's psychology; I suspect they were omitted because Pope felt they led the poem too far into levity, a reflection on the general satiric tone of the poem.

[16] The first edition read, 'Blame we for this the wise Almighty Cause? | "No ('tis reply'd) he acts by *gen'ral Laws*', and the substantive change led to the quotation marks being in the wrong place in editions up to the death-bed edition.

[17] Mack, perceptive as always, explains what the elements are, and wonders whether their presentation ridicules an opponent's speech. The link to the previous speech by Pride suggests it does, and a similar effect is achieved in lines 4. 157–62. Mack provides further helpful parallels of staccato speech.

into the world by man's abuse of free will; and seeing an argument by analogy for eternal rewards and punishments. Unfortunately for Warburton the actual focus of the poem up to this point is on evil of defect, into which Pope now surprisingly sweeps moral evil; the concern with free will is Warburton's own invention; and the argument by analogy for eternal life runs counter to Pope's declared intentions in the poem ('Thro' Worlds unnumber'd tho' the God be known, | 'Tis ours to trace him, only in our own').[18] Warburton's anxiety to supply Pope's missing articulation points up the rapid, sportive deployment of intellectual argument in the *Essay*. The speaker of *An Essay on Man* is impatient of argument, which tends to be redundant or even ridiculous. Pope introduces the lines about the Almighty Cause, not because they belong particularly to the position of anthropomorphic pride, but because they can be used, without direct endorsement, in an *ad hominem* argument against intellectual pride, which also sweeps aside the problem of moral evil. The conclusion of this section is characteristic:

> From Pride, from Pride, our very reas'ning springs;
> Account for moral, as for nat'ral things:
> Why charge we heav'n in those, in these acquit?
> In both, to reason right, is to submit. (1. 161–4)

There are, of course, problems in using the plenitude argument to account for moral evil, unless it is elaborated to take some account of human freedom, but Pope is less interested in the structure of argument than in ways of exposing the folly of complaint.

The third epistle begins with an echo of the lines on the 'Almighty Cause' we have just been examining.

> Learn Dulness, learn! 'The *Universal Cause*
> 'Acts to *one End*, but acts by various Laws.' (3. 1–2)

At one level this is straightforward, and Pope emends 'Learn Dulness, learn' to 'Here then we rest' for the *Works* of 1735, thereby suggesting that the position is already established. But though this motto clearly does echo the passage in the first epistle, 'the first Almighty Cause | Acts not by partial, but by gen'ral Laws', the sentiment is different and relatively uncontroversial: the issue now is not whether God makes exceptions to his laws (the Bible might suggest he does) but whether various systems serve one general purpose. The summary prose statement in the contents is more vital, 'The *whole Universe* one system of *Society*', and its connection to the discussion of the great chain of being that follows is much clearer. But the impression that there is a network of sayings about 'Cause' and 'Laws' is confirmed when the controversial

---

[18] Warburton, Pattison, and Mack (in his edition) would disagree (see, for example, *Twickenham*, 3 (1). 14 n.). Pope is willing to conclude from perceptible evidence of order that the universe is ordered, but he is not willing to reason about the operation of systems other than those he can observe. I see no reason why Pope should disagree with Bolingbroke, who was opposed to such analogical reasoning; see *Works* of Henry St John, Lord Viscount Bolingbroke, ed. David Mallet, 5 vols. (London, 1754), 5. 526, and Walter McIntosh Merrill, *From Statesman to Philosopher: A Study in Bolingbroke's Deism* (New York: Philosophical Library, 1949), 36.

passage about general laws is picked up in the fourth epistle and quoted as though its insight had been demonstrated:

> Remember Man! 'the Universal Cause
> 'Acts not by partial, but by gen'ral Laws . . . (4. 35–6)

Verbally this couplet combines elements from the two previous ones, but its sentiment is that of the first epistle. It is with some shock that we recollect that the idea was first formally introduced to the poem as an argument on the side of Pride.

The passages concerned with laws are merely one example of the layering of language and improvisation of argument within the poem, in contrast with Harte's straightforward intellectual debate. Quotation marks are used again in a debate about virtue and happiness in the final epistle (4. 149–65), again with some staccato ridicule, while some sentiments in the poem, for example, 'VERTUE alone is Happiness below' (4. 310) and those displayed by the use of capitals at the end of the poem, clearly have special status. The quotation marks emphasize the plethora of argument in the poem and Pope's refusal to present philosophical arguments in detail or with commitment. Such arguments tend to have an *ad hominem* air or to stand bracketed, serving to get an opponent to a position of submission to God. In this respect the quoted references to the Almighty Cause have some similarity with the poem's many rhetorical questions and those 'if' arguments that Mack and Solomon comment on; 'Of Systems possible, if 'tis confest | That Wisdom infinite must form the best' (1. 43–4): 'If to be perfect in a certain sphere, | What matter, soon or late, or here or there?' (1. 73–4); 'And if each System in gradation roll, | Alike essential to th' amazing Whole' (1. 247–8).[19] The arguments for hope of immortality have this equivocal quality throughout. In the social and psychological sections of the essay Pope does commit himself to particular ideas; but, in general, philosophical arguments are more like striking figures of speech that can aid him on his path to urge submission to divine order.

SIMILARITIES OF VISION

An Essay on Man and *An Essay on Reason*, therefore, differ in the status they give to philosophical arguments, even though their presentation as poetic essays is similar. They also share enough of a common approach to human questions and problems to make them suitable companion pieces. As one might expect, *An Essay on Reason* shows that Pope's influence on Harte is deeply felt, even though it may be impossible to identify the sections he actually revised. The two poems express a similar vision of the human condition, stressing the middle status of man between higher beings and the animals, and consequently condemning human pride and the desire for power

---

[19] See Mack, *Life*, 527–8, and Solomon, *The Rape of the Text*, 89–97 (containing some of the best modern criticism of Pope's *Essay*). The third of these is not usually discussed but seems to me a good example of Pope's refusal to make a direct claim about the general system. 'May, must be right' (1. 52) works in the same way. 'Or' is used to similar effect at 3. 227.

and knowledge. Both explore the limits of human reason, the topic advertised as the title of the first poem in Pope's planned second book of ethic epistles. These general positions, to some degree eighteenth-century commonplaces, are expressed in similar terms. Pope condemns the human propensity to

> Snatch from his hand the Balance and the Rod,
> Re-judge his Justice, Be the GOD of GOD! (1. 121–2)

Harte shows that in Eden

> No earth-born Pride had snatch'd th' Almighty's rod,
> O'erturn'd the balance, or blasphem'd the God . . . (19–20)

In portraying the Indian whose natural wisdom 'asks no Angel's wing, or Seraph's fire' (1. 110) and in assuring the reader that the life exists as fully in 'vile Man that mourns, | As the rapt Seraphim that sings and burns' (1. 277–8) Pope provides some sort of answer to Harte's rhetorical question, 'Shall Man, because he wants a Seraph's flame, | Not taste the Joys proportion'd to his frame?' (131–2). The advice Pope gives at the start of his first two epistles, 'Thro' Worlds unnumber'd tho' the God be known, | 'Tis ours to trace him, only in our own' (1. 21–2), 'Know then thy-self, presume not God to scan; | The proper study of mankind is *Man*' (2. 1–2), is echoed by Harte in a similar recommendation of humility:

> While Pride amidst the vast abrupt must soar—
> Alas, to fathom God is to be more!
>   Then dare be wise, into thy self descend,
> Sage to some purpose, studious to some end;
> Search thy own heart, the *Well* where knowledge lies;
> Thence (not from higher earth) we catch the skies . . . (167–72)

For both poets, Man has a middling status, which should produce contentment (*Essay on Reason*, 175–8; *Essay on Man*, 1. 174–92). The correct stance in relation to God is submission: Harte tells us that God is entitled to 'awe submissive' (73), and Pope that 'to reason right, is to submit' (1. 164). It follows from the poets' general stance on human pride that aspiration for happiness should be modest. The values in the prayer which closes Harte's poem echo those in Pope's epistle 4. Early in this epistle Pope instructs the reader:

> Know, all the Good that Individuals find,
> Or God and Nature meant to meer mankind,
> Reason's whole pleasures, all the joys of Sense,
> Lie in three words, *Health, Peace,* and *Competence.* (4. 75–8)

Harte's prayer begins:

> 'O Thou, the God, who high in Heav'n presides,
> 'Whose eye o'ersees me, and whose wisdom guides,
> 'Deal me that Portion of Content and Rest,
> 'That unknown Health, and Peace, which suit me best . . . (570–3; mislineated)

Both emphasize the importance of worthy action rather than powerful ideas. For Harte, wisdom should lead to action:

> The Question is not therefore, how much light
> God's Wisdom gives us, but t'exert it right . . . (153–4)

While for Pope, 'His can't be wrong whose Life is in the right' (3. 306). In this respect the poems move in the same imaginative world, making the same unsurprising choices of villain and hero. Nero is the representative tyrant (*Essay on Man*, 2. 198, *Essay on Reason*, 231); Socrates the good man (*Essay on Man*, 4. 236, *Essay on Reason*, 461).

HARTE ON HUMAN REASON AND REVELATION

Although the two poems have much in common, *An Essay on Reason* is more than a mere conventional companion to *An Essay on Man*. In its teaching on reason and revelation, it supplements and provides a counterweight to Pope's poem. Central to Harte's *Essay*—and to Pope's interest in it—is its Christian tone and its deliberate attempt to reconcile reason and revelation. This Christian emphasis persists in spite of the sense of classical decorum that pervades Harte's essay. The motto on his title page is from Manilius, and his footnote references are to Manilius, Cicero, Sallust, Lucretius, Seneca, and other Ancients. This is a poem in the classical tradition, overtly displaying the learning that commentators such as Mack and Solomon have seen underlying Pope's poem. But Harte uses this classical mode to put forward the general philosophy of life he shares with Pope in language that is frequently coloured by Christian belief. His poem is expressly directed against deist thinking, though in reality he does much to accommodate it. The contents makes the nature of his opposition clear: 'Hence is obviated an Error in Deism, of taking human Reason for the abstracted, or divine, V. 265 &c.'; 'Reason proved *de facto*, from the whole course of Antiquity, never to have done what the Deists imagine, but rather the contrary, V. 453 to 468'; 'Expostulation with the Deists, V. 554, &c.', the final expostulation being against pride and the refusal to trust revelation. Reason is an important topic for Harte because the deists exaggerate its human strength, while sceptics deny any correspondence between the order of the world and human understanding. A right understanding of reason helps us to accept a morally ordered universe and the revelation that will help us live morally in it. Harte's first paragraph presents a view of reason and a short conspectus of salvation history:

> FROM Time's vast Length, eternal and unknown,
> Essence of God, coeval REASON shone;
> Mark'd each recess of Providence and Fate,
> Weighing the present, past, and future state:
> 'Ere Earth to start from Nothing was decreed,
> 'Ere Man had fal'n or God vouchsaf'd to bleed! (1–6)

The final lines encompass the key events of the Christian narrative—the creation, the fall, and the atonement—and set up the expectation that they will fall within the discussion of reason. In fact the atonement is not addressed in this poem, and it was Harte's teaching on this topic that led to his revision of the sermon on *The Union and Harmony of Reason, Morality, and Revealed Religion* three years later. In the third edition of *An Essay on Reason* (1736), lines 3 to 6 are omitted, leaving an opening paragraph of two lines, but I suspect the omission was unrelated to the treatment of the atonement. Later in the poem Harte insists that to the thought of God

> No outward object e'er one image brought,
> The part, the whole, the see'er and the seen,
> No distance, inference, or act between . . . (310–12)

The 'weighing' in the introductory lines suggests no such unmediated perception, but rather the considering of alternatives. It was probably for that reason that Harte thought it better in his third edition to omit the operation of reason before the creation and move straight to the portrayal of Adam and Eve.

Harte's chief contention in the first half of his poem is that man in the fallen state has adequate reason. He does not doubt that we have enough knowledge for use (127 and 133), 'Clear, yet adapted to the mental Sight' (135). Like Pope he believes our reason is appropriate to our place in the scheme of things:

> REASON, like Virtue in a Medium lies;
> A hairs-breadth more might make us *mad* not *wise*,
> Out-know ev'n Knowledge, and out-polish Art,
> Till *Newton* drops down giddy—a *Descartes!*[20]

The deists are attacked for mistaking human for divine reason. As a power reason is always divine, always perfect, even though it is not always possessed fully or used rightly (367–72). Reason 'scarce appears' until we are 21, following on from sensation, perception, thought, imagination, fancy, experience, and knowledge:

> Bright Emanation of the Godhead hail,
> Fountain of living lustre, ne'er to fail;
> As none deceiving, so of none deceiv'd:
> Beheld, and in the act of *Sight* believ'd,
> In Truth, in Strength, in Majesty array'd,
> No Change to turn thee, and no Cloud to shade!
> Such in *her self* is REASON—: Deist say,
> What has thou here t'object, t'explain away?
> Think'st thou *thy Reason* this unerring Rule?
> Then live a madman—and yet die a fool! (259–68)

Harte gives man divine reason and then he takes it away. The reason that comes to us is divine (it is not just one of the human faculties) but we never enjoy it in its divinity.

---

[20] This passage makes very little sense, especially in view of Harte's elevated view of reason, which I discuss below; later in the paragraph he seems to be muddling reason and wit. I wonder whether he was helped with this passage.

From the discussion of the grandeur of reason, Harte moves to a defence of revelation. Reason in man's fallen state, Harte says, can be

> A faithful guide to comfort and to save,
> Till the mind floats, like *Peter* on the wave:
> Then bright-eyed *Hope* descends, of heav'nly birth,
> And *Faith*, our Immortality on Earth,
> A *Saviour* speaks! lo darkness low'rs no more,
> And the husht billows sleep against the shore. (275–80)

This link between hope and faith is an important element in the conclusion of *An Essay on Man* (4. 341–52), but here Harte relates it to revelation. The allusion is to Matthew 14, where Peter walks on the water in response to Jesus's call, but grows afraid and starts to sink. Jesus stretches out his hand and catches him, saying, 'O thou of little faith, wherefore didst thou doubt' (Matt. 14: 31). The episode echoes one earlier in the same Gospel when in a great tempest at sea the disciples wake Jesus, who rises and rebukes the winds and sea, creating a great calm (Matt. 8: 26). Harte is drawing an analogy between Jesus' intervention in this story and the role of revelation in the life of the believer, though, in contrast with much of his poem, the message is occluded. It may be that he takes the view that the truth of revelation is accepted through faith, not reason, but that he is unwilling to spell out his position. Instead, he moves immediately to deal with a technical objection to revelation, arguing that it does not impair human freedom. Harte believes that men are much more dependent on revelation than they realize. Revelation is the unacknowledged source of much human understanding, and, without it, the Ancients lacked moral system, sanctions, and authority (475–80). Half the knowledge the Ancients had came from revelation:

> Nay half the source of most the antients knew,
> From *Noah* they, as He from *Eden* drew . . . (489–90)

Harte finds the same debts in contemporary debates, where the Wits (Harte's usual name for those who challenge orthodoxy) fight religion with her own weapons:

> The truths they boast of, and the rules they know,
> Seen not, or own'd not, first from *Scripture* flow. (497–8)

The problem, as so often in these two poems, is pride: the Wits attribute to themselves what is given to them by God; they respond to guidance without acknowledging it.

Harte differs from Pope in insisting that man has been given guides through life. Reason itself is capable of being a guide, as it is not in *An Essay on Man*, and the work of reason is supplemented by revelation. The most famous of all Pope's textual changes concerns this very question of guidance. In the first edition of the first epistle of the *Essay* (Foxon P822), Lælius is invited to

> Expatiate free o'er all this scene of Man;
> A mighty maze! of walks without a plan . . . (5–6)

In the edition 'Corrected by the Author' a month later (Foxon P827), line 6 reads:

> A mighty maze! but not without a plan . . .

The correction exploits an ambiguity in both versions of the line: is the issue whether the maze has been designed by an artist or whether a map is available to the explorer? In the first edition the primary meaning of 'plan' is map—this is a maze without any guidance on how to move through it—and the implication is not that God did not design the world but that there is no revelation of what his design is. The corrected reading removes the implication that there is no revelation, but avoids positively claiming there is one by allowing the primary meaning of 'plan' to shift. The implication now is simply that someone did design the maze; if there is a map, there is no hint that Pope and Lælius are going to follow it. In the final version of the poem, of course, there is a guide, but it is not holy scripture; it turns out that Bolingbroke himself has been the poet's 'Guide, Philosopher, and Friend'. Harte responds to this aspect of Pope's poem by insisting on God's provision of a guide. Experience and reason are important, even if they are not fully adequate on their own. Experience is 'Nature's surest guide' (98); reason, rightly used, can be 'A faithful guide to comfort and to save' (275), a 'guide, and judge, and guardian of our ways' (447). But an examination of the ancient world shows the inadequacy of reason, specifically in relation to Pope's metaphor of the maze:

> Yet ah, how few ev'n antient times beheld,
> (When *Greece* and *Rome* in arms and arts excel'd)
> Who thro' life's maze the steps of Nature trod,
> Reason their guide, and Truth their unknown God? (453–6)

The Christian dispensation provides a contrast in man's capacity to 'thrid the Maze', though one less sharp than might be anticipated:

> For ah, ev'n *here* where life a journey runs,
> Blest with new day-light and with nearer suns:
> Virtue's dim lights by God's own hand supplied,
> With Sanction strengthen'd, honour'd with a Guide,
> How few (except instructed first and led)
> Can thrid the Maze, or touch the Fountain's head?
> Observe a Mean 'twixt Bigotry and Pride,
> Hit the strait way, or err not in the wide? (507–14)

I take it that the new light comes from the Christian dispensation; that sanction is strengthened by biblical teaching on eternal rewards and punishments; that the Guide is the gospel; and that the Church is responsible for instructing and leading. The same guide is alluded to after a discussion of the obstacles to the correct operation of reason:

> Then gracious God, how well dost thou provide
> For erring Reason an unerring Guide! (548–9)

Harte differs from Pope, therefore, in maintaining that we do have a guide through the maze: revelation complements and corrects what we can discover by the light of reason. In this respect Harte's poem differs from Pope's and supplements it. No writing on revelation could offer better support to the teaching of *An Essay on Man*.

POPE'S HOSTILITY TO HUMAN REASON

Where Harte and Pope differ is not over revelation, where Pope is merely silent, but over the nature of reason itself. At least, they differ when we are concerned with Pope's stance in *An Essay on Man*. In his lines 'From Boethius, de cons Philos.', first published in 1717, Pope is very close to Harte:

> O thou, whose all-creating hands sustain
> The radiant Heav'ns, and Earth, and ambient main!
> Eternal Reason! whose presiding soul
> Informs great nature and directs the whole!
> Who wert, e're time his rapid race begun,
> And bad'st the years in long procession run:
> Who fix't thy self amidst the rowling frame,
> Gav'st all things to be chang'd, yet ever art the same! (*Twickenham*, 6. 73)

Pope echoes lines 3–4 towards the climax of the first epistle of *An Essay on Man*:

> All are but parts of one stupendous Whole,
> Whose Body Nature is, and God the Soul . . . (1. 267–8)

In *An Essay on Man* reason is not identified with the deity and is not eternal; nature is again informed by a soul, but that soul is God. Harte's approach is in keeping with the lines from Boethius, as his grand opening suggests:

> FROM Time's vast Length, eternal and unknown,
> Essence of God, coeval REASON shone . . . (1–2)

The identification of reason with God is confirmed later in the poem, when Harte says of reason that the good man

> Sees *Thee, God, Nature,* (well explain'd) the same . . . (219)

Pope's attitude to reason is generally hostile. His conception of it is Lockean and entirely compatible with that of a moderate deist like Thomas Chubb, who begins his *Discourse Concerning Reason, with Regard to Religion and Divine Revelation* (1731), 'By *reason*, I understand that faculty or power of mind by which men *discern* and *judge* of right and wrong, of good and evil, and the like' (3). In the *Essay*, as opposed to his translation of Boethius, Pope does not regard reason as eternal or an emanation of the godhead. It has no qualities that are not apparent in human reasoning. The attack on reason begins early in the first epistle, though Pope reserves his explanation of its function for the second. Reason is intimately connected with pride 'In reas'ning Pride . . . our error lies' (1. 123), 'From Pride, from Pride, our very reas'ning springs' (1. 161), and Pope's aim is to limit and debunk its pretensions.

Reason has no power to lead us from the here and now (1. 18), though it draws us on pointless quests (1. 35). The best course is to use reason itself to abandon the promptings of reason: 'to reason right, is to submit' (1. 164).

The single tribute to reason in the first epistle is withdrawn crushingly in the third. In discussing the 'scale of sensual, mental pow'rs' (1. 208), Pope notes the links from the 'groveling swine', through the 'half-reas'ning Elephant', to human reason (1. 221–4). He concludes that reason is the source of man's authority over the animals:

> The pow'rs of all subdu'd by thee alone,
> Is not thy Reason all those pow'rs in one? (1. 231–2)

This is a traditional view of reason, but Pope overturns it in the comparison of reason and instinct in the third epistle. He begins even-handedly by saying that reason and instinct are given appropriately to creatures' needs. But the subsequent comparison is wholly to reason's detriment:

> Reason, however able, cool at best,
> Cares not for service, or but serves when prest,
> Stays till we call, and then not often near;
> But honest Instinct comes a Volunteer.
> This too serves always, Reason never long;
> One *must* go right, the other *may* go wrong.
> See then the *acting* and *comparing* pow'rs
> One in their nature, which are two in ours,
> And Reason raise o'er Instinct, as you can;
> In this 'tis *God* directs, in that 'tis *Man*. (3. 85–98)

This passage goes well beyond conventional praise of God's gift of instinct to animals; reason seems inadequate for its duties. Pope reinforces this view later when he gives an account of the formation of human societies. Nature supplies the guidance man needs:

> To copy *Instinct* then was *Reason*'s part (3. 170)

A curiosity of Pope's account is that, although his approach to reason is psychological, reason remains outside the human personality, a sort of calculator that cannot always be found and may not always work. As we shall see, this viewpoint follows from the conception manifested in the second epistle, which sees self-love as the active power and reason as the comparing one. This leads Pope to identify the human personality with self-love and ultimately with the ruling passion.

In his most thorough examination of reason, in the second epistle, Pope understands reason through its relation to its contrary, self-love, the two ruling human nature in dynamic opposition:

> Two Principles in human Nature reign;
> *Self-Love*, to urge; and *Reason*, to restrain . . . (2. 53–4)
>
> Self-Love, the Spring of motion, acts the soul;
> Reason's comparing Balance rules the whole . . . (2. 59–60)

> Most strength the *moving* Principle requires,
> Active its task, it prompts, impels, inspires:
> Sedate and quiet the *comparing* lies,
> Form'd but to check, delib'rate, and advise. (2. 67–70)

Pope's sympathy is with the energy of God's ordering, and throughout the poem he insists on the primacy of the propulsion of self-love. Both self-love and reason aim at pleasure; self-love responds to immediate stimulus but reason takes the long-term view, seeing 'the future, and the consequence' (64). This is an aspect of reason that worries Harte, 'T'attain the distant, we o'ershoot the near' (144), but for Pope the difficulty is obviated by the power of the passions. In particular, reason is unable to withstand the power of the ruling passion, which gathers all human energies to itself. In other ethical systems, including Harte's, the aim is for reason to attain dominion over the human personality, a process that Harte associates with maturity; until the age of majority the 'Master-Figure' reason hardly appears, 'While *Passion* . . . usurps the throne' (230). Pope, aware of this tradition, presents reason as a monarch under the control of a favourite:

> The ruling Passion, be it what it will,
> The ruling Passion conquers Reason still.
> We, wretched subjects tho' to lawful sway,
> In this weak Queen, some Fav'rite still obey. (2. 149–50)[21]

At best in these circumstances, reason may be 'no Guide, but still a Guard' (2. 162). Pope is keen to remove reason's traditional role as a guide or pilot. Harte tells us that in dangerous waters reason discovers the shelves and 'steers us right' (161). Pope offers us reason merely as a chart that might help us to steer:

> On Life's vast Ocean diversely we sail,
> Reason the Card, but Passion is the Gale:
> Nor GOD alone in the still Calm we find,
> He mounts the Storm, and *walks upon the Wind*. (2. 107–10)

Mack relates these lines to the passage from Matthew's Gospel I cited in relation to Harte's poem, but God is conceived here as uniting himself with the power of nature, not repressing it. The reference is to Psalm 104: 3 (Douai 103): 'Who layeth the beames of his chambers in the waters; who maketh the clouds his charet: who walketh upon the wings of the wind.'[22] Pope's God is a God of power, a power which is found in emphatically in human energies; the role of reason is subsidiary.[23]

---

[21] The first of these couplets appears in editions up to the quarto works of 1735 (Griffith 372). It is omitted from the octavo works of that year (Griffith 388) and from all subsequent editions. Pope may have rejected the statement as too crude, but he seems to have held by the sentiment.

[22] Elwin cites an identification by Bowles (Elwin–Courthope, 2. 384). The trope is familiar from Cowper's hymn 'God moves in a mysterious way', the subject of a valuable discussion in Vincent Newey, *Cowper's Poetry: A Critical Study and Reassessment* (Liverpool: Liverpool University Press, 1982), 297–303.

[23] The punctuation and interpretation of 2. 99 to give reason an altruistic force is mistaken, as Pope's drafts (Mack, *Last and Greatest Art*, 236) and revision (Foxon, *Pope and the Book Trade*, 229, fig. 105) show.

HARTE AND POPE ON THE REASON OF THINGS

Pope and Harte, therefore, share very similar approaches to human conduct while taking quite different views of reason. Harte takes it to be an emanation of the godhead; Pope takes it to be a comparing and restraining faculty of man. Their views can be accommodated by the doctrine of the fall which leaves Harte with a similar conception of human reason to Pope's (though a different psychology) and a similar wariness of placing too much confidence in it. There remains, however, one further complexity. Harte's view of the operation of human reason is closely tied in with his view of what contemporaries called 'the reason of things'. Harte sees an internal connection between human understanding and the structure of the world. They coincide, if human understanding operates properly, because they are both created by divine reason. This is the hinge on which his poem turns. The 'Wits', who, along with the deists, are his opponents throughout, listen to his account of fallen reason and revelation, but then ask what for them—and apparently for Harte—is the fundamental question:

> Yet still the Wits with partial voice exclaim,
> What art thou Truth? What Knowledge? but a name.
> In short, are Mortals free, or are they bound?
> Tell us, is Reason something, or a sound. (295–8)

The issue is whether the ethical and religious views that Harte has presented correspond to the external world of nature. Are the truths binding because the world enforces them? Douglas H. White outlines the issues in a helpful discussion:

'The reason of things' refers to relationships between and among the creatures and objects of creation that supply the reasons why some acts are good, or moral, and others are bad, or immoral. The least equivocal example would be the fact that pain is the result of certain actions, and pain is undesirable by the very nature of the sensible creatures of the world. Those actions that cause pain, therefore, are clearly bad, or immoral, not because God has made a commandment to that effect (though such a commandment might be a supplementary or enforcing measure), but because of the essential character of the creatures and the consequences of their actions.[24]

Harte summarizes a comprehensive theist point of view in response to the Wits:

> As *perfect* Reason from the Godhead springs,
> (And still unchang'd if *perfect*;) so from *Things*,
> *Truths, Actions*—in their kind and their degree
> Starts *real meaning, difference, harmony*.
> These all imply a *Reason, Reason* still
> A *Duty, good* if sought, if sought not, *ill*;
> Hence in the chain of causes, *Virtue, Vice*,
> And thence *Religion*, take their gen'ral rise. (315–22)

---

[24] *Pope and the Context of Controversy*, 114–15. Harte is also attracted to a related idea linking language and action, 'vice is only truth deny'd' (21), which he may have derived from the first chapter of William Wollaston's *The Religion of Nature Delineated* (New York: Garland, 1978), originally published 1724.

The relation between reason springing from the godhead and meaning starting from things is drawn out in the following lines. Both owe their existence to the same rational godhead:

> God first creates, the ref'rence, nature, force
> Of things created must result of course . . . (323–4)

It follows for Harte that the human mind can respond to art and history, because they bear the stamp of God's order:

> These truths congenial, nor devis'd tho' found,
> Live in each age, and shoot from ev'ry ground . . . (343–4)

Nature has laws independent of human perception of them, and the failure of earlier societies to obey the moral order does not mean that it does not exist. Ideally, though man now lives in a fallen state, human reason and the reason of things, emanations of eternal reason, coincide.

In *An Essay on Man* Pope does not mention the reason of things, and nothing is more revealing of his vision of the human condition than this omission. The link the concept provides between the human mind and the external world is severed. The omission is the more striking because typographically the great theme of *An Essay on Man* is order or rather 'ORDER'. Pope uses both capitals and small capitals (caps. and smalls) and italics for emphasis in *An Essay on Man*, but although he uses caps. and smalls regularly for personal names—for ST. JOHN in the first line, for example—he otherwise reserves them for key terms or for phrases he wishes to highlight more than he could through the use of quotation marks. There are some thirty-one lines using caps. and smalls at some point in their history, but, if the list is reduced to those instances found in both 1734 and 1743/4, the scores are as follows: Order 5; God 4; Love 2; Right 1; Heav'n 1; Nature 1; Virtue 1; Man 1.[25] The surprise is that ORDER occurs so often (six times in all) and REASON so rarely (once in the 1743/4 summing up of the poem). The poem is indeed a vision of order, but Pope's vision differs from that of Harte and others by seeing nature as power rather than primarily as structure. The problem for man is that although he has intuitions of order as structure, his experience is of order as power. Pope gives some idea of what an apprehension of structure might be like in the early lines of the first epistle:

> But of this frame the bearings, and the Ties,
> The strong connections, nice dependencies,
> Gradations just, has thy pervading soul
> Look'd thro'? or can a Part contain the Whole? (1. 29–32)

The point is that man has no comprehension of the structure of the whole universe, the 'Worlds unnumber'd' in which God may act, but the poem presents the reader with very little sense of the bearings, ties, connection, dependencies, and gradations of our 'Station here', our own world. The emphasis on ORDER draws attention

---

[25] I have counted a word only once when it appears capitalized in a line twice.

instead to the power manifest both as creative energy and as potential disruption and chaos. The most dramatic way of apprehending this energy is to conceive order as composed of strife:

> But *All* subsists by elemental strife;
> And Passions are the Elements of life. (1. 169–70)

The strife is internal and external, and, through this principle of conflict, neither the ordering of man nor the ordering of nature is readily intelligible. Man features as microcosm in the second epistle, with primacy given to self-love or passion: we sail on the ocean of life with passion as the gale, just as God 'mounts the Storm and *walks upon the Wind*' (2. 100). A 'MIGHTIER POW'R' than reason, and hence unintelligible to her, impels men according to its designs. With the status of reason so severely restricted in relation to the passions, and with those powers in the universe analogous to the passions so powerful, it is hardly surprising that Pope is unwilling to speak of the reason of things.

When the universal order does not express itself as conflict, it expresses itself as a comprehensive vitality rather than as a structure. The command, 'Cease then, nor ORDER *Imperfection* name', follows the magnificent passage in the first epistle in which Pope proclaims that all are parts of 'one stupendous Whole' that

> Warms in the Sun, refreshes in the Breeze,
> Glows in the Stars, and blossoms in the Trees,
> Lives thro' all Life, extends thro' all Extent,
> Spreads undivided, operates unspent . . . (1. 271–4)

In this passage the vitality of the universe is, quite simply, identified with its Order. The conception is developed in the third epistle, when Pope argues that happiness arises from mutual needs:

> So from the first Eternal ORDER ran,
> And Creature link'd to Creature, Man to Man.
> What'ere of Life all-quickening Æther keeps,
> Or breathes thro' Air, or shoots beneath the Deeps,
> Or pours profuse on Earth; one Nature feeds
> The vital flame, and swells the genial seeds. (3. 113–18)

Again the evocation of vitality, the sense of an 'all-preserving Soul' (3. 22), is the focus of Pope's writing, called in to support mutual dependency, but not actually illustrating it. More is achieved in that respect through deployment of the metaphor of the 'Chain of Love' at the start of the epistle. This time the presentation of 'plastic Nature' is more analytical, with analysis into components, though once again the emphasis is on the power of movement:

> See, plastic Nature working to this End,
> The single Atoms each to other tend,
> Attract, attracted to, the next in place,
> Form'd, and impell'd, its *Neighbour* to embrace. (3. 9–12)

This is the closest Pope comes to expressing the articulation of his structure, but even here the emphasis is on the energy that makes up the whole rather than its ordering. Atoms are 'impell'd' into undisclosed relations; forms are merely like bubbles on a vast 'Sea of Matter' (3. 19). It is the underlying power that counts.

Pope's apprehension of order, therefore, takes us to the heart of *An Essay on Man* as a philosophical poem. His vision springs from an attempt to unify his positive response to the natural world and his bewilderment at the power of human passion. The poem's deployment of argument is always in the service of that sense of 'All Matter quick, and bursting into birth' (1. 234) and of the human energies that belong to that same system. The poem is written out of the 'Chaos of Thought and Passion, all confus'd' that is the human lot. The consequent downgrading of the potentialities of human reason, so striking in the light of Harte's contemporary poem, made it difficult to write a conventional philosophical poem in which reasoning carried its usual authority. Any version of the imitative fallacy that would have Pope deliberately writing a confused essay in order to express the human condition must be avoided, but Pope's vision required the subordination of reasoning or philosophical argument to what he took to be self-evident truths. The intellectual energy of the poem is in the layering of intricate positions, often through the deployment of typography, and in exposing the folly of those who take reasoning to be capable of so much more. Nevertheless, there is a fundamental difficulty in comprehending the poem's teaching (a difficulty which is at the heart of debates about the poem involving Crousaz, Warburton, Nuttall, and others),[26] because Pope does take one philosophical position to be a matter of common sense. That position is the one fundamental to natural religion. Bolingbroke, otherwise so sceptical, believed that we could induce from the world that there was a God with the natural attributes of infinite power and wisdom (the moral attributes he took to be human projections), and from these natural attributes of infinite wisdom and power he believed we could deduce the quality of our own world. This view, taken to be common sense, Pope holds in place as a frame for the detailed reflections of his essay. It is this very view that Hume subjected to critique in his *Enquiry Concerning Human Understanding*, arguing through Epicurus:

If the cause be known only by the effect, we never ought to ascribe to it any qualities, beyond what are precisely requisite to produce the effect; Nor can we, by any rules of just reasoning, return back from the cause, and infer other effects from it, beyond those by which alone it is known to us.[27]

This argument is remarkably close to the position Pope takes at the start of *An Essay on Man*, warning us against letting reason pretend to deliver more on the basis of our

---

[26] A. D. Nuttall, *Pope's 'Essay on Man'* (London: George Allen & Unwin, 1984), gives a stimulating account of the philosophical and theological issues.

[27] David Hume, *An Enquiry Concerning Human Understanding*, ed. Tom L. Beauchamp (Oxford: Oxford University Press, 1999), 190 (para. 13). As Solomon points out, Hume presented a copy of *A Treatise of Human Nature* to Pope (Mack's List, *Collected in Himself*, Library, 97). Bolingbroke's position is clearest in 'Fragments or Minutes of Essays', 43, *Works*, 5. 331–2, 345–6.

experience than it possibly can. The opposition of the *Essay* to anthropomorphism is also echoed in Hume's pursuit of his argument.

> The great source of our mistake in this subject, and of the unbounded licence of conjecture, which we indulge, is, that we tacitly consider ourselves, as in the place of the Supreme Being, and conclude, that he will, on any occasion, observe the same conduct, which we ourselves, in his situation, would have embraced as reasonable and eligible. (*Enquiry*, 196, para. 27)

In Pope's reluctance to trust the powers of human reason and arguments by analogy, therefore, lurks a possible challenge to theism itself, and this generates a nervousness in commentators that goes beyond concerns over Pope's possible acceptance of deism. Pope leaves the fundamental claims of natural religion unargued and undisturbed, but his play within its bounds is potentially dangerous. The gulf that lies between his poem and *An Essay on Reason* springs from this sort of daring: his willingness to engage a much greater range of discourses, entertain a wider range of conceptions of the human condition, and write as part of that scene of man he was surveying.

HARTE RECANTS: *THE UNION AND HARMONY OF REASON, MORALITY, AND REVEALED RELIGION*

The story of *An Essay on Reason* has an interesting coda, which tells us something about Harte's attitude to Christian orthodoxy, and perhaps indirectly tells us what Pope's own position might be. The key document is a letter from Harte to the *Weekly Miscellany*, 274, of 24 March 1738, welcomed by the *Miscellany* because it is not a defence, or an explanation, but the retraction of a passage in a published sermon. On 27 February 1737 Harte had preached a sermon before the University of Oxford on *The Union and Harmony of Reason, Morality, and Revealed Religion*. Publication of the sermon, as was usual in these circumstances, was licensed by the Vice-Chancellor. The position taken by the sermon is apparent from its unconventional reading of its text: 'Ye believe in God, believe also in Me' (John 14: 1):

> These remarkable words of our Saviour contain as close an argument, and as pathetical an exhortation as can be conceived. We may infer from them in general terms, that as there is a plain gradation from NATURAL to REVEALED Religion, and as there is a real *union* and *harmony* between them, therefore every honest impartial mind, is, as it were, favourably pre-disposed and induced by the *former*, sincerely to believe and practice the *latter*. (1)

This is Christianity in defensive mode. Belief in God is not taken to be a matter of revelation; primacy is given to rational reflection on the natural order; the appeal of Christianity is that it does not conflict with natural religion and in some measure enhances it. As in *An Essay on Reason*, there is to be no appeal to miracles or prophecy in defence of revelation. In the first section of the sermon, on reason, Harte is relatively cautious. Like Thomas Chubb he argues that for heathens, '*Reason*, supposing it duly exerted and attended to, must be a sufficient guide in matters of religion' (4), but he says that for Christians, reason functions as a guide to revelation:

it judges it and then 'modestly submits to it'.[28] He praises reason as 'the prerogative and shining ornament of our nature' but, in Popeian fashion, he is wary of its getting too big for its boots. In the second section, on the Law of Nature and Natural Religion, he is less restrained. He is sure there is a rule for man in nature: 'those eternal immutable truths concerning good and evil, right and wrong, which result from the nature of things, as those things are supposed to affect and obligate rational agents' (12). He is less certain about out capacity to understand it, and its need to be supplemented by revelation, but at points he will go very far:

> In short, 'tis agreed on all hands, that men by natural conscience and the essential differences of things may know in the main what is good and evil. Some may proceed much further, and argue with great force and clearness upon the divine being and attributes, the works of creation and providence, the probable hopes of a future state, and God's mercy to penitent sinners. Yet still with confidence I affirm, no *subsidiary* system ought to be rejected, which ... confirms these hopes, strengthens these motives, advances and adds to these reasonings. (15)

The key word is *subsidiary*, and the italics are Harte's. It appears reason and nature can present a system of religion; revelation comes in as back-up and as a way of clarifying the message to the simple.

In his third and final section, Harte claims that revelation is valuable in providing societies with 'proper sanctions and authority from God' and 'a full satisfactory, and compleat character of God' (17–19). But even here, Harte is anxious to accommodate natural religion by showing that Christian doctrines would be accessible to reason. He wavers a little over the sanctions of heaven and hell, which he takes to be 'the main arguments and vital principle of the Gospel': either they are '*commentaries* in *so many words*, upon what was before implied in the nature of things' or they are 'subsidiary *motives* to virtue' (21). On God's forgiveness, or the vital topic of salvation, however, he has made up his mind:

> SOME indeed have attempted to prove, that man in a state of nature, could not possibly form to himself any moral assurance, that God would vouchsafe to pardon sin, even after repentance. But here I think they have lost ground by endeavouring to push the victory too far. *Reason* certainly assures us that God will forgive a frail erring creature after due repentance and submission; But *Revelation* graciously proceeds much further: It points out the very method whereby God has brought about this reconciliation, (matter of new incitement to devotion, of new obligation, and thankfulness:) It likewise tells us that God will not only pardon sin, (which questionless is *all* in strict justice he is obliged to do,) but that he will likewise confer upon us *immortality*. (19–20)

Harte then develops this idea of the free gift of immortality in a further paragraph, elaborating the contrast between this gift and God's pardon. Harte does emphasize

---

[28] Harte does not cite Thomas Chubb, *A Discourse Concerning Reason, with Regard to Religion and Divine Revelation* (1731), though he does cite Matthew Tindal, *Christianity as Old as the Creation* (1730) in order to criticize him. It is apparent that Harte has considerable sympathy with the basis of Chubb's argument 'That Reason is, or ought to be, a sufficient guide in matters of religion', though he believes reason can be supplemented by revelation.

that the life of Jesus adds new sanctions and motives to virtue, but his conclusion takes us back to the defensive stance with which he began: 'Let us first lay a firm and deep foundation in reason and morality, and then let us add the beautiful superstructure of Christianity' (28).

Crudely stated, Harte's difficulty is that, in showing Christianity to be rational in terms set by the deists, he risks making religion purely a matter of conduct and thus struggles to save revelation from redundancy. The recantation in the letter to the *Weekly Miscellany*, dealing with only one objection, nevertheless shows up the general point:

> The Rev. Mr. *Venn* . . . censures me for saying . . . that 'Reason certainly assures us that God will forgive a frail erring Creature after due Repentance and Submission.' And . . . that 'God in strict Justice is obliged to pardon Sin'. This he condemns as downright rank *Socinianism*, and entirely subversive of the Scripture Account of our Redemption.
>
> Having never entertained a Thought like this, I have reviewed the Passages attentively; and finding that I expressed myself stronger than in Reason I ought to have done, I take this Opportunity of owning it to the Public.

Harte explains that he was thinking of the position of heathens and the great hope that God would forgive sin after sincere repentance, but he says there could be no certainty on this point (no compelling argument from God's perfections) and that there is a potential conflict with scripture, which says that 'God forgives our sins for Christ's Sake'. He concludes,

> Upon this Account, and as a Consequence may be drawn from my Words, which never enter'd into my Thoughts, I readily and ingenuously *give up the Point*; and shall order (in case the Sermon be re-printed) those exceptionable Passages and the Paragraph herein they are contained, to be entirely left out.

The omission was duly made in the fifth edition, published, like the second, third, and fourth, by Lawton Gilliver. What is impressive about this retraction is the evidence of Harte's sincerity in his profession of orthodoxy. Although in the original sermon he was responsive to Tindal and Chubb and edged the province of reason and natural religion ever further forward, he believed that he was maintaining contact with Christian tradition and teaching, and clearing the way of unnecessary difficulties. When challenged, he could have attempted to argue against the orthodoxy of Venn's position or to refine his own, but instead he gave it up.

Harte's sermon, like his poem, is a valuable context for *An Essay on Man*. It shows how pious and orthodox Christians could be drawn so far towards deism as to strive to make Christian doctrine compatible with it. A view of religion, which, like Wollaston's, emphasized morality, allowed natural religion to become the test of revelation. Richard Venn, the critic of Harte's sermon, saw deism as pernicious, and predicted disastrous consequences to a comic extent:

> But it is the Business of the Civil Magistrate to curb this increasing Sect; for otherwise, Self-Murder will daily grow more common, unbounded Lust and Rapine range thro' all

Orders of Men, the Sacredness of Oaths and Contracts be despised, and Government at last dissolved.[29]

But his opposition did not lead him to take a harsh line with Harte, who is rebuked for the one mistake in a footnote. Harte's desire to be a faithful Christian was recognized. Given the climate of his times, Pope is entitled to a similar generosity. *An Essay on Man* is not a Christian poem, but Harte's directly syncretic work shows that it is a poem that could have been written by a Christian.

---

[29] *A Sermon Preached before the Right Honourable the Lord-Mayor, and Aldermen, in the Cathedral Church of St. Paul's, London, on Monday January 30, 1737. Being the Day Appointed to be Observed as the Day of the Martyrdom of King Charles I* (London, 1737), 19–20. Venn was also involved in attacks on Whiston's friend and erstwhile disciple Thomas Rundle, opposing his appointment to the bishopric of Gloucester on the grounds that he was a deist. Rundle is praised, briefly, in *Epilogue to the Satires*, 1. 71, while *An Essay on Man*'s evocation of nature has qualities that might be associated with Whiston.

6. The First and Second Satires of the Second Book of Horace *(1733–1734)*: Parallel Texts

After the *Dunciad Variorum* and its revisions, the imitations of Horace are Pope's most elaborate examples of bookmaking. Extravagant in print and paper, they presented the Latin (in italic) and English (in roman) on facing pages, and they employed a double apparatus to specify the relations between the two texts. A system of small superior numbers linked the texts by dividing them into equivalent sections, and key words in the Latin text were highlighted by being printed in roman. The striking physical appearance of the first imitation, *The First Satire of the Second Book of Horace*, provided the starting point for the swingeing attack on Pope by Lady Mary Wortley Montagu and/or Lord Hervey in *Verses Address'd to the Imitator of Horace*:

> In two large Columns, on thy motly Page,
> Where *Roman* Wit is stripe'd with *English Rage* . . .
> Whilst on one side we see how *Horace* thought;
> And on the other, how he never wrote:
> Who can believe, who view the bad and good,
> The dull Copi'st better understood
> That *Spirit*, he pretends to imitate,
> Than heretofore that *Greek* he did translate?[1]

The striping of Latin and English in the folio format invited this attack because it was a claim to status: the status enjoyed by Horace as an independent citizen with a voice of equal weight to those of the aristocrat and official, and also, implicitly, the status of the great poet. The elaborate apparatus tied Pope down to particular forms of engagement with the Horatian text—apparent fidelity to the original in all its parts and skill in rendering particular words and phrases—but the folio pamphlet as a whole presented Pope, with some élan, as the English Horace.[2] In this chapter I

---

[1] Lady Mary Wortley Montagu, *Essays and Poems*, ed. Robert Halsband and Isobel Grundy (Oxford: Clarendon Press, 1993), 265, lines 1–10 (first published 8 Mar. 1733).

[2] Butt's *Twickenham*, 4, represents the parallel texts but omits Pope's indices; I quote from his edition but restore the indices when they are relevant. There are several important studies of the Horatian poems. The most hotly debated question is the nature of Horace's reputation in 18th-century Britain, with Howard Weinbrot in *Augustus Caesar in 'Augustan' England* (Princeton: Princeton University Press, 1978) and *Alexander Pope and the Traditions of Formal Verse Satire* (Princeton: Princeton University Press, 1982) taking a largely negative view and Howard Erskine-Hill in *The Augustan Idea in English Literature* (London: Edward Arnold, 1983) taking a more positive one. My view of these early poems is in line with Erskine-Hill's (292), recognizing that new issues come into play with *The First Epistle of the Second Book*, published in 1737 and that Pope's poetry is sometimes influenced by Juvenal and Persius. Of the books

shall concentrate quite narrowly on the typographical presentation of Pope's relations with Horace. One consequence will be to highlight his pleasure—the typography invites applause—in achieving a fit between Horace's satire and his own targeted individuals and groups, but another will be to suggest the importance of the task of presenting the author himself as the basis for a critique of the pretensions of wealth and property.

THE RESPONSE TO THE *TO BURLINGTON* CONTROVERSY

The publication of the early Horatian poems was largely a response to aristocratic and court hostility to the *Epistle to Burlington*, a hostility expressed both in gossip and in print. Pope's resort to the Horatian imitations was a way of opposing the aristocratic authority of his critics, by claiming the right to a voice, a critical stance, and a stake in public affairs. Through Horace, Pope was able to urge the claims of self-expression, social independence, and moral frankness, and justify satirizing persons and institutions that the critics of *To Burlington* would have held taboo. The anti-aristocratic character of much of this satire has been obscured by biographers' emphasis on Pope's friendships with aristocrats and by the impossibility of associating him firmly with any opposing social grouping. Pope was unable to identify himself with the bourgeoisie, whose poet, on the basis of his background and business skills, he might have been, or with the country gentry, whose land and social position he did not share.[3] His identification with Horace was one way of negotiating this difficulty; another was to meet the challenges of his opponents by publicizing his own private life and personality. His first two imitations, published on 15 February 1733 and 4 July 1734, marked an acceleration in the process of Pope's self-definition as a personality, an acceleration which led to the publication of the second volume of his works in April 1735 and, more remarkably, of his letters in May that same year. Characteristically, the personality was revealed, not as a member of a particular social group or class, but as solitary and yet diffused over a wide range of external reflectors, as though Pope's spirit were restless like Coleridge's in 'Frost at Midnight', 'everywhere echo or mirror seeking of itself'.

The usual account of the origins of the Horatian imitations makes them seem casual, even accidental. Pope told Spence the idea came from Bolingbroke, on a visit to Pope when he was ill with a fever:

Lord Bolingbroke came to see me, happened to take up a Horace that lay on the table, and in turning it over dipped on the First Satire of the Second Book. He observed how well that

---

devoted to the Horatian imitations I have found Frank Stack's *Pope and Horace: Studies in Imitation* (Cambridge: Cambridge University Press, 1985) consistently helpful, and I have also gained from reading Jacob Fuchs's *Reading Pope's 'Imitations of Horace'* (Lewisburg, Pa.: Bucknell University Press, 1989), particularly the section on Horace's reputation, 42–52, and Richard Steiger's *The English and Latin Texts of Pope's 'Imitations of Horace'* (New York: Garland, 1988), a study written much earlier, on the evidence of the bibliography, which stops at 1969.

[3] No life can be free of these contexts, but I do not believe they are deeply implicated in Pope's self-presentation. For a different emphasis, see Maynard Mack, *The Garden and the City: Retirement and Politics in the Later Poetry of Pope* (Toronto: University of Toronto Press, 1969).

would hit my case, if I were to imitate it in English. After he was gone, I read it over, translated it in a morning or two, and sent it to the press in a week or fortnight after. And this was the occasion of my imitating some other of the Satires and Epistles afterwards.[4]

But although the first Horatian satire may have come out of a happy conjunction of circumstances—the speed of composition and the apparent casualness of publication are both remarkable—it can also be seen as the development of an existing plan. When Fenton wrote to Broome in June 1729 about the war the dunces were waging against Pope, he also recorded Pope's response: 'He told me that for the future he intended to write nothing but epistles in Horace's manner' (*Correspondence*, 3. 37). This is a reference to the moral essays and to writing in Horace's manner, not to the Horatian imitations, but it already gives a Horatian perspective to post-*Dunciad* plans. By the winter of 1731/2 these plans had foundered. Pope had pressing need for a mask or new mode of speech, and a defence of satire had gained a new urgency and appropriateness.

The publication of the *Epistle to Burlington* on 13 December 1731 and its hostile reception twisted Pope's career uncontrollably from the path he had set. The general publishing plan after the *Dunciad Variorum* was clear. Its literary programme is detailed in Miriam Leranbaum's *Alexander Pope's 'Opus Magnum'* and its publishing aspects dominate the third chapter of David Foxon's *Pope and the Early Eighteenth-Century Book Trade*. The plan was for a series of ethic epistles; the epistles were to fall into two books, the second illustrative of the general principles of the first. The two books were finally to be combined with *The Dunciad* in a second volume of collected works. The *Epistle to Burlington* was the first step in the programme, and it was followed by the signing of two agreements with the bookseller Lawton Gilliver, Pope's protégé and the holder of the *Dunciad* copyright, on 1 December 1732. In his contract with Pope Gilliver agreed to publish any essay or epistle Pope offered him, pay £50 for the rights for one year, enter the poem in the Stationers' Register, and then transfer any rights he had acquired back to Pope at the end of the year; in a consequent agreement Pope agreed to Gilliver's publishing the second volume of *Works*.[5] The gap of a year between the publication of the first element in the programme and the signing of the agreement may be an indication of how far the attack on Pope following *To Burlington* threatened to halt the project. The difficulty lay not so much in the number of attacks as in their source, and certainly in Pope's perception of their source.

The arguments over whether the portrait of Timon in *To Burlington* satirized the Duke of Chandos, as the campaign against Pope claimed, or whether it was aimed at Sir Robert Walpole, or twenty different aristocrats, have tended to obscure what was at stake and the real, if unintended, daring of Pope's poem.[6] Never previously or

---

[4] Spence, *Anecdotes*, 1. 143 (anecdote 321a). This and the surrounding anecdotes have useful information on unpublished imitations of Horace.

[5] The agreement is in the British Library, Egerton MS 1951, and is printed by Robert W. Rogers, *The Major Satires of Alexander Pope* (Urbana: University of Illinois Press, 1955), 116–19.

[6] Timon's Villa has stimulated a quantity of 20th-century debate. F. W. Bateson devoted appendix C to the Timon question in volume 3 (2) of the *Twickenham* Pope; Maynard Mack deals with the episode in his

subsequently in his career did Pope so deliberately fail in deference and affront a whole class.[7] In addition to general disparagement of the owners of great estates (5–6, 11–12, 17–18, 39–42), there were generic portraits offensive to the 'Quality' (Virro, Visto, Bubo, Villario, Sabinus, Timon), and even reflections on the King (195–6) and Queen (78). To put it very simply, a poem criticizing the abuse of wealth and land had offended the small and powerful group with wealth and land. The response was a campaign to put Pope in his place by accusing him of mocking the Duke of Chandos. The choice of Chandos as the poem's victim was almost incidental. A letter from Lord Hervey to Stephen Fox a week after the publication of *To Burlington* shows what Pope was up against: 'Everybody concurs in their opinion of Pope's last performance, and condemns it as dull and impertinent. I cannot but imagine, by the 18 lines in the last page but one, that he designed ridiculing Lord Burlington as much as he does the Duke of Chandois.'[8] With a poem so 'impertinent', any lord could serve as its victim. The attacks worked by insisting on an older conception of the writer's social position, representing Pope as a dependant in a system of patronage. His criticism was invalidated on that basis alone. The case against Pope was not that the picture of Timon's Villa misrepresented Cannons; it was that Pope's subordinate position forbade criticism. The clearest expression of this view is in the anonymous *Of Good Nature* (possibly by John Cowper):

> AND yet there lives (oh! Shame to human Race!)
> A Wretch who boasts within Your Heart a Place:
> Who like an Adder, swoln with cherishing,
> Darts at his Patron his relentless Sting:
> Well-treated, yet not pleas'd, caress'd, yet rude,
> And proud of the base Crime—INGRATITUDE.[9]

As he explained in a letter to Aaron Hill on 22 December 1731, Pope did not know how he could defend himself: 'It's an aukward Thing for a Man to print, in Defence of his own Work, against a Chimæra: You know not who, or what, you fight against: The Objections start up in a new Shape, like the Armies and Phantoms of Magicians, and no Weapon can cut a Mist, or a Shadow' (*Correspondence*, 3. 260). His fullest attempt to encounter and defeat his critics, the ironic 'A Master Key to Popery' (like the *A Key to the Lock*, a burlesque of misinterpretation) remained unpublished. In it Pope exposes the social prejudices of his opponents. He identifies Lord Fanny, Lady De-la-Wit, and Lady Knaves-acre among his critics and, and regrets that their gossip is not in print, where it can be examined.[10] Some of Pope's information is

---

Life, 495–501; and Rosemary Cowler gives an overview of the controversy in her edition of the *Prose Works*, 2. 405–8 and 421–30 (see particularly 408 n. 11 for a short bibliography).

[7] As I explain in the Introduction, I have used the term 'class' freely. In this case, there is undoubtedly a class-consciousness at work, though the 'Quality' would not have spoken of it in that way.

[8] Earl of Ilchester, *Lord Hervey and his Friends, 1726–1738* (London: John Murray, 1950), 124–5; cited by Cowler, *Prose Works*, 2. 405–6.

[9] See J. V. Guerinot, *Pamphlet Attacks on Alexander Pope, 1711–1744* (London: Methuen, 1969), 211–12.

[10] Cowler, *Prose Works*, 2. 410. Subsequent references are to this text.

surprisingly detailed, as when he claims Lord Fanny (Hervey) has identified Villario with Lord Bathurst (414). He claims ironically that the poet's design has been 'to *affront* all the *Nobility & Gentry*' (410) and neatly represents the satire of *To Burlington* as a trespass: 'I appeal to all my Superiors, if any thing can be more insolent than thus to Break (as I may say) into their Houses & Gardens, Not, as the Noble Owners might expect, to *Admire*, but to *laugh* at them?' (410). Most telling in the whole piece is the treatment of Pope's friend Charles Bridgman, the landscape gardener:

> The vast Parterres a thous^d hands shall make
> Lo Bridgeman comes, & floats them with a Lake.

As if he should have the Impudence, when a Gentleman has done a wrong thing at a great Expence, to come & pretend to make it a right one? Is it not his business to please Gentlemen? to execute Gentlemen's will and Pleasure, not his own? is he to set up his own Conceits & Inventions against Gentlemen's fine Taste & Superiour Genius? Yet is this what the Poet suggests, with intent (doubtless) to take the Bread out of his mouth, & ruin his Wife & Family. (415–16)

Bridgman was one of the new professional men, and the conception of him in the 'Master Key' is uncomfortably close to the aristocrats' conception of the poet. Only the wealth from Homer placed Pope in a different category from Bridgman, who helped him design his garden at Twickenham. And the accuracy of Pope's attack on the complaining Gentlemen in 'A Master Key to Popery' is confirmed by Bridgman's attitude. To Pope's distress, expressed in the draft of *To Arbuthnot* (*Last and Greatest Art*, 428, lines 42–3), he asked for his name to be removed from the poem. In the *Works* of 1735 it is replaced by Cobham's. Cobham was an aristocrat and was entitled to destroy others' parterres.[11]

'A Master Key' remained unpublished. The copy that survives at Chatsworth is in Lady Burlington's hand, an ironic index of the difficulty of transferring quarrels from the private to the public sphere. The danger of 'A Master Key', I suspect, was that it identified too accurately the question at issue without disqualifying the objection that had been made. Pope's only direct public response to the *To Burlington* affair was in the letter to Burlington prefixed to the third edition of the poem. The letter entertains the possibility of giving up writing, the interlocutor's starting point for the *First Satire of the Second Book of Horace*: 'Quiescas.' | 'I'd write no more'. Silence is a possibility because to speak is to risk being slandered. Pope argues that *To Burlington* ought to have avoided controversy by ridiculing follies in general pictures, with no offence personal or topical. The alternative, satire of vices, necessarily invites controversy: naming gives personal affront; non-naming invites misrepresentation. The difficulty was pressing because it threatened the *opus magnum*, where the aim of the second book was to illustrate the general principles of the first. The

---

[11] Pope's disappointment at Bridgman's reaction shows in the note Warburton added in 1751: 'This office, in the original plan of the poem, was given to another; who not having the SENSE to see a compliment was intended him, convinced the poet it did not belong to him.'

question for Pope was whether satirical writing of any sort which implicated his 'Betters', as he called them in his letter, was open to him.

The solution outlined in the *To Burlington* letter—to attack vices through named individuals rather than follies through allusions—was one put into practice in *To Bathurst*, which was held back and published in 15 January 1733. *To Bathurst* protected its author by satirizing those who were notorious villains or were dead. The Horatian poems, the first of which was published only a month after *To Bathurst*, were an attempt to find a voice, a cover, and a justification for returning with greater force to the social targets of *To Burlington*. In this respect they ran in parallel with the two imitations of Donne's satires that Pope also published during this period, finding confidence in having the example of the '*Freedom in so eminent a Divine as Dr. Donne*', which '*seem'd a proof with what Indignation and Contempt a Christian may treat Vice or Folly, in ever so low, or ever so high, a Station*'.[12] Horace, from a different philosophical perspective, inspired a similar confidence.

FIRST SATIRE OF THE SECOND BOOK OF HORACE

The First Satire of the Second Book of Horace, published 15 February 1733, imitates the second of Horace's defences of satire, and Pope's choice of the later poem for imitation and publication is significant. A section of the earlier poem, *Satire 1. 4*, was 'paraphrased' in the aftermath of the *Burlington* uproar and published in the *London Evening Post* on 22–5 January 1732:

> The *Fop*, whose Pride affects a *Patron*'s name,
> Yet *absent*, wounds an author's honest fame:
> That more abusive Fool, who calls me *Friend*,
> Yet wants the honour, injur'd to defend:
> Who spreads a *Tale*, a *Libel* hands about,
> Enjoys the *Jest*, and copies *Scandal* out:
> Who to the *Dean* and *Silver Bell* can swear,
> And sees at *C-n-ons* what was never there;
> Who tells you all I *mean*, and all I *say*;
> And, if he *lyes* not, must a least *betray*:
> —Tis not the *sober Satyrist* you should dread,
> But such a *babling Coxcomb* in his stead.[13]

These lines (cued in to the Latin) closely resemble some of the material in 'A Master Key to Popery', which must have been written around the same time. Pope has found a strikingly appropriate section of the poem to imitate, that where the poet denies that he takes pleasure in giving pain and reprobates backbiting and disloyalty in others (81–5), and develops it to fit his own situation. *Satire 1. 4* haunted Pope in this period, until he was able to use some central elements in *To Arbuthnot*. They include:

---

[12] From the advertisement to the imitations in the 1735 *Works* (*Twickenham*, 4. 3). For an admirable discussion of Pope and Donne, see Erskine-Hill, *The Augustan Idea in English Literature*, 74–98.
[13] The poem is printed with other occasional verse in *Twickenham*, 6. 339.

Horace's tribute to his father, who would warn him by examples; his sense of a tradition of satire; the contempt for inferior writers (including Fannius, who delivers his works unasked); and the idea that satire is an expression of personality. The *London Evening Post* lines found their place in *To Arbuthnot* as an introduction to the attack on Lord Hervey as Sporus. But other elements in *Satire 1. 4* were problematic: Horace says he writes very little; he does not seek publicity; he does not publish; and he does not think satire is poetry. This would have been enough to dissuade Pope from full imitation, but there are other reasons why it was *Satire 2. 1* that drew his attention for his first imitation.

There is a tendency, most ably and precisely expressed by Niall Rudd, to see Horace's *Satire 2. 1* as a particularly relaxed and non-serious defence of satire. Horace no longer needed to defend himself, the argument goes, and gives us here merely delicate and amusing shadow-boxing, a 'humorous charade'.[14] I suspect this is a false estimate; Pope would certainly have regarded it as one. If Horace and his distinguished friendly antagonist in the dialogue, the lawyer Trebatius, never lock horns, that is because they represent different systems of values. Horace is under no obligation to respond to Trebatius on his own terms. Part of the skill of Horace's response is its indirection and its insistence on taking each point thoroughly as it comes. This is in line with the advice in the 'Ars Poetica':

> ordinis haec virtus erit et venus, aut ego fallor,
> ut iam nunc dicat iam nunc debentia dici,
> pleraque differat et praesens in tempus omittat,
> hoc amet, hoc spernat promissi carminis auctor.

[Of order, this, if I mistake not, will be the excellence and charm that the author of the long-promised poem shall say at the moment what at that moment should be said, reserving and omitting much for the present, loving this point and scorning that.][15]

The argument in *Satire 2. 1* proceeds in this way, with a case being established slowly and the reader left to unite the strands into a single position. It defends satire as a natural activity and one to which Horace is suited, as a form of self-disclosure, as self-defence, as a stripping away of pretence, and as a vocation compatible with enjoying the favour of the great.[16] It also touches other issues in passing: the poles of lawlessness and triviality; the possibility of silence; the relation of the poet and ruler;

---

[14] Niall Rudd, *The Satires of Horace* (Cambridge: Cambridge University Press, 1966), 124–31, and 85. I have also learned much from Rudd's translation, *Horace: Satires and Epistles; Persius: Satires* (London: Penguin Books, 1979).

[15] *Horace: Satires, Epistles, and Ars Poetica*, trans. H. Rushton Fairclough (London: Heinemann, 1926, rev. 1929), lines 42–5. I have leant heavily on this edition and grown fond of its translation into that extraordinary dialect which is Classics English. When for a few lines of *Satire 1. 2* I have had to use my own translation, I have put it in italic.

[16] The first point, which is central to Pope's conception, is supported by the *Ars Poetica*: 'Sumite materiam vestris, qui scribitis, aequam | viribus et versate diu, quid ferre recusent, | quid valeant umeri' [Take a subject, ye writers, equal to your strength; and ponder long what your shoulders refuse, and what they are able to bear] (38–40). For a reading of Pope's poem which argues persuasively that Pope goes beyond Horace, 'a Horatian satire that nonetheless quarrels with Horace, his poem, and culture at several points' (239), see chapter 7 of Weinbrot's *Alexander Pope and the Traditions of Formal Verse Satire*.

and the nature of bad verses. All these were germane to Pope's situation, and if Pope's friends were out of power that added a poignancy to the parallel. Pope could also be said to be beginning a new book of satires. Horace wrote two books of satires and two books of epistles. Pope, who had written his *Ars Poetica*, *An Essay on Criticism*, early in his career, and was currently engaged in writing two books of epistles, had written one satirical book, the mock-epic *Dunciad*, and now really was beginning a second group of satires. Like Horace, he could look back on his earlier career and try to assess his position. His main aim in the first two of the imitations was to claim satiric dignity and freedom. As he says in the Advertisement: '*An Answer from Horace was both more full, and of more Dignity, than any I cou'd have made in my own person.*' The imitation of *Satire 2. 1* was to realize the satiric programme as well as defend it; the Horatian mask served equally for self-expression and for self-protection. Thomas Bentley was not far astray in his attack on Pope's *Sober Advice from Horace* (*Satire 1. 2*) in early March 1735: 'An admirable Expedient, and worthy of your Sagacity, *to get upon the Back of Horace*, that you may abuse every body you don't like, with Impunity!'[17] There were ambiguities in the poem to be exploited; and there were further ambiguities to be created through the relationship of the imitation to its original. The reader could be directed to these complex interrelations by the typographic resources at Pope's disposal.

The most striking aspect of the folio first edition of *The First Satire of the Second Book of Horace* is the prominence given to Horace's Latin, with Horace on the left balancing Pope on the right. The precedent in Pope's publishing career for this balancing of two authors is 'A Parallel of the Characters of Mr. Dryden and Mr. Pope', which formed appendix VI of *The Dunciad Variorum* (Fig. 9). Here Dryden appears on the left, Pope on the right. The pages are mirrored in their dropped heads, and a system of letters in parentheses connects the text with the footnotes. The fiddling intricacy of the parallel enforces the claim that these great writers are equivalent both in worth and in the vilification they provoked; the detail must be there, but its nature matters less than the overall visual impression. The early imitations of Horace have a layout very similar to this appendix. The original Horatian satire is not merely the base from which the English poem springs; it runs alongside it as equal and equivalent. Pope is presented as the modern Horace; the poets mirror one another. Significantly, when Pope later published the first edition of the *First Epistle of the Second Book of Horace* (1737), in which the differences between the poets became more marked than the similarities, he did so without parallel texts. The decision to choose parallel text in 1733 went in the face of contrary precedent. His printer, John Wright, and bookseller, Lawton Gilliver, had already started a little line of Horatian imitations with Pope's approval and it would have been possible for *The First Satire of the Second Book* to follow these. In a letter to John Caryll of 6 February 1731, Pope reveals something of his pleasure in dabbling in the bookselling business:

---

[17] Guerinot, *Pamphlet Attacks on Alexander Pope*, 251. I share Guerinot's view that this is in many ways a telling attack, while acknowledging *Sober Advice* to be one of the softer targets.

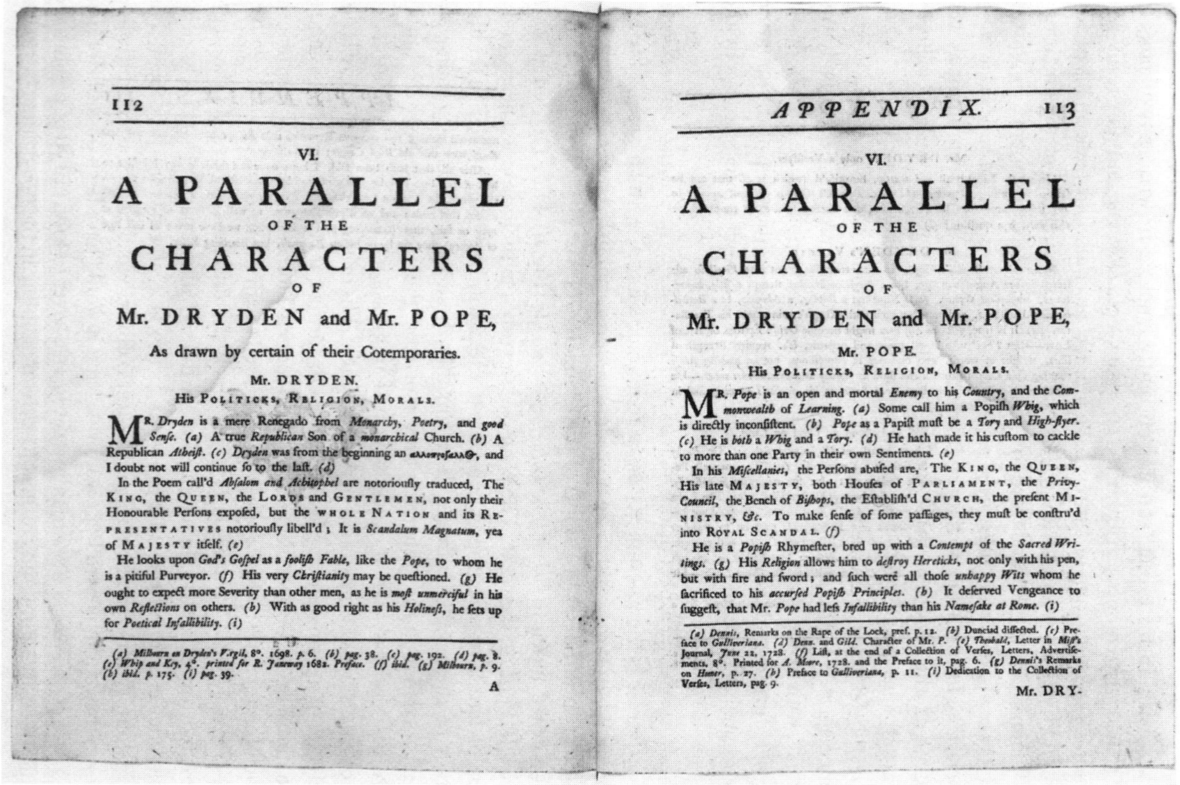

112

VI.

A PARALLEL
OF THE
CHARACTERS
OF
Mr. DRYDEN and Mr. POPE,

As drawn by certain of their Cotemporaries.

Mr. DRYDEN.

His POLITICKS, RELIGION, MORALS.

MR. *Dryden* is a mere Renegado from *Monarchy, Poetry,* and *good Senſe*. (*a*) A true *Republican* Son of a *monarchical* Church. (*b*) A Republican *Atheiſt*. (*c*) *Dryden* was from the beginning an ἀλλοπρόσαλλος, and I doubt not will continue ſo to the laſt. (*d*)

In the Poem call'd *Abſalom* and *Achitophel* are notoriouſly traduced, The KING, the QUEEN, the LORDS and GENTLEMEN, not only their Honourable Perſons expoſed, but the WHOLE NATION and its RA-PRESENTATIVES notoriouſly libell'd; It is *Scandalum Magnatum*, yea of MAJESTY itſelf. (*e*)

He looks upon God's Goſpel as a *fooliſh Fable*, like the *Pope*, to whom he is a pitiful Purveyor. (*f*) His very Chriſtianity may be queſtioned. (*g*) He ought to expect more Severity than other men, as he is *moſt unmerciful* in his own Reflections on others. (*h*) With as good right as his *Hodneß*, he ſets up for Poetical Infallibility. (*i*)

(*a*) Milbourn on Dryden's Virgil, 8°. 1698. p. 6. (*b*) pag. 38. (*c*) pag. 192. (*d*) pag. 8.
(*e*) Whip and Key, 4°. printed for R. Janeway 1682. Preface. (*f*) ibid. (*g*) Dedication, p. 9.
(*h*) ibid. p. 175. (*i*) pag. 39.

A

APPENDIX. 113

VI.

A PARALLEL
OF THE
CHARACTERS
OF
Mr. DRYDEN and Mr. POPE.

Mr. POPE.

His POLITICKS, RELIGION, MORALS.

MR. *Pope* is an open and mortal *Enemy* to his *Country*, and the Com-memorable of *Learning*. (*a*) Some call him a Popiſh *Whig*, which is directly inconſiſtent. (*b*) *Pope* as a Papiſt muſt be a *Tory* and *High-flyer*. (*c*) He is *bold* a *Whig* and a *Tory*. (*d*) He hath made it his cuſtom to cackle to more than one Party in their own Sentiments. (*e*)

In his *Miſcellanies*, the Perſons abuſed are, The KING, the QUEEN, His late MAJESTY, both Houſes of PARLIAMENT, the Privy-Council, the Bench of *Biſhops*, the Eſtabliſh'd CHURCH, the preſent MI-NISTRY, &c. To make ſenſe of ſome paſſages, they muſt be conſtrued into ROYAL SCANDAL. (*f*)

He is a *Popiſh* Rhymeſter, bred up with a *Contempt* of the *Sacred Wri-tings*. (*g*) His *Religion* allows him to *deſtroy Hereticks*, not only with his pen, but with fire and ſword; and ſuch were all thoſe *unhappy Wits* whom he ſacrificed to his *accurſed Popiſh Principle*. (*h*) It deſerved Vengeance to ſuggeſt, that Mr. *Pope* had leſs *Infallibility* than his *Nameſake* at Rome. (*i*)

(*a*) Dennis, Remarks on the Rape of the Lock, pref. p. 12. (*b*) Dunciad diſſected. (*c*) Pre-face to Gulliveriana. (*d*) Denn. and Gild. Character of Mr. P. (*e*) Theobald, Letter in Miſt's Journal, June 22, 1728. (*f*) Lift, at the end of a Collection of Verſes, Letters, &c. to the Author, printed for A. Moor, 1728, and the Preface to it, pag. 6. (*g*) Denn's Remarks on Homer, p. 27. (*h*) Preface to Gulliveriana, p. 11. (*i*) Dedication to the Collection of Verſes, Letters, pag. 9.

Mr. DRY-

FIG. 9. Parallel of the Characters of Dryden and Pope, *The Dunciad Variorum*, 1729 (Bodl. Vet. A4 d. 128 (2); 249 × 188 mm).

The Art of Politicks [by James Bramston] is pretty. I saw it before 'twas printed. There is just now come out another imitation of the same original, *Harlequin Horace* [by James Miller]: which has a good deal of humour. There is also a poem upon satire writ by Mr Harte of Oxford, a very valuable young man, but it compliments me too much: both printed for L. Gilliver in Fleet street. (*Correspondence*, 3. 173)

Harte's *Essay on Satire* has as an appendix Boileau's *A Discourse on Satires, Arraigning Persons by Name*, which has a strong Horatian emphasis, but the two imitations of the *Ars Poetica* are more directly relevant to Pope's turn to Horatian imitation. Both are expensively produced octavos, with engraved frontispieces and well-spaced texts. At the foot of each page of the poem is a rule right across the bottom of the page, with the Latin text running along below it in smaller type. Sections of the text, in English and Latin, are linked by preceding numbers in parentheses in *Harlequin-Horace* and by letters of the alphabet in *The Art of Politicks*. In this way each paragraph of the English poem is indexed to the appropriate section of the Latin. The imitations are generally loose. Miller's, which is largely an inversion of Horatian precepts, is the closer; Bramston's is ingenious rather than Horatian. The physical presentation of the Latin and the English is representative of their relationship: the Latin is a sort of base or springboard for the English; the trace of translation that belongs to imitation has almost disappeared.[18] It is clear from his letter that Pope knew and approved of these imitations. Some of their lines may have stayed in his mind. Peter Dixon has drawn attention to a parallel between *An Essay on Man* and *Harlequin-Horace*, and there may be some recollection of Miller's attack on Blackmore, 'How Arms meet Arms, Swords clash, and Cannons rattle', and of Bramston's 'Just such a Monster, Sirs, pray think before ye, | When you behold a Man both *Whig* and Tory' in *The First Satire of the Second Book*.[19]

Pope could have opted for this octavo format (as he had done for *The Rape of the Lock*), but he decided to follow the pattern set by *To Burlington* and *To Bathurst* and print in folio instead.[20] The crucial decision in planning publication was that of printing parallel texts. There were precedents. André Dacier's translation into French, which Pope knew, and which I believe, with Stack and Maresca, he drew on freely, presented the texts in parallel with italic type. But, of course, that was a translation, and parallel texts of translations, either in line-for-line equivalents or in

---

[18] I have nothing to add to the discussions of translation, paraphrase, and imitation to be found in the studies listed in note 1 and in the first two chapters of Howard D. Weinbrot's *The Formal Strain: Studies in Augustan Imitation and Satire* (Chicago: University of Chicago Press, 1969). Erskine-Hill's is a short lucid discussion (*The Augustan Idea in English Literature*, 291–2), while Steiger's attempt to shift the taxonomy seems to rest on the misconceived basis that to add a new distinction is to disqualify an old one (*The English and Latin Texts of Pope's 'Imitations of Horace'*, 1–9).

[19] Peter Dixon, 'Pope and James Miller', *Notes and Queries*, 215 (1970), 91–2. Blackmore is attacked by Pope in lines 23–6 and Tories call him Whig, and Whigs a Tory, in line 68.

[20] The decision over format was, by the terms of their agreement, strictly Gilliver's, but it seems unlikely that at this stage he would have disagreed with Pope. Pope was always concerned to help those who collected and bound his work in sets, and he would have had an eye to *Works* publication, which was initially in folio and quarto. The joint publication of *Satires 2. 1 and 2. 2* in quarto seems to have been aimed at incorporation in the *Works*.

prose, made easy work for the printer.[21] Imitations are a different matter; they are more loosely related to their originals, and frequently expand on them. Pope's certainly do: his version of *Satire 2. 1* has 156 lines to Horace's 86; *Satire 2. 2* has 180 to 116 (with omissions from Horace's poem). This creates a disparity between the two texts, which turning the Latin into footnotes disguises: the imbalance in the number of lines ceases to matter and yet the basis of the English in the Latin is readily apparent. Pope's decision to present parallel texts flies in the face of printing economy and convenience, but it did have some commercial advantages. It virtually doubled the cost of printing, but, on the other hand, it made a substantial pamphlet out of an English poem of 154 lines and enabled Gilliver to charge one shilling for it, as he had for *To Burlington* and *To Bathurst*, thereby making a satisfactory profit from the first year's rights (Foxon, *Pope and the Book Trade*, 119–20). From a book-trade point of view, then, the choice between footnotes and parallel texts was an evenly balanced one.[22] It was Pope's purpose in imitating Horace that determined the final mode of presentation.

The title page, perhaps the most interesting of all Pope's title pages, reveals a good deal of his conception of the poem. It draws immediate attention to the issues of status that underlay his current quarrels. The top half of the page uses large types with good space between the lines: THE [3.5] | FIRST SATIRE [8.5] | OF THE [3.5] | SECOND BOOK [6.5] | OF [3] | HORACE [11.5].[23] 'Horace' is much the most important word on the page, but it is followed by an immediate descent into social specifics. 'Imitated in a DIALOGUE between [5]' is still followed by a blank space, but then come these lines, not even, I think, leaded: '*ALEXANDER POPE*, of *Twickenham* in *Com.* | *Midd.* Esq; on the one Part, and his LEARNED | COUNCIL on the other [all 5].' The curious line-breaks in mid-phrase suggest the imitation of a different form of document, a legal one, but they cannot disguise the importance of the 'Esq;' lurking on the second of these lines. This setting up of the dialogue is followed, after a rule, by the line from the poem that Pope chose as a motto: 'Scilicet Uni Aequus Virtuti, atq; ejus Amicis [kindly in fact only to Virtue and her friends].' The capitals to begin the main words provide an emphasis similar to that in the body of the poem, where full capitals are used for the line.

---

[21] *Œuvres d'Horace en latin et en françois, avec des remarques critiques et historiques* (Paris: Ballard, 1709). I have not been able to decide which of the many editions of this work Pope used. Pope's praise of Dacier's remarks on Horace are in a letter to Buckingham of 1 September 1718 (*Correspondence*, 1. 492). I have also frequently consulted the Paris edition of 1697. I have not discovered a source for Pope's particular form of parallel text. Stack lists possibilities, *Pope and Horace*, 20–3, as do Steiger, *The English and Latin Texts of Popes 'Imitations of Horace'*, 17, and Weinbrot in *The Formal Strain*, 26–9. There were many precedents, including the fourth edition of Horace's *Satires, Epistles, and Art of Poetry*, ed. S. Dunster, 1729.

[22] David Foxon has emphasized the importance of the parallel texts in bulking out Pope's half share of the *Works*, taking them as a confession of Pope's failure to complete the *opus magnum* (*Pope and the Book Trade*, 123). The observation is characteristically acute, but Pope's agreement with Gilliver did not oblige him to contribute half the material for the *Works* and he and Jonathan Richardson could easily have filled up the volume with notes.

[23] Measurements are of the height of the type face in millimetres, from base on the line to the top of an ascender. Inking and paper shrinkage make accuracy unlikely; my aim is to give an impression of emphasis.

The most curious aspect of the title page is the 'Esq;'. Pope had until now been 'Mr Pope' and he was to remain 'Mr Pope' on his title pages until the octavo editions of his works were published in July 1735.[24] The odd situation in 1735 was that the expensive quarto and folio editions were called *The Works of Mr Alexander Pope. Volume II* (like the first volume in 1717), while the octavo series for a general readership were entitled *The Works of Alexander Pope, Esq.* Without the change in the title of the *Works*, the appearance of 'Esq;' on the title page of *Satire 2. 1* could be taken as mere legal colouring, but the two changes coincide with a change in Pope's title in his contracts and I think they represent some reassessment or redeclaration of Pope's social position. The title 'Esquire' already had a difficult status, as Steele explained in the *Tatler*, 19 (24 May 1709), recognizing that it was being used very generally in addressing letters and as a courtesy to all who were considered gentlemen. Others, on the other hand, were keen to maintain a distinction. The *OED* quotes Hearne in 1711 saying, 'I shall be glad to know . . . whether he be Esqr. that I may give him his true Title', and my impression is that the distinction was quite carefully maintained in subscription lists. The grounds on which the distinction was made are not at all clear; there were at least three different ways it could be done. The strictest definition is quoted by the *OED* from the *Encyclopaedia Britannica*:

Of esquires, legally so called, there are, according to some authorities, five classes: '(1) younger sons of peers and their eldest sons; (2) eldest sons of knights, and their eldest sons; (3) chiefs of ancient families (by prescription); (4) esquires by creation or office, as heralds and sergeants of arms, judges, officers of state, naval and military officers, justices of the peace, barristers-at-law; (5) esquires who attend the Knight of the Bath on his installation—usually two specially appointed.

Another definition entangles us in the question of 'armour-bearer' or alternatively 'one bearing heraldic arms'. This was a live question for the Lintots, who approached Humfrey Wanley in search of arms, but, as far as I am aware, it did not interest Pope.[25] The third qualification was to be simply 'a landed proprietor'. Pope falls short of esquire status by all these criteria. He was not descended from a titled family and he had no office. He was not a bearer of heraldic arms, though his choosing to wear a short sword was presumably a claim to gentility. Most importantly, he was not a landowner; he rented his villa at Twickenham from Thomas Vernon. This fact, a limitation imposed by his position as a Roman Catholic, is an issue in the Horatian imitations, considered at length at the close of the imitation of the *Second Satire of the Second Book*. The innocent-looking 'Esq;', therefore, is a claim to

---

[24] *Of the Use of Riches* continues to be 'By Mr. Pope', as is *To Cobham*; *Sober Advice* is in 'the Manner of Mr. Pope'; the epistle is 'From Mr. Pope to Dr. Arbuthnot'; *To a Lady* is 'By Mr. Pope'; the *Letters* are 'Of Mr. Pope'; Horace's *Ode to Venus* is 'By Mr. Pope', as are the other Horatian poems of 1737 and 1738, and eventually *Dunciad* 4 in 1742; but the deathbed edition of the ethic epistles was advertised as 'by Alexander Pope, Esq;' and Warburton's edition is of the *Works of Alexander Pope, Esq;*.

[25] *The Diary of Humfrey Wanley, 1715–1726*, ed. C. E. Wright and Ruth C. Wright, 2 vols. (London: Bibliographical Society, 1966), 2. 404.

dignity in defiance of the rules; it lurks ready to raise issues of property, just at a point when Pope had invaded and criticized the estates of the landed gentry.

The appearance of 'Esq;' on the title page of the first imitation lends a specific social colour to the dialogue. This is an English gentleman consulting his lawyer. Although Warburton gave the poem the title 'Epistle to Fortescue' thus drawing it into line with other epistles of the period, that title is misleading. Pope's interlocutor is certainly identified as W.F. in the manuscript, but the emphasis in the first edition of the poem is on the generic 'Learned Council'. *L.* is the speech prefix used and the speaker is not identified as '*F.*' until the second edition. Which is not to deny that the poem compliments William Fortescue, formerly Walpole's private secretary, later Master of the Rolls, and at the time of the poem attorney-general to the Prince of Wales, by implicitly comparing him to the distinguished jurist Trebatius. The style of the title page is that of legal documents and so draws the dialogue between poet and lawyer into the contemporary world. In particular, the language reflects Pope's most recent agreement, that with Gilliver for publishing the epistles. The title in the manuscript precisely reflects the language of a contract: 'Imitated in a Dialogue between W. F. Esq. A . . . . on ye one p^t. and A. P. of T . . . . . Esq. on y^e. other.' In the contracts for the *Iliad*, *Odyssey*, and *Works* 1717, Pope is initially named as 'Alexander Pope of Binfield [Twickenham] in the County of Berks [Middlesex] Gentleman', but in the contract of 1 December 1732 he becomes 'Alexander Pope of Twickenham in the County of Middlesex, Esqr.'[26] Fortescue had witnessed the *Odyssey* contract and some of the payments attached to the Bodleian *Iliad* indenture, but he was not connected with the epistles agreement as far as I know. The title page imitates the ordering of the legal document and introduces legal abbreviations. The change in Pope's designation in that agreement may not reflect a change in his relation to real property but I strongly suspect it does represent a change in his relation to what might be called literary property. In previous agreements Pope had sold the work he had created for perpetuity; but in the new agreement he merely allowed Lawton Gilliver the liberty of publishing it for a year. In this period Pope was accumulating literary rather than real property, and behaving like a gentleman, directing others in the cultivation of his land and then taking a share of the profits. What concerned him was the exclusive right to use his work; that was the right that was to return to him and the right that he defended through a series of court actions. In literary property he was an esquire.

The First Satire of the Second Book is notable in Pope's *œuvre* for the positive way it uses the language of bookselling. *P.* boldly announces his intention to:

> Publish the present Age, but where my Text
> Is Vice too high, reserve it for the next . . . (59–60)

[26] The agreement for the *Iliad* is BL, Egerton Charter 128, and Bodleian MS Don. a. 6 (transcribed in my 'The Contract for Pope's Translation of Homer's *Iliad*: An Introduction and Transcription', 206–25); the agreement for the *Odyssey* is BL, Egerton Charter 130 (transcribed by Sherburn, *The Early Career of Alexander Pope*, 313–16); the agreement for the *Works* is BL, Egerton Charter 129. Pope does not refer to himself as 'Esqr.' in the dispute with James Watson over the pirated *Letters*, but Watson does in the draft of his bond. See Mack, *Collected in Himself*, 500.

And 'Publish' has the dignity of a moral act. Later he says that, whatever the circumstances, he 'will Rhyme and Print', with the air of a man determined to do his duty. Both these passages point up a new commitment to the medium of print and publication as a means to carry out his vocation as poet. He concludes *The Second Satire of the Second Book* by saying,

> Let Lands and Houses have what Lords they will,
> Let Us be fix'd, and our own Masters still. (179–80)

The conflict over the *Epistle to Burlington* and subsequent contracts seem to represent a rethinking about property. As P. S. Atiyah points out in his *The Rise and Fall of Freedom of Contract*, the problematic nature of literary property does not mean that it is a marginal case in an otherwise simple conceptual field; rather literary property reveals the conventional nature of real property: 'The truth was laid bare for all who wanted to see it, though probably few did: property rights were not "natural" but artificial creations of the law, and it was the law, based on values and policies, which determined the extent of those rights.'[27] The first two of these Horatian poems suggest that Pope was one of the few who were prepared to recognize and draw attention to this truth.

The half-title of *Satire 2. 1* presents only the Horatian source, 'Q. Horatii Flacci Sermonum Libri Secundi Satira Prima', but the first two pages of text, with their dropped heads, restore the parallel (Fig. 10). The first pages of parallel text are striking, with the Latin dropped head suggesting a more ambitious programme than Pope ever realized: 'Sermonum Liber Secundus. Satira Prima'. From this heading we might be beginning the second book of satires, with this only the first of the eight. Pope may have intended to offer something of this promise—or threat.

The two names, Horatius and Trebatius, running across the page signal a major decision to formalize and dramatize the dialogue. Whereas the generality of Latin texts (Dacier in Latin and French is an exception) present the dialogue as a seamless whole, with the speeches undistinguished by quotation marks, let alone speech prefixes, Pope chose to present the dialogue as a Shakespeare play, even giving a new speaker a new line:

> *Quid faciam?* Præscribe.
>   *TREB. Quiescas.*
>   *HOR. Ne faciam inquis,*
> *Omnino versus?*
>   *TREB. Aio.*
>   *HOR. Peream . . .* (5–6)

The English dropped head has Horace's name printed more prominently than in the Latin. The speakers are not explained; they appear as *P*. and *L*. in the first edition. The latter became *F.* for fifteen lines of the second edition in quarto in 1734

---

[27] *The Rise and Fall of Freedom of Contract* (Oxford: Clarendon Press, 1979), 109.

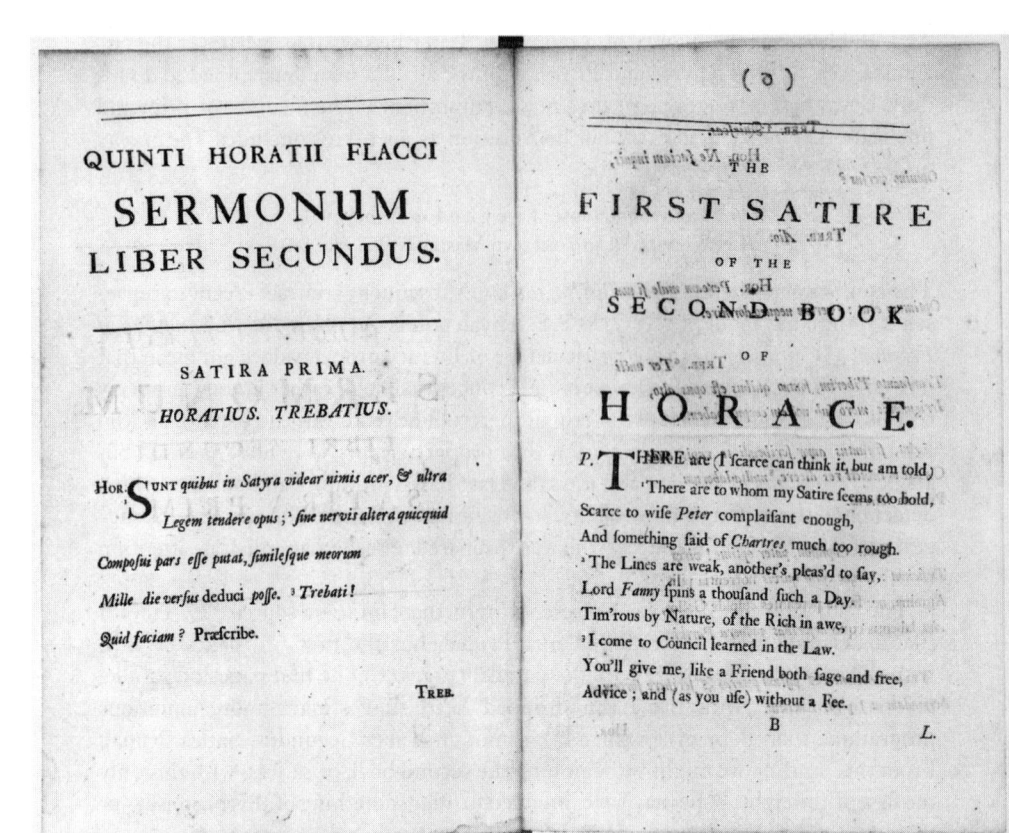

FIG. 10. First page of *The First Satire of the Second Book of Horace*, 1733 (Bodl. Fol. Δ 696; 320 × 205 mm).

but had to wait for the *Works* in 1735 for the change to be implemented consistently. The formalization is important for separating the speakers. Horace might be willing to allow some mingling of the two voices, just as there is some blurring of the border between Ofellus and Horace in the second satire, but Pope's text sharply distinguishes the two. They represent two different viewpoints, and two different personalities. This complements the insistence on the socialization of the satire in the English: for example, *L.* would normally be paid a fee, but it is a sign of their friendship and of *P.*'s cheek that he gets his advice free. At one point the creation of speakers seems to have created a deliberate difficulty. At the end of the English dialogue, *L.* has been completely won over. *P.* denies writing libels and satires:

> P. *Libels* and *Satires!* lawless Things indeed!
> But grave *Epistles*, bringing Vice to light,
> Such as a *King* might read, a *Bishop* write,
> Such as Sir *Robert* would approve—
>   *L.* Indeed?
> The Case is alter'd—you may then proceed.
> In such a Cause the Plaintiff will be hiss'd,
> My Lords the Judges laugh, and you're dismiss'd. (150–6)

But in Pope's Latin the equivalent words to those in *L.*'s speech, '*Solventur risu tabulæ; tu missus abibis*', are given to *HOR.* as the conclusion to his speech. Trebatius is not seen to be converted.[28] In separating the texts in two different ways (though he gives the Latin line its separate index number), Pope seems to be giving a victory to *P.* not enjoyed by his predecessor. The joke is probably at the expense of Fortescue's friend and colleague Walpole, who in 1729 had presented the *Dunciad Variorum* to the King. Pope expands the reference to Caesar in Horace into a reference to the King and first minister, Walpole's approval is delayed to the end of *P.*'s argument; but it settles the matter even for the wary lawyer.

Pope's parallel texts balance the general equivalence of Latin and English created by the typography with particular detailed comparisons. I have not been able to settle the question of Pope's Latin text. Lillian D. Bloom's account, based on a study of the editions of Horace known to have been in Pope's library, is still the best we have.[29] For one of the later imitations, *Sober Advice*, the situation is clear: this poem is accompanied by spoof Bentleian notes and uses Bentley's text. Spence says that it was Bentley's edition, specifically the Cambridge edition of 1711, that Pope was reading when Bolingbroke visited him and, by suggesting a reply from Horace, started off the whole series of imitations (*Anecdotes*, No. 321). *Sober Advice* follows this text very carefully in substantives and accidentals. The printer, John Hughs, probably worked from a printed copy rather than a transcription. He seems not to have had the accented characters Bentley used, and Pope created variation by paragraphing and highlighting certain words, but otherwise I count only six changes of punctuation—all could be errors. The first two *Satires of the Second Book* are more problematic. Bloom concluded that Pope's basic text came from Heinsius' Elzevier Horace published in Leiden in 1629, but that he had occasionally emended from Bentley, checking that Alexander Cunningham accepted the favoured reading in his edition of the *Poemata* (London, 1721) and his *Animadversiones* (The Hague, 1721). Only one of these Bentley readings was a conjecture rather than an emendation from a manuscript. The text Pope prints differs widely from other possible sources such as Cunningham's edition or Desprez's. Although R. W. Rogers upset the neatness of

---

[28] Stack notes that Pope follows Bentley's view in the English text but not in the Latin (*Pope and Horace*, 56), but the classical editors, including Bentley, do not allot speech prefixes in the way that Pope does; this is a special issue for him. Bentley may have given him the idea of contrasting endings.

[29] 'Pope as Textual Critic: A Bibliographical Study of his Horatian Text', *Journal of English and Germanic Philology*, 47 (1948), 150–5.

this solution by insisting on the need to consult editions not currently known to have been in Pope's library in case they presented the same combination of readings as Pope, he had not found such an edition and nor has anyone else.[30] I have not found Pope's source in any of these: John Pine's edition (London, 1733); William Baxter's (London, 1725; the Bodleian has the marked up copy of the 1701 edition, responding to Bentley and preparing for 1725); Michael Maittaire's (London, 1715); Cunningham's; Heinsius' emended by D. Emmanueli Caietano de Souza (1722), or Talbot's (Cambridge, 1699). A Heinsius edition was almost certainly the basis of the text used by Pope. To take one significant early feature, Heinsius punctuates *Satire 2. 1*, line 5, 'Quid faciam? Præscribe.' No one else does. But the accidentals of the 1629 edition we know to have been in Pope's library differ from Pope's too often for comfort. In *Satire 2. 1* I note around sixty-eight differences, only four of which are accounted for by the Bentley emendations.[31] In *Satire 2. 2* I note eighty-three, with three accounted for by Bloom's Bentley. Some readings have no apparent source: 'Irriguumve' (*Satire 2. 1*, line 9); 'lanceis' and 'istud' (*Satire 2. 2*, lines 4, 36). As the Pope library copy of Heinsius is unmarked, it is quite possible another copy, perhaps a modernized one, was sent to the printer. The British Library has a copy of the right sort of book: *Q. Horatius Flaccus ex Recensione Heinsii et Fabri, ac cum variis lectionibus Rich: Bentleii* (Amsterdam, 1719). This makes emendation easy by printing the Bentley readings at the foot of the page. It is influenced by Bentley in spelling and punctuation, and it agrees with Pope's text against the 1629 Heinsius on twenty-three occasions in *Satire 2. 1* and thirty-five in *Satire 2. 2*. It seems possible that Pope's reading comes from a text close to the 1719 Heinsius. Sometimes it agrees with the 1719 Heinsius against the others: for example, Pope and 1719 'Quinquennes'; 1629 'Quinquenneis'; Bentley 'Quinquennis' (*Satire 2. 2*, line 39). But I do not think 1719 was sent to the printer. It reads, 'Quid faciam, Præscribe', while the 1629 we know to have been in the library agrees with Pope's text. Nevertheless, I suspect Pope did have access to this or some intermediate text, which would have enabled him to make decisions without the extraordinary labours of collation the other solution seems to require.

The presence of 'Alexander Pope' on the title page of this poem and of *P.* in the dialogue clarifies one of the questions that dogs imitation: who is the speaker of the English? Is it Horace addressing modern conditions? Is it a reinterpreted Horace? Is it an inverted Horace? Pope makes it quite clear that he is speaking himself and taking responsibility for the views expressed. The main point of the parallel is that he speaks with an authority equivalent to Horace's, but the other two ideas, that this is what Horace says to the present or what he would say, exert their power. The general layout insists that the texts correspond, while the apparatus guides the reader and draws attention to local effects. The printer has attempted to equalize the texts

---

[30] R. W. Rogers, review of Lilian D. Bloom's article in *JEGP* 47 (1948), 150–5, in *Philological Quarterly*, 28 (1949), 397–8.
[31] I say 'around' 68 because of the difficulty of deciding which accidental differences should be counted and which ignored because Pope would have altered any text in that way.

as much as possible in order to suggest a balance between them. The English and Latin mirror one another, as Pope mirrors Horace (and Horace is made to mirror Pope). A possible arrangement of the text would be to begin sections at equivalent points on facing pages. This was the practice chosen for *Sober Advice*, printed by John Hughs, not at that point one of Pope's regular printers.[32] But the printings of *Satire 2. 1* and *2. 2* avoid the consequence, which is that the English runs on longer than the Latin in each section. For these first two Horatian imitations, the printer adjusts the spacing between lines and paragraphs to try to balance the texts.[33] Great flexibility was required, but on the whole the English is very lightly spaced at twenty-one to twenty-two lines per page, though the text, in the usual double pica, is leaded and elegant enough in appearance, while the Latin is much more loosely spaced at between eleven and fifteen lines per page. On the opening at pages 6–7, the lines of Latin and English are equally spaced at 7 mm between lines and 17 mm between paragraphs, but on pages 4–5 and 8–17 the Latin is spaced 12 mm to the English's 7 mm. Breaks between paragraphs afford the opportunity to increase diversity further. Some of the English paragraphs are not spaced, whereas the Latin expands the space to 25 mm at some points. On the opening at pages 10–11, the English goes down one line further on the page, while other opportunities for adjustment are afforded by the use and spacing of capitals in the Latin, as opposed to the English's use of caps. and smalls. The result of this, the free placing of rows of quads between the Latin lines, is a general appearance of equivalence that forms the basis of all subsequent manœuvres.

The *First Satire* was probably printed by John Huggonson, who had printed the first epistle of *An Essay on Man*, probably published a few days after *Satire 2. 1*, on 20 February 1733, and took over the printing of the *Grub-street Journal* in September 1733.[34] Any doubts as to whether this was Pope's preferred style of printing the parallel texts are removed by John Wright's printing of *Satires 2. 1* and *2. 2* together, with the aim of using that setting for individual publication and for the *Works*. Wright prints the English with a constant spacing of 6 mm between lines, but his spacing of the Latin varies greatly: 6, 10, 11, 13, 15, or 17 mm. The same pattern

---

[32] I have discussed Hughs's involvement with Pope in 'The First Printing and Publication of Pope's Letters'.

[33] Richard Steiger notes this practice and says that the layout in the *Twickenham* edition has the effect of making the imitation look as though it is using only a selection of the Latin text (205). This is not a mistake an informed reader would make, but Steiger is right in pointing out that the impact of the *Twickenham* is very different from that of the early editions, and the omission of Pope's indices is very difficult to justify.

[34] Once Pope had decided to go ahead after the *To Burlington* furore, there was a rash of publication in early 1733 (*To Bathurst* on 15 Jan.; *Satire. 1* on 15 Feb.; *Essay on Man* 1 on 20 Feb.) and again in the winter of 1733–4 (*The Impertinent* 5 Nov.; *To Cobham* 16 Jan.; *Essay on Man* 4 24 Jan.). The details are in table 12 of Foxon's *Pope and the Book Trade*. John Wright was probably already busy preparing the *Works*. The evidence for Huggonson's involvement is the ornament on p. 18 which is in the provisional collection of ornaments collected by Keith Maslen and J. C. Ross (information from Richard Goulden). Palmer died on 9 May 1732; when the administration of his estate was granted to his widow Elizabeth the securities were John Huggonson and John Pine. The business was to be sold, but it is unlikely that it had been before the printing of *Satire 2. 1*.

is found in the *Second Satire*. However, Pope's expansion of the final section of this second poem (when he himself becomes the substitute for Ofellus) defeats the plan and *Satire 2. 2* finishes with English only for ten lines on the verso of the pamphlet.

The typesetting cannot in itself settle the issue of Pope's intended relation to Horace in these poems, but it clarifies various sub-issues. Pope did wish to be compared with Horace, and the implied reader would be engaged in a double reading, a close reading of one text against the other—paragraph by paragraph, and very often sentence by sentence, phrase by phrase, and word by word. The relationship included differences as well as similarities, but the setting out of the poems suggests that both are working within the same framework, a grid on which both can be plotted. In close comparison Pope can often be seen using both poems to create an area of ambiguity between them in which two or more understandings of a word shimmer uncertainly, so that the reading of either text carries with it the colour it has not taken; understanding is worked out in the interstices of the two texts, where the Latin reawakens to etymologies and alternative readings and the English, while often amplifying one meaning, alludes to others.

Critics have paid some attention to Pope's use of indices and roman type to highlight words in the generally italic Latin, though the omission of the indices in the *Twickenham* text has placed an obstacle in their way. Frank Stack, whose enquiring and detailed analysis of the Horatian imitations is one of the few studies of Pope to draw on his interest in typography, calls the roman words 'key words' (22), which they are, but they are not key words in the sense that they could serve as headings for sections or paragraphs or provide a résumé of the poems.[35] They are connected with vital aspects of the imitations but their impact is local, if cumulative. They draw attention to technical skill and ingenuity, topical allusions and references, and central concepts. They generally, and this indicates the nature of Pope's programme of imitation, show some Horatian idea being attached to the contemporary and socially specific. Pope's aim is to saturate Horatian ideas and critiques in his own social world, with a sense of excitement at both take-up and resistance. The words that are highlighted by being printed in roman are sufficiently representative and important to be treated as a key to the *First Satire*, and I shall discuss them all, however briefly. The *Second Satire* uses them more loosely, but draws on the general pattern to challenge key concepts at the end.

The first highlighted word in *The First Satire of the Second Book of Horace* may fairly be treated as representative.

> [2]*sine nervis altera quicquid*
> *Composui pars esse putat, similesque meorum*
> *Mille die versus* deduci *posse.* (2–4)

---

[35] Aubrey L. Williams says that the key words point to similarities and differences, 'Pope and Horace: The Second Epistle of the Second Book', in Caroll Camden (ed.), *Restoration and Eighteenth Century Literature* (Chicago: Chicago University Press, 1965), 309–21, 311.

[The other half of them hold that all I have composed is 'nerveless', and that verses as good as mine could be turned out a thousand a day.]

Pope's imitation leads straight to the Court and the *To Burlington* scandal:

> ²The Lines are weak, another's pleas'd to say,
> Lord *Fanny* spins a thousand such a Day. (5–6)

This strikes a blow at an enemy, ridicules dilettante writing, and displays learning at the same time. 'Fannius' appears as a self-advertising poet in *Satire 1. 4*, and his counterpart, 'Lord Fanny', has already been encountered as one of the malicious gossips in 'A Master Key to Popery'. Now he is fixed as the type of the eager amateur poet and becomes identified with Lord Hervey. The driving force behind the identification is, as so often, unclear. If 'fanny' already had its modern English (as opposed to American) meaning of 'female private parts', it may be that Hervey's effeminacy combined with his poetry to identify him. The insult is gratuitous, because there is no Horatian figure to be paralleled, but it can be seen to grow out of the etymology of 'deduci'. Dacier serves as commentator:

Il faut bien remarquer icy *deduci* mis en mauvaise part, pour dire des vers foibles & décharnez, des vers filez si menu, qu'ils n'ont point de corps. C'est une metaphore prise du lin & de la laine qu'on file. (7. 21)

In utilizing this etymology Pope associates a charge made against him with a particular aristocratic stance and practice; he turns the attack round on the attackers. This process can be observed being worked out in the manuscript, where Hervey says the lines are weak, before being turned into the spinner of weak lines himself.[36] The development of 'deduci' is highly characteristic of the wordplay Pope creates through the interaction of Latin and English text. The imitation serves to enliven the Latin original and to cause certain words to reverberate with a semantic density that may not be present in Horace's Latin. The effect is not unlike that of Empson's explications in *Seven Types of Ambiguity*.[37] The positive and negative associations suggested by Dacier are picked up and explored by Pope, who nevertheless gives a single direction to his chosen treatment.

The second highlighting in the *First Satire* draws attention to the crucial social relationship that provides the character of the dialogue and small drama that is to be played out:

> Præscribe [give me advice]. (5)

---

[36] *The Last and Greatest Art*, 176. An impressive discussion of the manuscript is Julian Ferraro, 'The Satirist, the Text and "The World Beside": Pope's *First Satire of the Second Book of Horace Imitated*', *Translation and Literature*, 2 (1993), 37–63.

[37] *Seven Types of Ambiguity*, 3rd edn. repr. (Harmondsworth: Penguin, 1965). I have in mind particularly the famous comments on the axe and 'acies' in Marvell's 'Horatian Ode', but the whole of this chapter on 'the fifth type when the author is discovering his idea in the act of writing' is of potential relevance to Pope's imitations (155–75, esp. 166).

'Præscribe' is taken to be a medical term in Dacier's 1709 edition and a means of satirizing Trebatius' authority: he speaks as one to be obeyed, but he is politely disregarded. Pope's emphasis is on the details of contemporary life:

> ³I come to Council learned in the Law.
> You'll give me, like a Friend both sage and free,
> Advice; and (as you use) without a Fee. (8–10)

He sets up a contrast between friendship on the one hand (preparing for the declaration of friendship to Virtue and her friends in line 121) and the world of money and commercial transactions on the other. At the same time the 'as you use' suggests it may not be entirely fair of Pope to exploit friendship in this way. The emphasis is not on the authority or personality of the lawyer, as it is to some extent in Horace, but on social roles and their handling.[38]

The next three highlighted sections turn us to the political dimension of the poem, for they all concern Caesar. Both Horace and Pope needed to show tact here. Horace had enlisted under Brutus in the civil wars, though in Maecenas he had an influential patron. Pope had been a friend of the Jacobite conspirator Atterbury, who had fled to the exiled court; his equivalent of Maecenas was Bolingbroke, who had returned from that court and, professing loyalty to the house of Hanover, had assumed a leading role in the opposition to Walpole.[39] The first reference to Caesar enables Pope to introduce another of those social specificities that informed the beginning of the dialogue:

> *aude*
> CÆSARIS *invicti res dicere,* ⁸*multa laborum*
> Præmia *laturus.* (10–12)

[bravely tell of the feats of Caesar, the unvanquished. Many a reward for your pains will you gain.]

Pope frequently used caps. and smalls for proper names, but they are not used consistently here. Caesar himself has no caps at line 19, and neither do Scipio and Laelius at line 72. The capitals here serve to highlight the topic, even as TREB. and *L.* innocently introduce it. Pope gives *L.* very specific equivalents for 'Praemia':

> write CÆSAR's Praise:
> ⁸You'll gain at least a *Knighthood*, or the *Bays*. (21–2)

The rewards have become institutionalized and cheapened. This devaluation of poetry by contact with the court is elaborated in the following section. Horace's view is not easy to interpret:

---

[38] Trebatius was twenty years older than Horace, and his preoccupations and vocabulary seem to be reflected in this portrait, see Rudd, *The Satires of Horace*, 124–31.

[39] Howard Erskine-Hill's chapter in *Poetry of Opposition and Revolution* is the most influential account of Pope's politics. My uncertainty, anticipated by Erskine-Hill, is over how far political symbolism can be fully integrated into readings of the poems.

The First and Second Horatian Satires   163

> [9]*neque enim quivis* horrentia pilis
> Agmina, *nec* fracta pereuntes cuspide Gallos,
> *Aut* labentis equo *describat vulnera Parthi.* (13–15)

[Not everyone can paint ranks bristling with lances, or Gauls falling with spear-heads shattered, or wounded Parthian slipping from his horse.]

This is a good example of Horatian *recusatio*, refusing a request while accomplishing it. As Stack notes, Pope may have been helped by Dacier here. Dacier argues that although Lucilius had mocked Ennius for using 'horret', Horace had not followed him in doing so. Dacier refers to a fuller discussion of the question in *Satire 1. 10* (6. 644–5). Pope seized upon that precedent set by Lucilius, whether followed by Horace or not, and ridiculed Blackmore and Budgell in this passage.

> P. What? like Sir [9]*Richard*, rumbling, rough and fierce,
> With ARMS, and GEORGE, and BRUNSWICK crowd the Verse?
> Rend with tremendous Sound your ears asunder,
> With Gun, Drum, Trumpet, Blunderbuss & Thunder?
> Or nobly wild, with *Budgell*'s Fire and Force,
> Paint Angels trembling round his *falling Horse*? (23–8)

The middle couplet was added in the second edition to strengthen the element of parody.[40] What in Horace is ambiguous—the language is not manifestly absurd—is here ridiculous, and the typography has a role to play in the mockery. The capitals in the second line suggest obsequiousness; they literally 'crowd the Verse', pushing it to the very end of the measure (with no space after commas) in the first edition and over the line in subsequent ones. The type enacts what it denotes. A slyer effect may be aimed at in the reference to the '*falling horse*' highlighted by italic. The Latin 'labentis equo' is not a falling horse; it is the Parthian slipping, not his horse. Pope's reference is to the shooting of George II's horse from under him at the battle of Oudenarde. Perhaps he is suggesting that in the true translation George fell off. Certainly bathos is the effect of the final item of the group.

> [11]*Nisi* dextro tempore *Flacci*
> Verba per attentam non ibunt Cæsaris aurem . . . (18–19)

[Only at an auspicious moment will the words of a Flaccus find with Caesar entrance to an attentive ear.]

The auspicious moment is no longer a possibility; regulated tedium is the contemporary equivalent:

> P. Alas! few Verses touch their nicer Ear;
> They scarce can bear their *Laureate* twice a Year . . . (33–4)

Pope's attention is again to specific court practices. He was not daring enough to pick up the image in the following lines of Caesar as a sensitive horse ('Stroke the steed

---

[40] I do not know why Maynard Mack (*Life*, 903) thought this couplet (as opposed to lines 23–4) 'dangerous' but record the observation with respect.

clumsily and back he kicks') but these lines stayed in his mind to be used in *To Arbuthnot* in the dialogue that closes the section on King Midas:

> 'I'd never name Queens, Ministers, or Kings;
> 'Keep close to Ears, and those let Asses prick,
> 'Tis nothing'—Nothing? if they bite and kick? (76–8)

The attack was too good to waste.

The next group of highlights engages the poem's second picture of Lucilius, but a curious highlighting intervenes. Pope puts 'ovo prognatus eodem' (26) in roman, and it is not clear from the printed edition why.

> [15]*Castor gaudet equis*; ovo prognatus eodem
> *Pugnis* . . . (26–7)

[Castor finds joy in horses; his brother, born from the same egg, in boxing.]

The imitation,

> [15]*F*— loves the *Senate*, *Hockley-Hole* his Brother
> Like in all else, as one Egg to another . . . (49–50)

is not very powerful or economical, with an awkward sense that Pope has missed an old proverb without making a new one. The reference is to the Fox brothers, and Stephen Fox was Hervey's close friend, probably his lover. The manuscript shows that this connection is what Pope was aiming at. The original line was 'A Boy Lord Fanny loves, a Wench his Brother'. The highlighting is directed at the ghostly presence of the original line.[41]

The appropriation of the portraits of Lucilius to Pope himself is the most surprising and important move in this imitation.[42] Horace's poem is a complex exploration of his own relation to Lucilius, which had seemed much simpler in *Satire 1. 4*. As Rudd suggests, Horace avoids a central position which offers itself, 'I write scathing satires like Lucilius', and investigates instead other aspects of his predecessor: dedication to writing metrically; frankness and self-revelation; fearless exposure of pretension and vice, with the support, and even intimate friendship, of those in power. Pope's projection of himself as Lucilius as well as Horace is aesthetically and morally problematic. It presents another aspect of the colonizing impulse that informs much of his later work: the ordering and colouring of critical voices, especially Cleland's, in the *Dunciad Variorum*; the setting up of the *Grub-street Journal* and the encouragement, and rewriting, of younger authors' work; the publication, possibly, of *Verses Address'd to the Imitator of Horace* and then the redevelopment of that material in *To Arbuthnot*; the indifference to the claims of actual addressees in

---

[41] For Hervey and Fox, see Robert Halsband, *Lord Hervey: Eighteenth-Century Courtier* (Oxford: Clarendon Press, 1973). For the original line see Mack, *The Last and Greatest Art*, 174.

[42] Pope puts lines 35–9 in square brackets, presumably to show that the material, on the Venusian settlers, is not applicable. This is part of his faithfulness to Horace's text in this poem. In treating *Satire 2. 2* he makes various excisions, some of them silent.

publishing his correspondence. Pope's tendency is to cover other identities with his own. *Satire 2. 2* shows precisely the same development as *Satire 2. 1*. Having translated Ofellus into Bethel for the majority of the satire, Pope places himself in Ofellus' position at the close. As these figures, Lucilius and Ofellus, are presented for admiration, a recognition of the egoism, albeit courageous egoism, of the manœuvre is inescapable. Pope's is not the romantic egoism of Wordsworth, where solipsism and the relation of the mind to nature become a philosophical problem; this is social egoism.[43] The self does not diffuse itself over creation, it invades, occupies, polarizes, and simplifies the social world. Paradoxically, it weakens its hold on its own identity as a result. Horace finds himself as one in a complex set of persons, criss-crossed by rank, history, and ability. Pope's world falls rapidly into Pope and not-Pope, with a consequentially reduced set of values, Virtue and not-Virtue. In individual works, the recourse to a moral reductionism is not, of course, as rapid as this summary suggests, and Virtue has its powerful political resonance, but the technique is central to the high-risk strategy Pope was adopting in response to his opponents. It had its success in the myth-making Maynard Mack has detailed in *The Garden and the City*, but it remains a major problem in evaluating Pope's achievement.

The first of the Lucilius passages is without highlighting, but it forms a necessary background to Pope's appropriation of Lucilius. Horace describes Lucilius as 'nostrum melioris utroque' [a better man than either of us] (29), a humility made impossible by Pope's self-identification with the figure. Horace says that the poet's 'life is open to view'; Pope gives the passage a more personal and religious application, 'The Soul stood forth' (54). But Pope then turns the issue of self-revelation (in Horace something admired for its own sake) into the basis for a threat:

> In me what Spots (for Spots I have) appear,
> Will prove at least the Medium must be clear.
> In this impartial Glass, my Muse intends
> Fair to expose myself, my Foes, my Friends;
> Publish the present Age, but where my Text
> Is Vice too high, reserve it for the next . . . (55–60)

What in Horace is the discovery of a new conception of Lucilius's writing is now reinterpreted as a strategy of making self-exposure the ground for satire. The response to the criticism of *To Burlington* is to accept the personal nature of the attacks and to launch a personalized poetry that while setting Pope's own reputation at stake frees him to attack those of others. 'Reserve' enters the passage only with the concept 'Publish'. The glass is 'impartial'; it shows all. But not all will be shown. The artist intends disclosure; the publisher-artist manœuvres in response to the times.

The second passage on Lucilius is used to justify the attack on the 'Quality' that seemed necessary after *To Burlington*.

---

[43] For a pioneering discussion of the topic, see G. K. Hunter, 'The "Romanticism" of Pope's Horace', *Essays in Criticism* (1960), 390–4.

> [29]*Quid? cum est Lucilius ausus*
> *Primus in hunc operis componere carmina morem,*
> [30]Detrahere & pellem, nitidus *qua quisque per ora*
> *Cederet*, introrsum turpis . . . (62–5)

[What! when Lucilius first dared to compose poems after this kind, and to strip off the skin with which each strutted all bedecked before the eyes of men, though foul within]

Jacob Fuchs argues provocatively here that Pope weakens the force and violence of the Horatian metaphor, elevating his practice above the apparent cruelty of the original (70). This is one of Pope's more elaborate developments of a highlighting. The index 30 points to a simple equivalent, 'And I not [30]strip the Gilding off a Knave' (115), picking up and extending 'nitidus', but the preceding lines have seen a typically social development of the same idea:

> Brand the bold Front of shameless, guilty Men,
> Dash the proud Gamester in his gilded Car,
> Bare the mean Heart that lurks beneath a Star;
> Can there be wanting to defend Her Cause,
> Lights of the Church, or Guardians of the Laws? (106–10)

The metaphors of branding and of baring the heart replicate the cruelty of the original, but, unlike Horace, Pope specifies the tenors of his metaphors and engages directly with social signs—the garter, the carriage, gold braid or lace—forcing social specification on the reader. The individuals Pope will expose represent institutional power—aristocracy, the court, the Church, the law.

The next section continues the concern with the powerful:

> *Atqui*
> Primores *populi arripuit*, populumque *tributim*;
> *Scilicet* [31]UNI ÆQUUS VIRTUTI ATQUÆ EJUS AMICIS. (68–70)

[Yet he laid hold upon the leaders of the people, and upon the people in their tribes, kindly in fact only to Virtue and her friends.]

Although they are highlighted, Pope has nothing to say about the 'populus', though he does pick the idea up at line 140: 'Scriblers or Peers, alike are *Mob* to me.' His concern is with the 'primores', and his defiance is explicit:

> Yes, while I live, no rich or noble knave
> Shall walk the World, in credit, to his grave.
> TO VIRTUE ONLY and HER FRIENDS, A FRIEND . . . (119–21)

The effect in the original folio is very striking, with large caps. for the Latin, making it exceptionally emphatic, and small caps. for the English, which suggests a relative modesty. Nevertheless, this typographical shouting risks absurdity and forms a significant departure in tone from the Horatian original. Dacier, himself a moralizing commentator, finds nothing remarkable in these words; they form a generous tribute in Horace, but they have nothing of a manifesto about them. The sentiment that interests Pope finds stronger expression in Juvenal's eighth satire and Boileau's fifth.

As Claude Brossette, Boileau's authorized commentator, says, 'L'Auteur y fait voir que la veritable Noblesse consiste dans la Vertue, indépendamment de la Naissance.'[14] From this viewpoint, Pope finds in Horace's words an admirable monolithic quality, and in his translation they acquire an additional significance from their ordering and the consequent wavering between the abstract and the personal. Pope is not in this passage criticizing the aristocracy from a populist, bourgeois, or gentry viewpoint; the capitalized line is a bid for transcendence. By reserving self-reference to the end of the line he defines the group abstractly: Virtue's friends might be moral qualities—patience, humility, integrity—but the initial metaphorical status of the friends is transformed by Pope's appearance as friend at the end of the line, which simultaneously places him as virtuous and transforms the group into one with a personal existence. 'Æquus' (fair, friendly, kind), a term that might be attached to justice, is domesticated. That this effect is intended is clear from a letter to Jonathan Richardson quoted in the *Twickenham* edition, which says, 'I think I have made a panegyric of you all in one line.' A certain moral vacuity in the claim is masked by the rhetorical skill.

The final element in the picture of Lucilius continues the theme of friendship by complimenting St John and Peterborough as virtuous Scipio and wise Laelius.

> [32]*Quin ubi se a* Vulgo & Scena, *in* Secreta *remorant*
> Virtus Scipiadæ, & *mitis* Sapientia Læli ... (71–2)

[Nay, when virtuous Scipio and the wise and gentle Laelius withdrew into privacy from the throng and theatre of life]

Here there is more territory for the ego to occupy, and the retirement of Scipio and Laelius becomes 'my *Grotto*' and 'my *Retreat*'. The highlights that follow pick up the idea of friendship with the great.

>   *tamen me*
> [34]*Cum* magnis vixisse *invita fatebitur usque*
> Invidia, & *fragili quærens illidere dentem,*
> Offendet solido;—(75–8)

[yet Envy, in spite of herself, will ever admit that I have lived with the great, and, while trying to strike her tooth on something soft, will dash upon what is solid.]

The first element is translated, '[34]*Envy* must own, I live among the Great' (133), but there is no specific equivalent for the solid against which she strikes her tooth. The victory over Envy must come from friends who tell the truth: 'This, all who know me, know; who love me, tell' (138). The repetitions enact an enclosure similar to that of the earlier line on Virtue and her friends.

The remaining highlights pertain to the friendly lawyer and the law. The address, 'docte Trebati', becomes 'Council learned in the Laws', and Trebatius' citing of the law is caught in a burst of jargon from *L*.

---

[44] For the influence on Pope of the Geneva *Œuvres de Mr. Boileau Despréaux. Avec des éclaircissemens historiques, donnez par lui-même* (1716), see Chapter 4 on the *Dunciad Variorum*.

> ³⁷Consult the Statute: *quart.* I think it is,
> *Edwardi Sext.* or *prim. & quint Eliz* . . . (147–8)

Characteristically Pope uses language culled from abbreviations in books. He cannot exactly parallel the play on 'mala' and 'bene' in the lines that follow.

> ³⁷ 'Si mala condiderit in quem quis carmina jus est Judiciumque.'
> HOR. *Esto, siquis* ³⁸*mala; sed* bona *siquis Judice condiderit laudatur* CÆSARE . . . (82–4)

[If a man write ill verses against another, there is a right of action and redress by law. HOR. To be sure, in the case of ill verses. But what if a man compose good verses, and Caesar's judgement approve?]

There is redress against wicked/bad verses, but Horace says he writes good ones. Pope specifies a contrast between '*Libels* and *Satires!*' on the one hand and 'grave *Epistles*' on the other. He seems to have his publishing plans in mind again. The reference to Caesar is subtly expanded:

> Such as a *King* might read, a *Bishop* write,
> Such as Sir *Robert* would approve—
>   *L.* Indeed?
> The Case is alter'd . . . (152–4)

The chief minister rather than the King settles the matter. Pope not only adopts Bentley's reading by giving the final lines to *L.*, he uses the speech prefixes layout he has adopted from the drama to interrupt *P.* at the mention of Walpole's name. There are private jokes here—Walpole was Fortescue's friend; he had presented the *Dunciad Variorum* to the King—but the public impact is clear. The last word is neither *P.*'s, nor *L.*'s, nor the King's, but Walpole's.[45]

THE SECOND SATIRE OF THE SECOND BOOK OF HORACE

Pope's imitation of *Satire 2. 1* deals with the question of the satirist's relation to the powerful. The imitation of *Satire 2. 2* moves through the topic of plain living to question one of the bases of that power, the notion of property. The publication of the *Second Satire of the Second Book* along with the *First Satire* in folio and quarto on 4 July 1734 as part of the lead up to *Works* II seemed confirmation that Pope intended a series of imitations of the *Second Book*, though in fact this was the last, except for the completion of Swift's imitation of the *Sixth Satire*. This time the printer was Pope's regular printer since the *Dunciad Variorum*, John Wright. The printing generally conforms to the pattern set by the first edition of the *First Book*, with the Latin spaced out to suggest an equivalence. This time, however, the use of

---

[45] *Twickenham*, 5, pp. xxvii and 60. Pope seems to have first mentioned this presentation, which took place on 12 Mar. 1729, in the *Works* of 1735. The anti-Hanoverian implications identified by Erskine-Hill are present in the poem, but they cannot have been as unmistakable as he maintains in *Poetry of Opposition and Revolution*, 102.

roman type to vary the pattern in the Latin italic is much less interesting. Many of the words highlighted are the names of foods, and in these cases the reader's attention is being drawn to what is really a form of translation. The effect is sometimes even simpler, as with '*you*' for 'teipso' (line 16/14). The most interesting of the highlights are used to draw attention to specific social practices and personalities. This satire engages very closely with the contemporary sophistications and indulgences it rejects.

Pope's imitation of the *First Satire* began in a scholarly fashion by exploiting Dacier's note on 'deduci'. The *Second Satire* has similarly learned treatment of the first Latin words highlighted by roman type. Here Ofellus is described as '*Rusticus*, [4]abnormis *sapiens*, crassaque Minerva' [a peasant, a philosopher unschooled and of rough mother-wit] (3). The description interests Pope but he delays his imitation of these lines in order to clarify the speakers in his poem. Horace presents Ofellus' views through indirect speech, but Pope imposes a clear demarcation: the speech is (from line 11) or is not (at the beginning) Bethel's; it never hovers in between poet and character. The index numbers show that Pope has arranged the text so that the description of Bethel immediately preceded his speech:

> Hear Bethel's Sermon, one not vers'd in schools,
> [4]But strong in sense, and wise without the rules. (9–10)

The description again coincides with Dacier's conception:

Abnormis sapiens] Mot à mot: Philosophe sans regle, c'est-à-dire, Philosophe qui ne suit point de Maître, & qui n'a été ni dans les Ecoles des Stoïciens, ni dans celles des Epicuriens . . . (7. 85)

Pope's imitation draws on Dacier's gloss, demonstrating a refined understanding of the original. The case parallels that of the first satire, except that this time a person is required by Horace's text and a friend is engaged rather than an enemy. 'Rusticus [a peasant]' is left out by the placing of the index, and Pope is able to compliment Bethel by praising a wisdom without academic training.

Dacier's influence continues in the remainder of the first paragraph and generates an example of the exotic vocabulary that Pope uses to pin the satire to his own social milieu.

> *Discite non inter lances*, mensasque nitentis,
> *Cum stupet* insanis acies fulgoribus, & *cum*
> *Acclinis falsis animus meliora recusat*;
> *Verum hic* impransi *mecum disquirite*. (4–7)

[learn, I say, not amid the tables' shining dishes, when the eye is dazed by senseless splendour and the mind, turning to vanities, rejects the better part; but here, before we dine let us discuss the point together.]

Dacier explains with the usual elements of paraphrase: 'Il appelle *insanos fulgores*, le trop grand éclat qui vient de la folle magnificence de la table, & de la trop grande somptuosité du buffet' (7. 86). Pope seizes on 'buffet', adopting the French word, but characteristically tying it directly to the values discussed.

> Lets talk, my friends, but talk before we dine:
> Not when a gilt Buffet's reflected pride
> Turns you from sound Philosophy aside . . . (4–6)

The word 'buffet' is itself exotic and fashionable (there is only one previous instance in *OED*, from 1718), and the vocabulary has an up-to-date air, while expressing the deluded sophistication of those satirized, quite simply, as proud. A further example of Pope's knowledge of contemporary taste soon follows. Horace is concerned with gluttony:

> [12]Porrectum magno magnum spectare catino
> Vellem (*ait* Harpyiis *gula digna rapacibus*) (23–4)

['But a big fish on a big dish outstretched! That's what I'd like to see!' cries a gullet worthy of the greedy Harpies.]

For the dish appropriate to gluttony Pope supplies something wonderfully specific:

> [12]*Oldfield*, with more than Harpy throat endu'd,
> Cries, 'Send me, Gods! a whole Hog *barbecu'd!*'

This time Pope recognized that the word might be too obscure for his readers and glosses '*barbecu'd*' for the *Works* and subsequently: 'A *West-Indian* Term of Gluttony, a Hog roasted whole, stuff'd with Spice, and basted with *Madera* Wine.' *OED* confirms the definition and gives only one earlier example, from 1690. The *Twickenham* edition tentatively identifies *Oldfield* as Richard Oldfield, but 'Harpy throat' points to a woman, perhaps, in a characteristic blurring of reference, Anne Oldfield, shortly to figure in the first paragraph of *Sober Advice*. Certainly the harpy image lingered to be reflected in the portrait of Constantia Phillips in the third paragraph of that poem, 'And Lands and Tenements go down her Throat' (14).

The miserliness which contrasts with the gluttony of the opening is similarly specific, with Pope developing the portrait of the miser Avidienus into a caricature of the Wortley Montagus.

> [19]*Avidienus*
> [20](*Cui* Canis *ex vero ductum cognomen adhæret*)
> *Quinquennes oleas est, & sylvestria corna.*
> [21]*Ac nisi* mutatum *parcit defundere* vinum, *&*
> *Cujus oderem olei nequeas perferre* (*licebit*
> *Ille* repotia, natales, *aliosque dierum*
> [22]Festus *albatus celebret*) *cornu ipse bilibri*
> Caulibus *instillat*;[23]*veteris non parcus aceti.* (37–44)

[Avidienus, to whom the nickname 'Dog' quite rightly clings, eats his olives five years old with cornels from the wood, and is chary of drawing his wine till it has soured; as to his oil, you couldn't bear its smell, yet even if in his whitened garb he keeps a wedding or a birthday feast or some other holiday, he drops it on the salad from a two-pound horn with his own hands, though his old vinegar he does not stint.]

The purpose of the highlighting is partly to show the integrity of the application of this portrait to the Wortley Montagus and partly to advertise its ingenuity. Lady Mary featured in the *First Satire* and she appears again in *Sober Advice*. One of the pleasures of the imitations is the reappearance of familiar characters in the series, but, as Horace has no such continuity, regular readers were made aware of the calculation and skill that went into maintaining it. As the indices suggest, Pope follows his original very closely:

> [19]*Avidien* or his Wife (no matter which,
> For him you'll call a [20]dog, and her a bitch)
> Sell their presented Partridges, and Fruits,
> And humbly live on rabbits and on roots:
> [21]One half-pint bottle serves them both to dine,
> And is at once their vinegar and wine.
> But on some [22]lucky day (as when they found
> A lost Bank-bill, or heard their Son was drown'd)
> At such a feast [23]old vinegar to spare,
> Is what two souls so gen'rous cannot bear;
> Oyl, tho' it stink, they drop by drop impart,
> But sowse the Cabbidge with a bounteous heart. (49–60)

In spite of its cruelty, this is a passage which gives its victims a rude comic energy, partly through its close attention to Horace's details of diet. Pope uses his indices differently here, as a supplementary highlighting. The section begins with Avidienus; there is no need for index 20 except as a means of further highlighting 'Canis', which enables Pope to abuse Lady Mary by calling her 'a bitch'. This is an example of the conscious misogyny which here and in *Sober Advice* leads Pope to substitute or supplement male victims with female ones. The stale wine which follows is translated straightforwardly to the English, with the half-bottle adding a telling domestic detail, but it is the feast that affords Pope his major opportunity. Horace's text is highlighted to emphasize values of religion, custom, and family. The 'festum'—holiday, festival, festal banquet, feast (the word has some religious associations)—becomes merely a 'lucky day', unconnected with tradition or custom. The family celebration, the wedding reception (*repotia*) or birthday (*natalis*), is inverted into rejoicing at the death of a son. The Wortley Montagus' son had tried to run away to sea, and Pope's imaginings probably come close to the facts of the case. The general tendency of this passage, which is to identify meanness at table with meanness of heart, is finely realized in its close. The translation of 'caulis' as 'Cabbidge', rather than the Loeb's refined 'salad' (Pope is the great poet of vegetables in these imitations) provides social realism, while 'with a bounteous heart' clinches the irony of the portrait.

A different form of social specificity is found in the handling of religion. There are two important references to the Church. The first comes in the famous passage which associates feasting with the clogging of the spirit. In Horace the satire is general:

> [30]*Vides, ut pallidus omnis*
> *Cena desurgat dubia? quin corpus onustum*
> *Hesternis vitiis*, animum *quoque prægravat una,*
> *Atque affigit humo* divinæ particulam auræ. (58–61)

[Do you see how pale rises each guest from his 'puzzle feast'? Nay more, clogged with yesterday's excess, the body drags down with itself the mind as well, and fastens to earth a fragment of the divine spirit.]

Pope is successful in rendering the philosophical aspects of the passage in a language acceptable to Christian theology, but he narrows the whole attack to focus on clerical worldliness.

> [30]How pale, each Worshipful and rev'rend Guest
> Rise from a Clergy, or a City, feast!
> What life in all that ample Body, say,
> What heav'nly Particle inspires the clay?
> The Soul subsides; and wickedly inclines
> To seem but mortal, ev'n in sound Divines. (75–80)

The clergy are a social group like the merchants. 'How unfair it is of the soul', Pope seems to say, 'to disguise its nature, when it might show itself plainly through the clergy.' The passage satirizes the Anglican clergy's complacency; the soundness of the Divines is made a reflection of their failure of spirit. The second reference to religion picks up the attack on inadequate public works at the close of *To Burlington*.

> *quare*
> [42]Templa *ruunt* antiqua *Deum? cur* improbe! *caræ*
> *Non aliquid* patriæ *tanto emetiris acervo?* (85–7)

[Why are the ancient temples of the gods in ruin? Why, shameless man, do you not measure out something from that great heap for your dear country?]

This gives Pope the ideal opportunity to substitute new for old:

> Shall half the [42]new-built Churches round thee fall?
> Make Keys, build Bridges, or repair White-hall:
> Or to thy Country let that heap be lent,
> As M**o's was, but not at five *per Cent*. (119–22)

The reference is to the churches built under Acts passed in the reigns of Anne and George I. Pope's view was that the foundations were unsound, a case of natural symbolism. As in *To Burlington*, he connects the failure of public works to the abuses of private wealth. The case of the Duchess of Marlborough with which this passage closes focuses a more general exploitation of the nation by its ruling class.

The Second Satire of the Second Book ends with Ofellus' speech on his current circumstances: his land has been assigned to others; he continues his life of modest fare and self-sufficiency; he is hospitable and pious; he has not suffered materially from the arrival of a landlord; nature gives no one complete dominion over the earth; villainy, the law, death take property away; the use of the land will pass from one to another. Pope himself takes the place of Ofellus in his imitation, introducing a short dialogue with Swift to emphasize the point about property. The highlighted vocabulary shows up the aspects of Ofellus' position that most closely relate to Pope's. Ofellus' wealth, his land, has been cut down ('accisus'); Pope is now '*Excis'd*', a reference to Walpole's bill then being considered by Parliament, with strong opposition from, among others, John Wright's former master, John Barber. The index 46 is placed next to '*Excis'd*', which is italicized to mark the wordplay. Ofellus' land is 'metatus', measured off for confiscation, which Pope picks up later, 'My lands are sold, my Father's house is gone' (155). The topic of dispossession is sustained by the continued highlighting of words for food, now seen as the produce of the land: one's relation to food expresses one's relation to the land. Pope piddles on broccoli ('holus'); his guests do not eat turbots ('pisces') but chicks ('pullus'), mutton ('haedus' [kid]), walnuts ('nux'), dried grapes ('pensilis uva'), and figs ('ficus'). Pope cannot claim to produce all these foods himself. Some—the chicks, the fruits—are his own; over the others he has a claim as a native of the place. He belongs there as does the animal life:

> But gudgeons, flounders, what my Thames affords.
> To Hounslow-heath I point, and Bansted-down,
> Thence comes your mutton, and these chicks my own . . . (142–4)

This is not a contrived sense of connection, as anyone who speaks of Stoke FC as 'my team', or recommends Staffordshire oatcakes to strangers, will agree. Pope expresses a sense of belonging, a local loyalty, a knowledge of place and product, that is independent of proprietorship. The reference in Horace to 'venerata Ceres' gives this relation to the land a religious turn, which Pope is able to echo with an additional force because his religion has lost him ownership of land: 'And, what's more rare, a Poet shall say *Grace*' (150).

Pope's dispossession is fully politicized in lines 154 to 174. The '[52]novus Incola' [new landlord] is represented by '[52]Standing Armies'; '[54]Nequities' [villainy] is '[54]Peter Walter'; and '[55]vafri inscitia juris' [ignorance of the quirks of law] is '[55]Equity'. The key challenge in this conclusion, however, is to the concept of property. Ofellus makes a bold claim:

> *Nam* propriæ telluris *herum natura neque illum*
> *Nec me, aut quemquam statuit* (109–10)

[Nature, in truth, makes neither him nor me nor anyone else lord of the soil as his own]

Pope dramatizes his own position in this respect by creating a formal dialogue with Swift, who says, 'I wish to God this house had been your own':

> Well, if the Use be mine, can it concern one
> Whether the Name belong to Pope or Vernon?
> What's [53]*Property* dear Swift! you see it alter
> From you to me, from me to [54]Peter Walter . . . (165–8)

Pope is following Horace closely here. Fairclough points out that the distinction is between *ususfructus*, the right of using and enjoying property, and *dominium*, owning it. I have suggested earlier that such a distinction might recently have shaped Pope's sense of literary property. In questioning the concept of property, Pope is challenging the system that might deny his right to style himself 'Esq;' on the title page of the *First Satire*. The same system had been criticized in *To Burlington*, where a historical perspective had denied the importance of ownership by any single individual:

> Another age shall see the golden Ear
> Imbrown the Slope, and nod on the Parterre,
> Deep Harvests bury all his pride has plann'd . . . (173–5)

Questioning the very notion of property further subjects great landlords to the possibility of critique. If they are merely users, they may be judged by the criteria of effective use; if Bridgman has the use of an estate, he is effectively master of it. In this imitation Pope has expanded and developed the implications of the rule he advances in *To Burlington*: ''Tis Use alone that sanctifies Expence' (179). This is the radical position with which the poem closes:

> Let Lands and Houses have what Lords they will,
> Let Us be fix'd, and our own Masters still.

It is probably his most radical couplet, but its challenge to a society in which property in land was of crucial importance went unanswered. Interest and controversy focused rather on the personal attacks of *Satire 2. 1*.

These early imitations of Horace exemplify a characteristically Popeian defiance, one that mingles boldness with a catch-me-if-you-can avoidance of being caught in the expected place. Pope dramatized the reception of *To Burlington* until it left him stark choices: silence, acquiescence, defiance. The defiance chosen involved both guile and an unusual measure of self-risk. Pope stayed in the game of public satire by staking his own reputation and personality on the result. The first two imitations involve a double operation: the identification with Horace himself, which makes a claim for authority and freedom, and the identifications with Lucilius and Ofellus, which set up the poet's own life as a guarantee of his integrity. The typography of the parallel texts shows Pope's overwhelming concern to be one of fit, with a determination to draw attention to the ingenuity and subtlety as well as the neatness of the parallels. The imitations are rather like a jigsaw and the highlighted words draw attention to the pieces that almost fail to fit. Over and again, institutions and personalities are slotted into surprising places and made to lie there. A condition of their fitting is that Pope fits too—which is the central risk of the parallel on which he embarks. The same tactic of engaging with opponents through self-exposure will be seen in his handling of the response to *Satire 2. 1*, *Verses to the Imitator of Horace*, in the *Epistle to Dr Arbuthnot* and *Sober Advice from Horace*.

7. To Arbuthnot *and* Sober Advice: *Textual Variation, Sexuality, and the Public Sphere*

An Epistle to Dr Arbuthnot and *Sober Advice from Horace* were designed as twin publications, contrasting and complementary, responding in different ways to that most vital of all attacks on Pope, *Verses to the Imitator of Horace*.[1] Pope hoped to publish them both in late December 1734. As it turned out, *Sober Advice*, which had passed out of his own control, appeared successfully on 21 December 1734, but *To Arbuthnot*, whose disturbed lineation suggests Pope's own late interference, was published late, on 2 January 1735.[2] Both poems were published in the knowledge that the second volume of Pope's works was soon to appear (23 April 1735). But whereas *Sober Advice* was scurrilous and anonymous and did not appear in a volume of works until 1738, the first edition of *To Arbuthnot* provided a censored text, acknowledged its authorship in its full title, *An Epistle from Mr. Pope to Dr. Arbuthnot*, and presented itself as a 'House of Pope' publication, 'Printed by J. Wright for Lawton Gilliver'. Pope's plan was to reserve a fuller text of *To Arbuthnot*, containing an attack on Lord Hervey's bisexuality, for the *Works*, when his enemies would have had the chance to respond to *Sober Advice*. The first edition of *To Arbuthnot* attempts to fuse two elements: a critique of the institutions of contemporary literary culture (court and aristocratic patronage, the journal and the coffee house, the exploitation of major writers) and an account of Pope's private life and values. The fusion presents the poet and his poetry as part of the intimate sphere of domestic life and filial duty; Pope is both the poet-victim of the public sphere and the virtuous embodiment of private life. *Sober Advice*, on the other hand, is, strictly, anonymous, with a sly wink to the alert reader, 'Imitated in the Manner of Mr. Pope', and it was published by booksellers remote from Pope's circle. It satirizes the corruption of private life by castigating the sexual mores of contemporary society, largely through an attack on women, but it differs from *To Arbuthnot* in presenting no binary opposition between satirist and satirized. Its own stance is prurient, and its parodic notes are licentious and sometimes pederastic in their focus. Pope is absent from this poem, except as a possible victim of his own satire. The two poems together contrast virtuous and vicious domestic life in the persons of Pope and his critics, Lord Hervey and Lady Mary Wortley Montagu. *To Arbuthnot* and *Sober Advice* are separate poems, but they

---

[1] The two editions of the poem have slightly different titles. When I am speaking indifferently of the versions, I call the poem *Verses to the Imitator of Horace*, which is a title used by Pope.

[2] For Pope's plans for *To Arbuthnot*, see the letters to Swift and Oxford, *Correspondence*, 3. 444–6. Sherburn suggests both poems were completed at Bevis Mount that August. Quotations from the first edition and from the *Works* version are taken from David Foxon's facsimile edition (Menston: Scolar Press, 1970).

are necessary to one another. They intersect in the portrait of Hervey in *To Arbuthnot*, where Pope attempts to unify his critique into one figure, the anti-Pope, Sporus. But *Sober Advice* shows that the contrast between Sporus and Pope is more complex than criticism usually suggests. Pope uses *Sober Advice* to suggest what *To Arbuthnot* cannot openly declare, that Pope's own sexuality, private and possibly pederastic in orientation, presents a purer form of masculinity than Hervey's. The textual instability in the early editions of *To Arbuthnot* comes not only from the fear of prosecution for libel (real enough) and from the strain of uniting variant strands into one poem, but also from the difficulty of providing a satisfactorily complex and truthful account of the self.

POPE AND *VERSES TO THE IMITATOR OF HORACE*

The two different forms of *To Dr Arbuthnot*, in a separately printed folio of 2 January and in *Works* II of 23 April 1735, mark a further stage in Pope's desire for self-representation. The advertisement, or introductory note, to the *Works* of 1735 presents the issue with clarity. 'ALL I had to say of my *Writings* is contained in my Preface to the first of these Volumes . . . And all I have to say of *Myself* will be found in my last Epistle [To Dr Arbuthnot].' As I argued in Chapter 2, the *Works* of 1717 were concerned with the projection of Pope in his writings. The creation was a public Pope, which in some way embodied Pope the man. But it was understood there was a remainder: the *Works* gave no access to Pope's domestic and family life, nor was the poetry to be read as confessional. The Pope we encountered was the equivalent of Pope the society figure of the frontispiece. Although the preface showed some niggling anxiety about the relation between *Pope* (the writings) and Pope (the person)—was Pope entirely happy to leave *Pope* to his fate at the hands of the reading public?—it was generally relaxed about the separateness of the public and private spheres. The *Works* was a monument in the public sphere; it no more involved the private man than a statue in Westminster Abbey would. But in 1735 Pope used his preliminary advertisement to announce not only the discussion of his writings (which seems entirely appropriate) but also the discussion of himself. *To Arbuthnot* is both that discussion of '*Myself*' and the account of the breakdown in the integrity of the public sphere that makes the—on the face of it, rather shocking—disclosures necessary.

The first paragraph of the advertisement to the first independent edition of the poem, which preceded the *Works*, accounts for the genesis of the poem and is worth quoting in full.

This Paper is a Sort of Bill of Complaint, begun many years since, and drawn up by snatches, as the several Occasions offer'd. I had no thoughts of publishing it, till it please'd some Persons of Rank and Fortune [the Authors of *Verses to the Imitator of Horace*, and of an *Epistle to a Doctor of Divinity from a Nobleman at Hampton Court*,] to attack in a very extraordinary manner, not only my Writings (of which being publick the Publick judge) but my *Person*, *Morals*, and *Family*, whereof to those who know me not, a truer Information may be requisite. Being divided between the Necessity to say something of *Myself*, and my own Laziness to undertake so awkward a Task, I thought it the shortest way to put the last hand to this Epistle. If it have any thing pleasing, it will be That by which I am most desirous to please, the *Truth* and the *Sentiment*; and if any thing offensive, it will be only to those I am least sorry to offend, the *Vicious* or *the Ungenerous*. (Italics reversed; square brackets Pope's.)

The bill of complaint, which amounts to nothing less than Pope's rejection of the public sphere of letters, was precipitated into the present poem, Pope claims, by the attack by persons of rank and fortune in *Verses to the Imitator of Horace*, that is from the sphere of the court and aristocratic regulation or patronage.[3] The two spheres, at least in theory, combine together awkwardly in opposition to Pope. The institutions of the literary public sphere—the coffee house, the club, the journal, the newspaper—should constitute an alternative to the court, one of independent persons judging and communicating freely. Yet Pope suggests that both spheres merit the same rejection, which depends on self-investigation and revelation. I shall consider the impetus given to Pope's disclosure by the aristocratic attack before turning to the general critique of the public sphere.

Pope's engagement with the *Verses to the Imitator of Horace*, a cruel, vigorous, and entertaining attack on Pope, probably written by Lord Hervey in association with Lady Mary Wortley Montagu, represents a final stage in his march towards total control of the context of print and publication. Publishing explanations, justifications, and defences of your work is one stage; publishing pseudo-attacks, as in the *Key to the Lock*, pushes a front further; but publishing the work of your enemies when they have decided to keep it private, or within the world of the court, which is what I suspect Pope did in this case, is a bid to take over the whole territory. In the British Library is a four-volume collection of pamphlet attacks on Pope, consisting of twenty-four items in all, that Pope had owned and had bound together. On the flyleaf is a curious inscription from the Book of Job: 'Behold it is my desire, that mine Adversary had written a Book. Surely I would take it on my Shoulder, and bind it as a crown unto me' (Job 31: 35–6).[4] The verses come from Job's response to his comforters, developing the plea, 'Let me bee weighed in an even ballance, that God may know mine integritie' (31: 6), a response that concludes, 'So these three men ceased to answere Job, because he was righteous in his owne eyes' (32: 1). The pamphlet attacks range in date from 1711 to 1733. They do not include *Verses to the Imitator of Horace*, the first of the attacks by 'Persons of Rank and Fortune' to which we owe *To Arbuthnot*; the collection was closed just after its publication. Pope stopped collecting the attacks in 1733, I would suggest, because in *Verses to the Imitator of Horace* he had found the book that advanced the charges he wanted to answer. In *To Arbuthnot* he bound it to him as a crown by asserting his integrity and righteousness. The publication of the *Verses* was odd. Two different versions of the work appeared

---

[3] I shall have nothing to say about *An Epistle to a Dr of Divinity*, which is a feeble piece in comparison with the *Verses*. I discuss its publication in '"Of Which Being Publick the Publick Judge": Pope and the Publication of *Verses Address'd to the Imitator of Horace*', *Studies in Bibliography*, 51 (1998), 183–204.

[4] The volumes, which once belonged to John Wilson Croker, are BL C.116.b.1–4. They are noted by Mack, *Collected in Himself*, Library, 3 etc. and Guerinot, *Pamphlet Attacks on Alexander Pope*, p. li. The quotation is discussed by Helen Deutsch in *Resemblance & Disgrace*, p. viii.

on the same day, 8 March 1733, with slightly different titles: *Verses Address'd to the Imitator of the First Satire of the Second Book of Horace* (published by Anne Dodd) and *To the Imitator of the Satire of the Second Book of Horace* (published by James Roberts). As the poem is now generally thought to have been the work of Lady Mary Wortley Montagu in collaboration with Lord Hervey, the double publication is sometimes explained in relation to the joint authorship, but in a recent article I have suggested that the most likely explanation is that one version (*Verses*) was published by Pope himself, forcing calumny shrouded by the court out into the public, while the other (*To the Imitator*) was published by Lord Hervey, acting for both authors, in response.[5]

The suspicion that Pope was involved in publication of the *Verses* arises for three reasons: he had a powerful motive for drawing his quarrel with his aristocratic critics into the public arena; he had a record of clandestine publication in this period, culminating in the trick he played on Curll to get him to publish his letters in May 1735: and he had the opportunity to publish secretly through his contacts with the book trade, in this case Anne Dodd, the publisher of the *Dunciad*, and Henry Woodfall, the printer of much of Pope's poetry in the middle of his career.[6] Pope's motivation sprang from his relations with Lady Mary. In the 1730s his attacks on her seem to have provided a focus for uneasiness at court about his dutifulness as a subject. In his defence of his satiric attacks on persons of position and reputation, the *First Satire of the Second Book of Horace*, published on 15 February 1733, Pope had attacked both Lady Mary and Lord Hervey. He had sneered in passing at the effeminacy and pretensions to be a poet of Hervey, the court Vice-Chamberlain and ally of Walpole:

> The Lines are weak, another's pleas'd to say,
> Lord *Fanny* spins a thousand such a Day. (5–6)

And he had grossly insulted Lady Mary, another Walpole ally and former court favourite, causing serious offence:

> Slander or Poyson, dread from *Delia*'s Rage,
> Hard Words or Hanging, if your Judge be *Page*.
> From furious *Sappho* scarce a milder Fate,
> P—x'd by her Love, or libell'd by her Hate (81–4)[7]

Pope always denied the identification of Sappho as Lady Mary; because denial, however preposterous, was necessary in order to avoid a charge of libel himself. He could

---

[5] 'Of Which Being Publick the Publick Judge', cited in note 3. There is an excellent account of the episode by Isobel Grundy, '*Verses Address'd to the Imitator of Horace*: A Skirmish between Pope and Some Persons of Rank and Fortune', *Studies in Bibliography*, 30 (1977), 96–119. Grundy notes that Lady Mary and Lord Hervey both had dealings with Roberts, and that the superiority of the Dodd version, at least in marketing, is something of a mystery (103).

[6] I am a touch uneasy about relying on Pope's contacts with Woodfall. He may have been Lintot's choice for printer of the octavo *Works*, rather than Pope's. According to John Nichols, Pope was responsible for giving Woodfall his start in business (*Anecdotes*, 1. 300 n.), but I am not sure I believe the story.

[7] For an account of Lady Mary's reaction, see Isobel Grundy, *Lady Mary Wortley Montagu: Comet of the Enlightenment* (Oxford: Oxford University Press, 1999), 334–40.

neither demonstrate that Lady Mary had attacked him in print, nor, alternatively, that she gave her lovers the pox, nor that the inoculation against smallpox that she had introduced was dangerous. Libel was a serious matter, and the close to *Satire 2. 1*, where Fortescue consults the Statutes, shows that Pope had investigated the legal provisions. According to John Butt, the maximum penalty was 'imprisonment for life and loss of goods on the second offence' (*Twickenham*, 4. 19 n.). Libel of peers was particularly dangerous. Lady Mary herself protested vigorously about the attack in *Satire 2. 1*, speaking to the Earl of Peterborough, who responded by relaying Pope's equivocating denial, but it seems the court and the Prime Minister also took an interest. Letters from Pope to Fortescue of 8 and 18 March 1733 show that Pope was under some pressure from Walpole to moderate his treatment of Lady Mary:

> I wish you would take an opportunity to represent to the Person who spoke to you about that Lady, that Her Conduct no way deserves *Encouragement* from him, or any other Great persons: & that the Good name of a Private Subject ought to be as sacred even to the Highest, as His Behavior toward them is irreproachable, loyal, & respectfull.—What you writ of his Intimation on that head shall never pass my lips. (*Correspondence*, 3. 357, 18 Mar. 1733)

It is a pity that Walpole's intimation was not recorded in the letter, but the implication is that Pope's attacks on Lady Mary had led the court to question whether his behaviour was 'irreproachable, loyal, & respectfull'. Although Lady Mary was no longer an important figure at court, she was the daughter of a duke and the wife of an MP and former ambassador. When one considers the authorities ranged against him—aristocrats, Prime Minister, King and Queen ('Great persons')—Pope's response, insisting on reciprocal responsibilities, is daring: 'the Good name of a Private Subject ought to be as sacred even to the Highest, as His Behavior . . .' The use of the word 'Private' is particularly interesting. Pope's poem was published, but he did not speak as a public man, a person of rank or a holder of office, but as a 'Private Subject'. His attitude anticipates the theorization of the bourgeois public sphere as 'the sphere of private people come together as a public'.[8] A man's private life merits respect, Pope claims, even from those set in authority over him.

The affront to Pope's good name referred to in his letter of 18 March is *Verses to the Imitator of Horace*, and Pope explains in his *Letter to a Noble Lord* that he believes his reputation was damaged when the poem was read out to the King and Queen at court by Lord Hervey:

> Your Lordship so well knows (and the whole Court and town thro' your means so well know) how far the resentment was carried upon that imagination [that Lord Hervey had been attacked in the imitation of Horace], not only in the *Nature* of the *Libel* you propagated against me, but in the extraordinary *manner*, *place*, and *presence* in which it was propagated; that I shall only say, it seem'd to me to exceed the bounds of justice, common sense, and decency. (*Prose Works*, 2. 444)

Later in the *Letter*, when referring to his position as a Catholic under penal laws, Pope reminds Hervey of his vulnerability to the disapproval of those in power,

---

[8] Habermas, *The Structural Transformation of the Public Sphere*, 27.

saying, 'you inadvertently went a little *too far* when you recommended to THEIR perusal, and strengthened by the weight of your Approbation, a *Libel*' (*Prose Works*, 2. 455). Pope's complaint is that he has been endangered by a libel that had been given publicity at court, and possibly in the town as well, without being placed in the public arena of print, to which all men had access and were therefore equals. Indeed, the *Letter to a Noble Lord* is itself a sharp illustration of Pope's difficulty, for the *Letter* was printed but circulated only privately. Pope had no certain knowledge that Hervey was the author of either the *Verses* or the *Epistle from a Nobleman* and he had to avoid a public libel, but until these works were published he had no opportunity to reply at all. The *Letter* was worth printing, even though it could not be published, because a printed work could be sent to the Queen, and, unlike Hervey, Pope had no easy access to her. Warburton, who was the first to publish Pope's *Letter*, says, 'It was for this reason [the original propagation of the libel] that this Letter, as soon as it was printed, was communicated to the Q.'

Pope initially thought the *Verses to the Imitator of Horace* were written by Lady Mary alone, and if, as I suspect, he published the Dodd version of the poem, *Verses Address'd to the Imitator of Horace*, that is why it was advertised as 'By a LADY of QUALITY', with its title page confirming that the poem was '*By a* LADY'. I take it that Pope's claim in *A Letter to a Noble Lord* that he 'took it for a *Lady*'s (on the printer's word in the title page)' is a characteristically cheeky laying of a false trail. The appearance in print of the poem with this identification gave Pope the opportunity to refuse to end his satirical campaign against Lady Mary. The very day of publication, 8 March 1733, he wrote the first of his letters to Fortescue about the court's attempt to silence him. Pope tells Fortescue:

Your most kind Letter was a Sensible pleasure to me: & the Friendship & Concern shown in it, to suggest what you thought might be agreable to a Person whom you know I would not disoblige, I take particularly kindly. But the affair in question of any alteration is now at an end, by that Lady's having taken her own Satisfaction in an avowed Libell, so fulfilling the veracity of my prophecy. (*Correspondence*, 3. 354)

Pope presumably says the poem is 'an avowed Libell' on the basis that the title page says it is 'By a Lady'. He claims that Lady Mary has declared herself the author, but only one version of the poem does that and the likelihood that Lady Mary should wish to identify herself as author in that way is slight. The episode is suspiciously shaped to Pope's convenience. That the poem should be published was essential to justify a reply. That Lady Mary should be identified as the author was necessary to justify Pope's attacks. That Pope should know immediately of the publication of the *Verses* and immediately be prepared to exploit it ('the . . . question of any alteration is now at an end') requires some special explanation. When we further note that the *Epistle to Arbuthnot* used its apparatus of footnotes to refer the reader to the *Verses* as a source text, and that the *Verses* were specially reprinted in 1735, again by Henry Woodfall, as a companion text to *To Arbuthnot*, with footnotes added to reciprocate those in *To Arbuthnot*, the possibility strengthens that Pope was behind the publication of these convenient editions. Indeed, the likeliest explanation for the double publication of the poem on 8 March is that Pope threatened Hervey with publication, which Pope then went ahead with, and that Hervey took 'publish and be damned' one stage further by also publishing himself.[9]

J. V. Guerinot points out that although *Verses to the Imitator of Horace* repeats some of the charges made familiar by his bibliography of pamphlet attacks, a summary of the charges falsifies the 'total impact of the poem'. The charges are '1. Pope ignorant of Greek; 2. Pope ungrateful to Chandos; 3. Pope incapable of loving; 4. Pope deformed.'[10] The poem goes beyond these stale complaints because Lord Hervey and Lady Mary pick up the concern with self-representation in the *First Satire of the Second Book of Horace*, treat it with a measure of moral seriousness, and translate it into the grotesque. Their response builds from elements that are within the poem, transforming them as they respond. They begin with the typography, correctly taking the printed text as one of Pope's modes of self-representation and transforming Pope's parallel texts into contrasting texts. Pope is not the modern representative of Horace, they say; the two poets are antithetical. Then, in the poem's central and dazzling manœuvre, they deal with Pope as a representative not of Horace but of humanity:

> Thine is just such an Image of *his* Pen,
> As thou thy self art of the Sons of Men:
> Where our own Species in Burlesque we trace,
> A Sign-Post Likeness of the noble Race;
> That is at once Resemblance and Disgrace. (11–15)

These lines respond to the concern in *Satire 2. 1* with satire as the expression of self:

> I love to pour out all myself, as plain
> As downright *Shippen*, or as old *Montagne*.
> In them, as certain to be lov'd as seen,
> The Soul stood forth, nor kept a Thought within;
> In me what Spots (for Spots I have) appear,
> Will prove at least the Medium must be clear. (51–6)

If, as Pope says, his head and heart flow through his quill (63), the consequence, his critics claim, is violence and ugliness. Pope's body may be a burlesque of humanity, but his body fully represents his spiritual nature, allowing his soul to stand forth:

> It was the Equity of righteous Heav'n,
> That such a Soul to such a Form was giv'n . . . (50–1)

Through their attention to Pope's declared intention in *Satire 2. 1*, the *Verses* are liberated into easy movement between man and verse. There are two main failures of Pope as a human being: he is not a gentleman and he is sexually inadequate. The first charge is established fairly systematically. Pope's numbers are 'Hard as thy Heart,

---

[9] Hervey made revisions for an unpublished second edition himself. For the campaign run by the *Craftsman* to try to get Hervey to admit authorship of *Epistle to a Nobleman*, and several other relevant but not overwhelmingly persuasive arguments, see McLaverty, 'Of Which Being Publick the Publick Judge'.

[10] *Pamphlet Attacks on Alexander Pope*, 225. The summary is, nevertheless, a good one.

and as thy Birth obscure'; his poems are like weeds, though they 'seem produc'd by Toil'; his satire is like an oyster-knife; his books are equivalent to his person, which would not be allowed in the house. The second charge emerges through metaphor as well as direct statement. Pope's hatred in his satire is like love in a brothel:

> 'Tis the gross *Lust* of Hate, that still annoys,
> Without Distinction, as gross Love enjoys . . . (30–1)

He is unresponsive to beauty, youth, and charms, because he is 'No more for loving made, than to be lov'd'; his antipathy to mankind is like that of the snake to Eve; his physical weakness is like the female scold's. Binding the two elements together is the imagery of impotence: the hacking and hewing with the oyster-knife; the scolding; the harmless quill shooting from the back of the '*fretful Porcupine*':

> Thus 'tis with thee:—whilst impotently safe,
> You strike unwounding, we unhurt can laugh. (77–8)

Ridiculing the claim in *Satire 2. 1* to punish wrongdoing, 'Hear this, and tremble! you, who 'scape the Laws', they present the poet as a self-destructive insect:

> like thy self-blown Praise, thy Scandal flies;
> And, as we're told of Wasps, it stings and dies. (87–8)

In *To Arbuthnot* Pope was able to appropriate and redirect much of this material, working hard to integrate it with what he had already prepared. His writing could serve as its own defence and redirect the metaphors of unnaturalness, animal life, and exclusion; his family was already the central topic of a draft poem; but the 'Pope incapable of loving | Pope sexually inadequate' material was difficult to rebut. It required *Sober Advice* as well as *To Arbuthnot* to encounter it properly, and that in turn led to some of the variation in *To Arbuthnot*, with the most daring material omitted from the first edition and reserved for the *Works*.

THE PRINTED VERSIONS OF *TO ARBUTHNOT*

The precise relation between the text of *To Arbuthnot* in the first edition of 2 January 1735 and that in the *Works* of 23 April is perplexing. Although they were printed around the same time by John Wright, in a shop Pope supervised very closely, there are many differences between them. Most striking are the twenty-three lines they do not share. The first edition has five couplets not found in the *Works*, while the *Works* has five couplets and one triplet not found in the first edition. They also differ in punctuation forty-seven times, and there are over fifty differences in wording, though I have counted these somewhat subjectively, sometimes counting several changes in a line independently and sometimes regarding them as one revision. John Butt and David Foxon have made valuable attempts to clarify the relations between the texts, but both found the manuscript evidence inconclusive. As Butt remarks, there is a danger of frolicking in conjecture, but I think we can gain a general picture without too much frolicking. The texts have different styling. The first edition has proper names in italic and uses footnotes; the *Works* aims at classical simplicity, with roman names and endnotes. Moreover, the first edition makes a policy of not naming its victims. These general decisions led to different settings of type. As printers generally prefer to set from printed copy, it would have been convenient to use marked-up copy of the first printing, whichever it was, to print the second. But if that had been the case, the coincidence of accidentals should have been higher; it is reasonable to suppose, as Butt and Foxon seem to, that both editions were set from manuscript. A sensible explanation of the failure to use printed copy is that it was not ready when serious work began on the second printing. So that while the first printing was proceeding from one manuscript, Pope provided a second manuscript from the foul papers that had been the basis of the first. Further revision might then take place in the second manuscript before it went to press, and both printings would be subject to revision in proof. It remains to decide which was the first printing. Some examples are suggestive. Butt proposes that the *Works* text is the earlier because in lines 289–304 it follows the text published in the *London Evening Post* on 22–5 January 1732, but a collation of the two texts with the *London Evening Post* yields mixed results. In major features, the *Works* follows the newspaper: it retains the same order of the couplets at lines 297–300; it lacks the new couplet at 301–2; and it concludes, 'Let never honest Man my satire dread', rather than 'A Lash like mine no honest man shall dread'. In all these cases also, the octavo *Works* of 1735, which determined the subsequent textual history of the poem (henceforth called simply the octavo, to distinguish it from the quarto and folio, which I continue to call *Works*), agrees with the revisions present in the first edition but not in the *Works*, and in the case of 'Let never honest Man' we can see the *Works* line is very close to the early manuscript draft 'Let never modest men my satire dread' (*Last and Greatest Art*, 442. 68).[11] It does look from the substantive readings as if the *Works* is the earlier text. A couple of trivial readings, however, suggest the opposite. The *London Evening Post* lines begin 'The Fop' and so does the first edition; the *Works* and subsequent editions read 'That'. The newspaper reads 'if he lies not' and so does the first edition ('lyes'), but not the *Works* and subsequent editions, which read 'lye'. I think this fits well with my initial hypothesis and fleshes it out: Pope's most polished manuscript was used to set the *Works* (hence the superiority of its text in these small matters), but Pope then went back to a slightly earlier manuscript to make the major revisions for the first edition, in which he was trying to respond more fully to the *Verses'* attack. This interpretation is confirmed by the introduction to the Atticus portrait, where no revision to respond to the *Verses* was needed and the first edition is consequently clearly the earlier text. The manuscript reads, 'Well might they swear, not Congreve's self was safe!' (426. 70); the first edition has 'How did they swear, not *Addison* was safe'; and the *Works* has 'And swear, not Addison himself was safe', which is adopted by subsequent editions. A final example suggests that Pope sometimes went back to his

---

[11] In citations from Mack's reproductions of the manuscript, I have given the page number he allocates to the relevant page of the manuscript and the line number from his transcription from the facing page. There are often variants in a line, but when the variants are not important to my discussion I have selected a particular version of the text.

first thoughts, but that provides no difficulty for the current hypothesis. The present lines 169–70 appear in the manuscript with alternatives written above and alongside them. The first line is either 'The Thing, we know, is neither rich nor rare' or 'The Things, we know, are either rich or rare'. The second line is either 'But wonder, how the devil it got there?' or 'But all ye wonder is how the devil they got there?' (426. 20–2). The first edition combines the two:

> The things, we know, are neither rich nor rare,
> But wonder how the Devil they got there?

The *Works* produces something different:

> Not that the things are either rich or rare,
> But all the wonder is, how they got there?

This is another of the cases where we can see that the first edition is closer to the manuscript, but this time its reading is the one chosen by the octavo. It seems likely that in revising the couplet for the *Works*, Pope eliminated any hint of blasphemy, but that he later reconsidered and restored the more colloquial original.

If the relation between the printed versions of *To Arbuthnot* is complicated, that between the two earliest editions and the octavo is easier to work out. Analysis of the variants suggests that up to and including line 151 the first edition was followed as copy text; after that Pope used the *Works* text.[12] If, as seems likely, Pope was marking up unbound folio sheets, he marked up B and C of the first edition and then switched to P, Q, and R of the *Works*. As this was an edition for the popular market, Pope chose to print the names in full and in italic, something that could be achieved only by combining the practices of the two editions, but with these exceptions he initially followed the first edition in dozens of details, with only two cases of punctuation (probably revisions in proof) agreeing with the *Works*. If, as I have suggested, the text was throughly revised for the first edition, the choice would be a simple one. But in sheet D of the first edition Pope knew he would encounter two of the poem's major satirical portraits, and that the second, Bufo, had lines missing in the first edition that were supplied in the *Works*. I think that explains the switch. Initially Pope followed his new copy text faithfully, with only one first edition reading added on sheet P (lines 171–2), but Q1^r has heavy revision of one couplet and the addition of two more; Q1^v–Q2^r have couplets reversed in order, a new couplet, and another line substantively revised, and Q2v has one line revised. In R things have settled down again, but there is a sudden burst of petty revision restoring the readings of the first edition around lines 350–4. I think the text that emerged in the octavo was one Pope wanted—he was obviously comparing both editions and making choices—but accidentals certainly follow the copy text, with rare exceptions, and it is highly likely

---

[12] I use 'copy text' in a fairly orthodox way, though I do not suppose Pope had developed a rationale. Whichever text was marked up determined readings Pope regarded as indifferent. My analysis leaves the present lines 157–62 unaccounted for. As the octavo follows the *Works* in line 157, it seems likely this passage was taken from the *Works*.

that some revisions were not reinstated. Certainly following copy text resulted in some departures from the general styling that had been adopted. Line 60 still reads 'The Play'rs and I are, luckily, no friends' from the first edition, when I suspect the specific 'Cibber' should have been inserted; and at line 198 the octavo, following *Works*, does not italicize 'Turk'.[13] But without more manuscript evidence, it would be difficult to carry the analysis of the switch in copy text further.

THE DEVELOPMENT OF THE MANUSCRIPT OF
TO ARBUTHNOT INTO PRINT

The tight timetable that led to the complex variations between the first two texts of *To Arbuthnot* may in part have derived from the desire to publish the poem before Arbuthnot's death (he was seriously ill by 17 July 1734 and died on 27 February 1735), but I suspect Pope also wanted an interval between the first publication of the poem and the appearance of the *Works* in order to give his enemies an opportunity to respond to it. Some variations arose from the need to use a fairly short time to adapt the draft poem Pope had lying by into a full and publishable response to the *Verses*; but others arose from an unwillingness to print virulent personal attacks in the first edition. Important sections of the characters of Bufo and Sporus were left out of the first edition, thereby lessening the offence these sections would cause but also reserving some material for a response to consequent attacks. The first manuscript version of *To Arbuthnot*, which did not have that title and was not addressed to Arbuthnot, already began, 'And of myself too something must I say?' (424. 2). The impulse to a poem overtly egoistic, even something quasi-confessional, was already there, but the decorum to which Pope adhered constrained it and made it false to itself, just as it did in the contemporary subterfuge over his letters. The line introduces a 'must' without motivating it, but the excuse for self-analysis was fortuitously (or deliberately, depending on Pope's involvement) provided by the publication of the *Verses*. The acerbity of the *Verses* opened up the opportunity for a full self-defence, and even provided some of the materials on which that defence could be based.

In a British Academy lecture in 1954, John Butt launched the study of Pope's manuscripts with a fascinating analysis of the development of *To Arbuthnot*, an analysis that has subsequently been elaborated by Maynard Mack and Julian Ferraro, who has disputed some of Butt's conclusions.[14] There is a consensus that there is an

---

[13] In his letter to Pope, Cibber remarks on the change from 'Cibber' in what he calls 'your Folio Edition', that is, the *Works*, to 'The Play'rs' in 'your smaller edition', saying, 'This is so uncommon an Instance of your checking your Temper and taking a little Shame to yourself, that I could not in Justice omit my Notice of it' (*A Letter from Mr Cibber to Mr Pope*, introd. Helene Koon, Augustan Reprint Society 159 (Los Angeles: William Andrews Clark Library, 1973), 44. This is touching, but the first edition is full of such instances and I think Cibber's name was left out of the small edition by mistake.

[14] John Butt, 'Pope's Poetical Manuscripts', Warton Lecture on English Poetry, *Proceedings of the British Academy*, 40 (1954), 23–39; Julian Ferraro, ' "Rising into Light": The Evolution of Pope's Poems in Manuscript and Print' (unpublished doctoral dissertation, University of Cambridge, 1993). There is also a fine introduction to these manuscripts in Mack's *Last and Greatest Art*, 419–23, including a dazzling paragraph on the poem (422), the best criticism of it I know.

early draft of a poem about Pope, poetry, and his parents, and that there are further manuscript fragments elaborating or adding to these materials; my concern is not with the detailed relations of the manuscripts but with the way Pope was able to develop an early draft that already privileged the interior life over the public life by suggesting that the life of a writer was altogether a mistaken course. This is an odd stance in a poet who had made his fortune from his pen and was just about to bring out his second volume of works, but none the less it emerges as sincerely felt. It repeats the sentiment of the prefatory letter to the third edition of *To Burlington*: 'Critics of *this Sort* can intimidate me, nay half incline me to write no more.' The deleted opening lines evoke a figure in contrast to Pope himself, 'The Man of Learning, yet too wise to write', and immediately start him on the section of the poem which now begins at line 125:

> Why did I write? what Sin to me unknown
> Dipt me in Ink? my Parents, or my own? (424. 13–14)

In a deft account of his life, Pope makes poetry a natural activity emerging from family life and the circle of friendship. He insists in the draft, as he does in the published poem, that his life as a writer is compatible with family duty. In the manuscript poem Pope defends himself by presenting himself as the product of his mother and father; he is, above all, a filial poet. His writing is not his own sin ('No duty broke, no Father disobey'd') and it cannot be the sin of his parents either, for the conclusion of the poem declares their virtue. He does not, therefore, merit punishment. In the original draft the flatterers' praise of his body leads directly to the source of that body, his parents, and to the praise of their lives that is to conclude the poem.

> But Friend! this Shape wch you & Curl admire
> Came not from Ammon's Son, but from my Sire;
> And for my Headake (you'l the truth excuse)
> I had it from my Mother, not the Muse:
> Happy! thrice happy, had I heir'd as well
> The Christian Kernel, as the crazy Shell . . . (434. 37–47)

In the manuscript version, Pope may not be confident that he embodies Christian virtue to the extent that his parents do, but he claims for himself a version of their lives ('I pay my debts, nor blush to say my prayers', 440. 23–4), and in the final revision dedicates himself to the care of his mother.

The treatment of the life of the writer as a public figure, in contrast to that of the private man, is negative. The assessment differs from that of the *Dunciad* by ignoring the growth of the market, but, in a conversational and seemingly casual way, it analyses the institutions of literary culture and finds them wanting. *To Arbuthnot* is in many ways a new essay on criticism, written from the point of view of the established writer and concerned with the social realization and distortion of principles of criticism. Whereas *An Essay on Criticism* idealized a public sphere that was neither a circle of friends, nor public institutions, nor the product of the market, *To Arbuthnot* represents a range of public or semi-public behaviour that renders criticism vicious.

Pope's own place in the public sphere is represented as a natural, if unfortunate, development from the private circulation of his poems. It is the product of friendship, one of the poem's key concepts:

> But why then publish? Granville the polite,
> And knowing Walsh, would tell me I c^d write ... (424. 25–7)[15]

The list of those who encouraged his early writing concludes with a strong declaration of the superiority of friendship to even the best-informed criticism:

> Happy my Studies, when by these approv'd!
> Happier their Author, when by these belov'd! (424. 34–5)

Pope is propelled into the public sphere by public men who are valued for personal affection rather than public virtues. There is, I suppose, an implicit explanation of publication in this paragraph—good writers publish—but there is no appeal to the public good or to the social function of literature. Such social and political issues are left blank throughout this version of the poem, except in the discussion headed, 'But why write Satyr?' whose materials were largely transferred to *Satire 2. 1*.

Public criticism is vitiated by the personal emotion of anger. Pope's early draft moves through various cases until it reaches a temporary halt with the account of Addison as Atticus. The first examples, Gildon and Dennis (raving 'in furious fret'), connect anger with venality—you can earn a living through being angry—and these figures tie in neatly with those from Addison's circle who resent the exposure of their incompetence. In between them Pope places verbal critics, whose endeavours preserve their names, linked forever with Milton's or Shakespeare's, in a practice parasitic on the literary values that it negates. Pope's critique so far reiterates briefly some of the positions taken by the *Dunciad*. When he moves to the portrait of Atticus, however, he invokes a different sphere, that of the ideal critic:

> Blest with each <u>Talent</u> & each <u>Art</u> to please,
> And born to write, converse, & live with ease ... (428. 4–5)

Atticus presents the tarnished ideal of the private man functioning to the public good. His abilities move freely between private and public—living, conversing, writing—the easy transition from life to literature echoes the progress claimed for Pope himself. In this section of the poem on Addison's little senate, Pope evokes the world of the coffee house—community, intelligence, evaluation, criticism, civilization—but it is vitiated by jealousy, a barbaric hostility to a 'brother'. The barbarism, however, has no direct expression. Atticus attacks his potential rival by means so sly that they evade resistance, and in this passage Pope starts up that network of metaphor that characterizes his society by its failure of connection even in aggression: a structure of indirection, impotence, and slime:

---

[15] Dustin Griffin notes the importance of 'friend' in this poem (it appears twenty times) in his valuable chapter on Pope in *Literary Patronage in England, 1650–1800* (Cambridge: Cambridge University Press, 1996), 123–54 (151).

> Willing to wound, and yet afraid to strike;
> Just hint a Fault, & hesitate Dislike . . . (428. 12–13)

The idea is picked up in the revision of the poem, especially in the Sporus character, where 'well-bred Spaniels civilly delight | In mumbling of the Game they dare not bite' (*Works*, 288–9). In the little senate indirection performs its malicious work through egoism and sycophancy. The ideal critical order involves the exchange of rational views among equals, but the little senate is not an assembly of equals; it receives laws from its master and responds with applause. In the manuscript the abuse of power in the assembly of the coffee house is followed by abuse of power by the patron. Atticus is succeeded by Bubo, at this stage of the poem's development clearly a satirical sketch of George Bubb Doddington. Bubo is satirized for his ostentation but also, like Atticus, for his corruption of critical judgement. In his draft Pope was playing with their relation to Apollo, the god of poetry who figures so prominently in the illustrations of *Works* I. The master-critic and the patron function in a context of idolatry. Originally the Atticus portrait began with the query 'were there One, whose fires | Apollo kindles, and fair Fame inspires' (428. 1–3), but 'Apollo' is crossed out and down the page at line 57 the new portrait starts:

> Proud, as Apollo on the forked Hill,
> Sate fullblown <u>Bubo</u>, puft by ev'ry Quill . . . (428. 57–9)

Bubo has such power over his admirers that he takes the place of Apollo himself. His life is sustained by praise—dedication, admiration, flattery—the corruption of criticism by wealth being expressed in a characteristic zeugma:

> Receivd of wits and undistinguishd race,
> Who first his Judgment askd, and then a Place. (428. 69–71)

This critique of aristocratic patronage was originally picked up in lines that did not reach the published version of the poem:

> And Peers for Flattry make such large
> [demands, They take a hundred Dunces off my hands (432. 69–71)

Pope tried a variant on the same page, 'Or Lordly Pride for flattry makes demands' (432. 84–5), but discretion changed the published line to 'Or simple Pride for flatt'ry makes demands' (*Works*, 242).

Already in the draft version of his poem, Pope turned from the abuse of power by great men, Atticus and Bubo, to his own role as a sort of patron. The picture is, of course, a generalized one, but it springs in part from his role as an entrepreneur after 1729, when he took on his own printer, John Wright, and bookseller, Lawton Gilliver.[16] The depiction is, once again, negative:

> But Authors, Authors are a heavier Curse!
> Their anger dreadful! but their Friendship worse! (430. 31–8)

---

[16] See Foxon, *Pope and the Book Trade*, 102–8, and its references.

Twickenham is invaded not by authors we know Pope helped—Gilbert West, George Lyttelton, Walter Harte, James Miller, James Bramston, George Jeffreys, Henry Brooke, David Mallet, Aaron Hill—or those who wrote complimentary verses to be prefixed to his works, but by maudlin poetesses and parsons 'much bemus'd in beer'.[17] The invasion of Pope's home vividly expresses the radical confusion of public and private, and in the final version it was promoted to the start of the poem. The private life of family, which Pope has denied abandoning in order to pursue the life of the poet, is now invaded from the outside. The would-be writers break through Pope's hedges, assail him as he travels, accost him in church, and interrupt his meals. Domestic life at Twickenham is delicately evoked through the details of its destruction. Pope's response is portrayed as sad bewilderment. The demands made of him draw in various sources of financial power: the patron ('<u>You know his Grace . . . Ask him for a Place</u>', 432. 29–31); the theatre ('<u>Commend it to the Stage</u>', 432.53); and the bookseller ('Your Int'rest Sir with <u>Lintot</u>', 432. 62–3). A new topic is collaboration. In the case of his collaboration with Fenton and Broome over the *Odyssey* Pope initially took an older view of collaboration, thinking it unnecessary to specify the contribution of each individual; he had been driven to be more specific by the responses of his critics. In *To Arbuthnot* the request for collaboration is an invasion of the poet's privacy. One invader demands a prologue, as well as friendship and ten pounds, another, driven from the stage to publication and told that he expects too much for the copy, replies, '<u>Not Sir, if you revise it, & retouch</u>' (432. 66). When he offers to share the profits by way of persuasion, Pope can dismiss him, 'Glad of a Quarrel, strait I clap the door' (432. 72). The privacy which in this poem is the poet's necessary context is restored. From the representation of these supplicants, with their demands, Pope moves to their flattery, and through that to his body and his parents, who bring the poem to a close. In this early draft, *To Arbuthnot* is successful in treating poetry as a private matter, belonging to the circle of family and friends. Pope's misfortune, we understand, was to stray beyond that boundary, into a public sphere where the duties of family and friendship are not observed and no compensatory virtues are to be found; the rational critique that should characterize the public sphere is vitiated by the abuse of power by those acting from personal vanity and jealousy. In a final dramatization of the innocent poet's fate, the intimate sphere of the home is invaded by the mad, selfish, and incompetent, all seeking to appropriate the poet's art and influence to their own ends.

In many respects the materials in the manuscript draft of *To Arbuthnot* found the perfect opportunity for publication in *Verses to the Imitator of Horace*. For example, the passage beginning 'Shut, shut the door, good John!' (432. 2) now emerges as a response to

> And to thy Books shall ope their Eyes no more,
> Than to thy Person they wou'd do their Door. (99–100)

---

[17] I discuss Pope's possible role in the publication of these authors in 'Lawton Gilliver: Pope's Bookseller', 116–18. I also, a touch recklessly perhaps, suggest that Burscough, Young, Dodsley, Lockman, and Brownsword might also have been published by Gilliver through Pope's influence.

To such a casting out, Pope's response is a Coriolanus-like 'I banish you.' The position of the outsider is one the poem happily claims, but whereas in the draft he is outside a corrupt literary culture, in the published poem he is also outside a corrupt elite. The portrayal of Pope in the *Verses* as a monstrous outcast gives him the necessary cue for self-analysis, 'And of myself too something must I say?', and enables him to recast his emphasis on the family as a critique of his aristocratic opponents. The advertisement and annotation to the first edition protect the author against any charge of egoism, and his acknowledged adaptation of their figurative language liberates his attack into a new ferocity. There are six significant new sections in the published version of *To Arbuthnot*, which can best be identified by reference to the Twickenham text. A brief address to Arbuthnot himself early in the poem (27–30); a discussion of the objects of Pope's satire, beginning 'You think this cruel?' (83–106); a short comparison of himself with his monarch (221–4); a reflection on the loss of Gay (255–60); and the portrait of Paris/Sporus, present in manuscript fragments (305–33), followed by a contrasting section on the poet himself (334–81). These sections are common to the two first publications of the poem, though they are subject to varied revision.

The address to Arbuthnot serves to establish the poem's addressee at an early point in its progress (though the poem begins, confusingly for contemporaries, by addressing John Serle, Pope's gardener), it introduces the theme of friendship ('Friend to my Life'), and it strengthens the poem's concern with moral and physical health ('What *Drop* or *Nostrum* can this Plague remove?'). The draft of the poem already contained a reference to 'this long Disease, my Life' (424. 24), the contrasting praise of Pope's father, 'Healthy, by Temp'rance & by Exercise' (434. 72), and reflections on the death of the poet and his mother. Making Arbuthnot the poem's addressee highlighted these aspects of the poem and made another link between public and private life.

The section of the poem on the insensibility of Pope's satirical targets affords an opportunity to develop analogies between human and animal life in response to the attack in the *Verses*, where Pope is a snake, a porcupine, 'A puny Insect shiv'ring at a Breeze' (quotation from *To Burlington*), and a wasp that stings and dies. The appropriation of these images was to culminate in the representation of Hervey as Sporus, but Pope, conscious of these motifs, adds to this early section of the poem a supportive simile:

> Scriblers like Spiders, break one cobweb thro',
> Still spin the slight, self-pleasing thread anew,
> Thron'd in the centre of their thin designs!
> Proud of a vast extent of flimzy lines! (*Works*, 87–90)

Possibly because it was new, this passage was subjected to heavy revision in the first edition, which also added a further couplet at the last minute, developing the analogy and anticipating the disgust at Sporus:

>   Destroy his Fib, or Sophistry; in vain,
>   The Creature's at his dirty work again . . . (First edition, 89–90, lineation disturbed)

The couplet is picked up in the *Works* Variations—too good to lose but too late to be incorporated into the poem—and forms part of the poem in the octavo. Pope follows up his simile of the dirty spider by increasing the range of persons attacked in the poem. The contempt for Pope's satire described in *Verses* ('You strike unwounding, we unhurt can laugh', 78) is translated into brazen indifference to criticism, and the critique of aristocratic arrogance, already present in the portrayal of Bubo, is given an extra thrust:

>   Whom have I hurt? has Poet yet, or Peer,
>   Lost the arch'd eye-brow, or *Parnassian* sneer? (First edition, 91–2)

Further targets are then named, or in the first edition of the poem, referred to but not named: Colly Cibber, Orator Henley, James Moore-Smythe, William Arnall, Ambrose Philips, Lady Mary Wortley Montagu—but then Arbuthnot speaks to halt the poem:

>   Still *Sapho*—'Hold! nay see you, you'll offend:
>   'No Names—be calm—learn Prudence of a Friend:
>   'I too could write, and I am twice as tall,
>   'But Foes like these!' One Flatt'rer's worse than all . . . (First edition, 97–100)

The reaction to Sapho's name mimics the court's response to Pope's criticism of her in his satires; Fortescue had taken a similar, though less explicit, line in *Satire 2. 1*, reflecting his correspondence with the poet. Arbuthnot held a similar position. He was, after all, the person Lady Mary had complained to of Pope's treatment of her in two letters in October 1729 (*Correspondence*, 3. 59–60), while a further letter to him denied her authorship of the *Verses* in response to *To Arbuthnot* itself (*Correspondence*, 3. 448–9), and a letter from Lord Hervey tells Henry Fox that Arbuthnot had asked him why he had been so severe on Pope.[18] Arbuthnot intervenes immediately to suggest that these enemies are too powerful to risk offending, but at this point in the poem Pope simply deflects the issue, by moving on to material he had prepared on his flatterers. In the first edition, however, he added a couplet that prepares for both the Atticus and Sporus portraits.

>   Of all mad Creatures, if the Learn'd are right,
>   It is the Slaver kills, and not the Bite. (First edition, 101–2)

This couplet also is picked up in the *Works* Variations. It defines the offensiveness of Pope's enemies, whose attacks are the more loathsome for their indirection. The couplet completes, for this early section, a major element in Pope's own web of image and analysis. It is noteworthy that a couplet added to the *Works* at the last minute ('Time, Praise, or Money, is the least they crave, | Yet each declares the other, fool or

---

[18] Earl of Ilchester, *Lord Hervey and his Friends*, 189.

knave', 109–10, lineation disturbed), which contributes to the poem's analysis but not to its imaginative patterning, is omitted from subsequent editions.

The short sections Pope added on the King and on Gay are part of a development of the theme of corrupt patronage already present in the manuscript draft. They show some hesitancy in Pope, who was himself willing to wound and yet afraid to strike, as far as the monarch was concerned. The printed versions of the poem have an earlier feint at the King in the lines on Midas (69–82), where the lineation looks as though it has been disturbed by the addition of an extra couplet. A comparison with the manuscript (432. 105) suggests that the extra lines are

> 'Good friend forbear! you deal in dang'rous things!
> 'I'd never name Queens, Ministers, or Kings . . . (First edition, 75–6)[19]

That is, Arbuthnot's warning serves to point up the possible application of the Midas story to the present monarch, but it is not clear what that application would be. The new reference to the King comes immediately before the portrait of Bubo (now protectively changed to Bufo), with Pope making some gesture at tackling the topic of patronage from the top:

> Poems I heeded (now be-rym'd so long)
> No more than Thou, great GEORGE! a Birth-day Song. (First edition, 216–17)

The couplet anticipates *Epistle 2. 1* of 1737 in its playful irreverence, but it mounts no attack on the culture of the court, and the poem moves on to Bufo instead. At the end of that section we find one of the major disparities between the *Works* and the separate edition. After the first edition's conclusion of the portrait with Bufo's cruelty in paying some poets in kind, the *Works* continues for four more lines:

> Dryden alone (what wonder?) came not nigh,
> Dryden alone escap'd this judging eye:
> But still the Great have kindness in reserve,
> He help'd to bury him he help'd to starve. (*Works*, 238–41, lineation disturbed)

Because line 244 is actually labelled 240, it looks as though these lines were added to the *Works* at the last minute, perhaps in proof, without a complementary change being made to the first edition. However, the final couplet was present in some form in the manuscript (the end of the page is now torn off) at the very end of the Bubo portrait (428. 96–7), and the lineation of the first edition has been disrupted, with line 242 being labelled 240. The *Works* Variations records the presence of these lines in one edition but not the other. Although the explanation of a last minute revision is entirely plausible, I wonder whether Pope might have dithered about whether or not to include these lines, perhaps excising them from both texts and then restoring them to the *Works*. The Earl of Oxford has interesting notes on this passage in his copy of the first edition, now in the Bodleian (M 3.19 Art (17)). First he writes, 'This character made fit many'; but, probably later, adds, 'but I think it is cheifly the right

---

[19] Butt disagrees (*Twickenham*, 4. 93), suggesting instead that 73–4 are added, but a form of this couplet appears in the manuscript.

of Mr Bubb Doddington'; and finally, in red pencil, 'it would also fit the late earl of Halifax'.[20] I suspect the information in red pencil comes from Pope himself and that this is a characteristic example of Pope's double reference: the passage points distinctly to two incompatible figures. Doddington is engaged by 'Bufo' and the references to his seat; Halifax is engaged by the reference to Dryden's funeral—he planned to erect a monument to Dryden in Westminster Abbey. The problem raised for Pope by the four lines on Dryden, and implicitly Halifax, is that although Pope regarded Halifax as 'rather a pretender to taste than really possessed of it' and told a good story of his foolish behaviour towards Pope over the *Iliad* translation, Halifax had subscribed for ten sets of the *Iliad* and had been named in the preface as 'one of the first to favour me' (Spence, *Anecdotes*, 204). He had also offered Pope a pension (*Correspondence*, 1. 271). He was a good example to use, and yet in satirizing him Pope would risk a recurrence of the charge of ingratitude that followed the Timon portrait in *To Burlington*. Finally, Pope decided to take the risk of including these lines in the *Works* but not in the first edition, which, appearing on its own, was bound to attract more attention. The concluding couplet, of course, generalized the attack to 'the Great' in any case.

The lines on Gay follow on shortly after the section on Doddington and appear directly after the couplet whose reference to 'Lordly Pride' was censored. They act as a detailed substitute for it:

> Blest be the *Great!* for those they take away,
> And those they leave me—For they left me GAY,
> Left me to see neglected Genius bloom,
> Neglected die! and tell it on his Tomb;
> Of all thy blameless Life the sole Return
> My Verse, and QUEENSB'RY weeping o'er thy Urn! (First edition, 244–9)

The lines relate the issue of patronage to Pope's own circle without directly involving his own position. Again, the Earl of Oxford has an interesting note:

Gay] he was neglected by the court & had no place though often promised, He lived with the Duke of Queensberry & died at his House Dec. 4. 1732. He was buried at the Duks expence and will set up a monument for him.

As so often, Oxford seems to endorse Pope's attack even when it is directed at 'the *Great*', but latterly these lines, recognized as powerful myth-making, have been taken to misfire. David Nokes points out with justice that Gay was far from neglected: he died leaving an estate of more than £6,000 and was given a funeral suited to a peer of the realm.[21] I suspect the desire to find a place for Gay in the pattern of the poem produced a distortion. Gay clearly was not one of those neglected by aristocratic patrons; both the Duchess of Marlborough and the Queensberrys were strong in his support. His neglect was quite specifically, as

---

[20] See James McLaverty, 'Pope in the Private and Public Spheres: Annotations in the Second Earl of Oxford's Volume of Folio Poems, 1731–1736', *Studies in Bibliography*, 48 (1995), 33–59 (58).
[21] *John Gay: A Profession of Friendship* (Oxford: Oxford University Press, 1995), 6–7.

Oxford says, by the court, whose support of Walpole in banning *Polly* was a rare act of open hostility to Pope's circle. It was also supported by Hervey, prominent in the court's opposition to the *Craftsman* and its campaigns, of which Gay's plays could be seen as part. Gay therefore belongs in a potential version of the poem, but a failure of political nerve has him in the wrong place. Two additional couplets, appearing in the first edition only, but recorded in the *Works* Variations, follow on from the memorial of Gay by attempting to summarize the point of these passages in a picture of Pope himself. The first defines his relation to patronage:

> Above a Patron, tho' I condescend
> Sometimes to call a Minister my Friend . . . (First edition, 254–5)

The second draws in those values that are used to contrast with the world of patronage:

> Has Life no Joys for me? or (to be grave)
> Have I no Friend to serve, no Soul to save? (First edition, 261–2)

Gay and Dryden function in the poem as examples of neglect but their most powerful role is as contrasts with Pope himself.

SOBER ADVICE, SAPHO, AND SPORUS

The revisions of *To Arbuthnot* that remain to be considered all concern the representation of Lord Hervey as Sporus and the contrasting sketch of the poet himself, beginning 'Oh keep me what I am'. Collation immediately shows that there is something unusual going on in this section. The texts of the *Works* and the first edition differ quite sharply in these lines, more than anywhere else in the poem. The character being sketched has a different name: in the first edition he is Paris; in the *Works* Sporus. In the *Works* a central passage concerns Sporus' ambiguous sexuality and occupies seven lines:

> His Wit all see-saw between *that* and *this*,
> Now high, now low, now Master up, now Miss,
> And he himself one vile Antithesis:
> Amphibious Thing! that acting either part,
> The trifling head, or the corrupted heart,
> Fop at the Toilet, Flatt'rer at the Board,
> Now trips a Lady, and now struts a Lord. (*Works*, 315–21, lineation confused)

But of all this passage, only one modified couplet appears in the first edition, the sexual ambiguity suppressed:

> Did ever Smock-face act so vile a Part?
> A trifling Head, and a corrupted Heart! (First edition, 312–13, lineation confused)

Even more surprisingly, these differences are not recorded in the *Works* Variations, though all the other variants I have discussed so far, and some slighter ones, are. It is not difficult to surmise that Pope was nervous about this central element in his attack and suppressed it in the first edition. Such an interpretation is borne out by the manuscript fragments of the poem reproduced by Mack, which include an early version of the Sporus portrait and show that all the suppressed material must have been available for the first edition. The decision to omit it can best be understood in the light of *To Arbuthnot*'s companion *Sober Advice*, into which other material of this sort was projected and published anonymously.

Sober Advice (Horace's *Satire 1. 2*) complements *To Arbuthnot* by satirizing the intimate sphere in contemporary society. It responds to the 'Pope incapable of loving/Pope sexually inadequate' charge of *Verses to the Imitator of Horace* by attacking the sexual morality of the court and Quality, largely through ridicule of women, including Lady Mary Wortley Montagu. Horace's poem was notorious because its initial topic, the importance of the golden mean, soon gives way to that of adultery, and its controversial nature had been enhanced by André Dacier's edition, which had focused its discussion on a Christian condemnation of a passage recommending pederasty. Not surprisingly, Pope published the poem anonymously and showed circumspection in imitating the sexually explicit passages in Horace. The circumspection was balanced by the ingenious, and unfair, typographical device of adding notes purporting to be by 'the Rev^d. R. Bentley', the distinguished editor of Horace and Milton, which make dirty jokes, point up the timidity of the imitation, reintroduce the interest in pederasty, and underline the poem's misogynistic turn. I shall call this invented commentator *Bentley* to distinguish him from the Master of Trinity. As in *Satires 2. 1* and *2. 2*, Pope prints parallel texts and uses highlighting to point up the social embedding of his material, but this time these features take second place to the capitals that are used to draw attention to physiological details and the *Bentley* notes.

In order to maintain anonymity, Pope avoided publishing this poem through his normal bookseller and printer. According to Curll, he received sixty guineas for the copyright from J. Brindley, O. Payne, T. Boreman, and C. Corbet; Boreman's is the name on the title page (Griffith 347 and 355). This seems to have been a payment to print and sell the poem for one year, because Pope, astonishingly in view of its content, printed it in his *Works* in 1738, 1740, and 1743. If Gilliver had sold the poem on to Boreman for sixty guineas because Pope had asked him not to publish it himself, he would have received the £50 he needed to pay Pope and gained a little more than the £10 that David Foxon calculated would have been his profit on the first 2,000 copies—a confirmation of the value of the epistles agreement he signed with Pope on 1 December 1732.[22] On the evidence of the ornaments, the printer was John Hughs, who helped with the printing of the *Letters*, published in May 1735. The presence of a new printer may explain the absence of indices, and the regularity of the spacing in both English and Latin. But the closeness of the English to the Latin in number of lines (178 to 134) means that the passages can be kept roughly parallel without any major adjustment of space.

[22] *Pope and the Book Trade*, 117–20. Warburton did not print *Sober Advice*, nor did Courthope.

The title page sets up the playfulness that characterizes the whole production: 'Sober Advice from Horace, to the Young Gentlemen about Town. As Deliver'd in His Second Sermon. Imitated in the Manner of Mr. Pope. Together with the Original Text, as restored by the Revd. R. Bentley, Doctor of Divinity. And some Remarks on the Version. London: Printed for T. Boreman, at the Cock on Ludgate-Hill; and sold by the Booksellers of London and Westminster. [Price One Shilling.]'. Pope picks up a quality of the original—its recommending the golden mean—by calling it sober advice, and plays on the dispute about the meaning of 'sermones' by calling it a sermon.[23] The nature of Horace's advice—avoid adultery, but indulge in sexual congress with prostitutes, maids, or boys instead—is sober in the sense that it is prudential, but its morality is not that of a Christian sermon, nor is *Bentley*'s commentary that of a Doctor of Divinity. Even Boreman's address is apposite; I wonder whether Pope asked for it specially.

Pope dedicated the poem to himself, equivocating genteelly: 'this Imitation . . . whose Birth I may truly say is owing to you'; 'tho' perhaps, I may not profess myself your Admirer so much as some others, I cannot but be, with as much inward Respect, Good-will, and Zeal as any Man, *Dear Sir, Your most Affectionate AND Faithful Servant.*' The limited admiration is an even better clue to Pope's involvement than the paternity claim. The dropped head, as well as the title page, emphasizes poor Bentley's role, eliding genuine editing and spoof commentary: 'Textum *Recensuit* V. R. RICARDUS BENTLEIUS, S.T.P.' The text is indeed Bentley's, but the presence of 'j' in the first word 'Ambubajarum' makes me wonder if there is not the residual influence of another text, perhaps a copy of Heinsius used for Pope's first efforts at imitation.

The first *Bentley* note refers to the title and to the first highlighted word in the Latin text, 'Benignus', one of those Empson-like ambiguities:

[NOTÆ BENTLEIANÆ.] *Imitated*. Why Imitated? Why not translated? *Odi Imitatores!* A Metaphrast had not turned *Tigellius*, and *Fufidius*, *Malchinus* and *Gargonius* (for I say *Malchinus*, not *Malthinus*, and *Gargonius*, not *Gorgonius*) into so many LADIES. *Benignus*, *hic*, *hunc*, &c. all of the Masculine Gender: Every School-boy knows more than our Imitator.[24]

The note picks up a vital issue: Horace's male targets become female. The change is reflected in generalization as well as particular portraits: 'Dum vitant stulti vitia, in contraria currunt' [in avoiding a vice, fools run into its opposite] (24) becomes 'Women and Fools are always in Extreme' (28). Pope changes the opening satire on masculine extravagance and miserliness into an attack on female predation. The point is no longer that all human beings run into extremes and that consequently men ruin themselves by pursuing married women or prostitutes. Instead we are

[23] For the problem of names, see Rudd's discussion in *The Satires of Horace*, 124–31.
[24] A metaphrast is a literal translator. Pope was fully aware of these distinctions (the *Second Satire* was referred to as paraphrased or 'praprhased' as John Wright has it in a notable misprint), and is drawing attention to his own freedom and invention in this imitation.

shown that women are violently contrary and rapacious. An early development applies what is originally a simile directly to Lady Mary Wortley Montagu.

> *Vix credere possis*
> *Quam sibi non sit amicus: ita ut* Pater *ille, Terenti*
> *Fabula quem miserum* gnato *vixisse* fugato
> *Inducit, non se pejus cruciaverit atque hic.* (19–22)

[You would hardly believe how poor a friend he [the usurer] is to himself, so that the father whom Terence's play pictures as having lived in misery after banishing his son, never tortured himself worse than he.]

Lady Mary replaces the usurer Fufidius:

> With all a Woman's Virtues but the P—x,
> *Fufidia* thrives in Money, Land, and Stocks:
> For Int'rest, ten *per Cent.* her constant Rate is;
> Her Body? hopeful Heirs may have it *gratis.*
> She turns her very Sister to a Job,
> And, in the Happy Minute, picks your Fob:
> Yet starves herself, so little her own Friend,
> And thirsts and hungers only at one End:
> A Self-Tormentor, worse than (in the Play)
> The Wretch, whose Av'rice drove his *Son* away. (17–26)

As with 'benignus' the highlighting draws attention to the change in gender. The reference to the pox makes a link with the controversial lines in *Satire 2. 1*, but the new name separates this attack from the series on Sapho. The attack is outrageous in its adaptation of scripture to convey the strength of Fufidia's sexual appetite ('Blessed are they which do hunger and thirst after righteousness: for they shall be filled', Matt. 5: 6) and in its exploitation of the misfortunes of Lady Mary's family (the madness of her sister and the running away of her son). The moral point is that avarice corrupts all family relations and is only counteracted by sexual appetite (*Son* is italicized to make sure the parallel is not missed). *To Arbuthnot* will bring out the implicit contrast with Pope's own life.

The portrait of Lady Mary as Fufidia launches Pope on a romp through sexual corruption in Georgian society, with an emphasis on those in positions of social authority. The interest is largely in sensational and deviant behaviour, with the notes compensating for reticence in the poetry. Horace's man of fashion who wears his garments pulled up to the groin ('Inguen *ad* obscaenum') is translated into bashful *Jenny*, who at morning prayer 'Spreads her Fore-Buttocks to the Navel bare' (34). A few lines later the 'notus homo' leaving the brothel becomes a '*noted Dean*' (40) fornicating in the park. Both these examples are neat thrusts at the established Church and at a disturbed relation between public and private. Pope then moves on to what was to become a notorious aristocratic divorce case. 'CUNNI CUPIEN-NIUS ALBI', capitalized in case we should miss it (36), is translated as '*Hi-sb-n*'s *hoary Shrine*'. *Bentley*'s note, making great play of *Thing* as a translation of 'cunnus',

198   To Arbuthnot *and* Sober Advice

points out that 'albi' applies to the garment; this is the sort of mistranslation Pope's readers are usually left to work out for themselves. The effect of the note is to point up the insult to the Viscountess Hillsborough. As the Viscount was to start proceedings for divorce only two months later, Pope could hardly be commenting on a more immediate aristocratic scandal. He follows it up with a rare example of male deviance; the detail has an obscure contemporary application, but its class comment is general.

> *Hunc perminxerunt calones; quin etiam illud*
> *Accidit, ut*cuidam* TESTIS, CAUDAMQUE SALACEM
> *Demeterent ferro.* (44–6)

[another [has] been abused by stable-boys; nay, once it so befell that a man mowed down with the sword the testicles and lustful member.]

The poet's translation is in some ways even more delicate than Fairclough's:

> K— of his Footman's borrow'd Livery stript,
> By worthier Footmen pist upon and whipt!
> Plunder'd by Thieves, or Lawyers which is worse,
> One bleeds in Person, and one bleeds in Purse . . . (55–8)

Pope develops social aspects of these punishments that are merely shadowed by Horace. A 'calo' is a soldier's servant or drudge. By changing this to 'footman' Pope makes the humiliation more domestic, and the attempt at disguise degrades the adulterer below the level of a servant. *Bentley* claims the translation of '*Perminxerunt*' fails to enter into the 'deep Meaning of the Author'; presumably Pope has sodomy in mind. What seems frank because of its vocabulary ('pist upon') is shown to be euphemistic. The implication is not so much that Horace's world is coarser than this one, as that the contemporary social world is coarser than the poem. In two later notes, *Bentley* first objects to 'Thing' as a translation for 'CUNNUM', having used it earlier himself, and then asks for a literal translation of 'dum futuo'. In the second case Pope has prepared the poem for him:

> Oh Love! be deep Tranquility my Luck! (175)

Bentley can achieve his favoured reading by his favourite change, that of a single letter.

There remained for Pope the opportunity of dealing with Horace's treatment of homosexuality, and possibly of moving from criticism of Lady Mary to criticism of Lord Hervey.[25] The suppressed elements in the contemporary first edition of *To Arbuthnot* might find a comfortable place here within the protection of anonymity. For Dacier at least the most shocking passage was one which occurs in the final paragraph of Pope's poem:[26]

---

[25] I am aware that there are serious problems of vocabulary here, and that 'homosexuality' is largely a 20th-century concept. But there are no good choices, and I have tried to be historically aware, while being fairly relaxed about vocabulary.

[26] Dacier is disturbed by the way in which the condemnation of adultery is linked to approval of activity condemned in Jewish and Christian teaching, *Œuvres* (1709), 6. 93–6, 157–8.

> *num esuriens fastidis omnia praeter*
> **Pavonem, rhombumque? tument tibi cum inguina, num, si*
> *Ancilla aut verna est praesto puer, impetus in quem*
> *Continuo fiat, malis tentigine rumpi?*
> *Non ego: namque* parabilem amo venerem, facilemque. (115–19)

[when hungry, do you disdain everything save peacock and turbot? When your passions prove unruly, †would you rather be torn with desire? I should not, for the pleasures I love are those easy to attain.]

Fairclough still leaves out the section (†) which offended Dacier over 200 years before: 'if a handmaid or a boy-slave is ready, and you can make an immediate attack.' Pope is at his craftiest here. He omits reference to the boy, leaving the asterisk as a route whereby *Bentley* can reintroduce him in a footnote, and he creates an independent play on 'praesto' [at hand] to supplement Horace's routes to sexual pleasure.

> When sharp with Hunger, scorn you to be fed,
> Except on *Pea-Chicks*, at the *Bedford-head?*
> Or, when a tight, neat Girl, will serve the Turn,
> In errant Pride continue stiff, and burn?
> I'm a plain Man, whose Maxim is profest,
> 'The Thing at hand is of all Things the *best*. (149–54)

In the note to '*Pavonem' *Bentley* rakes through the poem for homosexual references, without discussing the reference to the boy Pope has omitted:

PAVONEM, Pea-Chicks] *Not ill-render'd, meaning a* young *or* soft Piece, *Anglice* a Tid-bit: *such as that Delicate Youth* Cerinthus, *whose Flesh, our* Horace *expressly says, was as* tender as a Lady's, *and our Imitator turn'd*

> Such Nicety, as Lady or Lord F—

not amiss truly; it agrees with My own Reading of tuo femore, *instead of* tuum femur, *and savours of the true Taste of Antiquity.*

The note supplies the poem's omissions. The initial play is on the missing 'cock' in 'peacock', and the note's admiration for boys stands substitute for the 'puer' of the passage. The note then moves into parody of Bentley's long note on Cerinthus, in which he famously resolved a crux, while insisting on the homoerotic nature of the reference.[27] Pope had passed over the reference to Cerinthus in his lines 106 to 107, but he retrieves it now. The earlier reference to Lady or Lord Fanny, picked up by *Bentley* in this note, comes in a section of the poem concerned with class prejudice in sexual selection, as a question whether 'that honest Part that rules us all' asks for 'Such Nicety, as Lady or Lord Fanny' (92). The note, I suspect, is designed to make it clear that this is a reference to Lord Hervey on his own; he is being depicted as the master/miss of *To Arbuthnot*, though very briefly.

---

[27] For a discussion of the crux and Bentley's solution, see Eduard Fraenkel, *Horace* (Oxford: Clarendon Press, 1957), 84–6.

Pope's handling of Hervey in the poem is much lighter than one might have anticipated from the attack on Lady Mary, and surprisingly the 'praesto puer' passage raises the question of Pope's self-portayal in this indecent poem. At this point he chooses to speak of himself. The maxim he puts forward as a plain man, 'The Thing at hand is of all Things the *best*', is ambiguous, but the strongest suggestion is not that he takes advantage of any maid that is around but that his chief pleasure is private and masturbatory. Earlier in the poem he introduces an even more important line that may be bluff or double bluff. The Bishop of London, seeing the noted Dean 'busy'd in the Park', cries,

> "Tis *Fornicatio simplex*, and no other:
> 'Better than lust for Boys, with *Pope* and *Turk* . . . (42–3)

Fuchs notes (155 n. 17) that these words are echoed in *Epistle 2. 1* of 1737:

> Verse chears their leisure, Verse assists their work,
> Verse prays for Peace, or sings down Pope and Turk. (235–6)

And the *Twickenham* edition points out, in relation to this line, but not to that in *Sober Advice*, that Pope comically identified himself as the Pope prayed against with the Turks in the Hopkins and Sternhold psalms in a letter to Swift in 1725 (*Correspondence*, 2. 334). I have no doubt a potential self-reference is present in *Epistle 2. 1* and in *Sober Advice*; the problem is in deciding what weight to attach to it.

Horace was believed to have mocked his friends in *Satire 1. 2*, and it may be that Pope was playing with the idea that the *Sober Advice* imitator was mocking Pope, just as he mocked Bolingbroke in lines 63 to 66. But this takes no account of the nature of the mockery, or of Pope's willingness to print the poem unrevised in his works. I think an undoubted motive was to create some controversy around the homoerotic material in this poem that would then justify the full-scale portrait of Sporus in the *Works*, where Pope would be acknowledged as author. *Satire 1. 2* was chosen for imitation because it afforded the opportunity to insult Lady Mary and to raise a pother about homosexuality. Any accusation against Pope himself could then be turned against Lord Hervey. However, that intention does not rule out the possibility that the accusation that Pope lusted after boys was a double bluff, a saying something of himself that could not be incorporated in *To Arbuthnot* and a confession of truth in line with the promise of *Satire 2. 1*:

> In this impartial Glass, my Muse intends
> Fair to expose myself, my Foes, my Friends . . . (57–8)

This interpretation of the line in *Sober Advice*, which need not, of course, indicate an exclusive sexual preference, is made more plausible by Pope's next two Horatian imitations, which both have homoerotic concerns or shadows.

The first Horatian imitation after *Sober Advice* is *Horace his Ode to Venus* (*Ode 1. 4*), published 9 March 1737, an address to the goddess of love:

> Again? new Tumults in my Breast?
> Ah spare me, Venus! let me, let me rest!

Horace's poem is about being disturbed late in his life with a new love, for the boy Ligurinus; it also compliments Horace's friend the lawyer Paulus Maximus. Pope's friend, the lawyer William Murray, takes the place of Paulus Maximus in the poem, and there can be little doubt that it was intended chiefly as a compliment to him. The homoeroticism of the original poem rests unused, and Mack emphasizes Pope's heterosexuality, making a good deal of the close of the poem, which in a newspaper version was addressed to 'Patty', that is, Martha Blount: 'the whole quatrain, beautifully catching the crosscurrents of sheepish tenderness and confusion that every man who has ever loved a woman knows, assumes an endearing intimacy' (*Life*, 674). But if Pope's aim was to write a poem to a woman he had known all his life, Horace's ode to a boy he had just met was a strange one to imitate. Nor was Pope's tone towards Martha Blount in his later years notably erotic. The compliment to her at the end of *To a Lady* as a 'softer Man' (272) is in keeping with his rejection of John Caryll's suggestion that he should marry her ('I have no tie to your God-daughter but a good opinion, which has grown into a friendship with experience that she deserved it').[28] The decision to imitate this poem remains perplexing, particularly in the light of *Sober Advice* and *To Arbuthnot*; the answer to the puzzle may lie in Pope's relationship with Murray, about which we know so little.

Pope's next imitation, *Epistle 2. 2*, published on 28 April 1737 also begins oddly, especially if the 'si Ancilla aut verna est praesto puer' passage is remembered:

> Dear Col'nel! *Cobham*'s and your Country's Friend!
> You love a Verse, take such as I can send.
> A Frenchman comes, presents you with his Boy,
> Bows and begins.—'This Lad, Sir, is of Blois:
> 'Observe his Shape how clean! his Locks how curl'd!
> 'My only Son, I'd have him see the World:
> 'His French is pure; his Voice too—you shall hear—
> 'Sir, he's your Slave, for twenty pound a year. (1–8)

The point of the story is that if the father says the boy is a thief you must not be surprised if he steals, but the set-up seems extravagant. John Butt in the Twickenham edition, following Warburton, writes a long note on Blois as the place for the purest French, but that seems only one of the boy's attractions in the eyes of his father, and not the most important. Blois does, however, seem to afford the most promising clue to the meaning of the passage. The city was the centre for the French court in the Renaissance, and it was central to the activities of Henri III (1551–89) and his

---

[28] *Correspondence*, 3. 75. He adds, 'you will never see me change my condition any more than my religion, because I think them both best for me'. The poem was published in the *Whitehall Evening Post*, 26 Feb. to 1 Mar. 1737. When John Wright entered the poem in the Stationers' Register, he said he had an assignment from Pope dated 25 Feb. 1737. It is quite likely that Pope was responding to an unauthorized publication.

'mignons', his favourites and, in some interpretations, catamites. The reputation of that court for homosexuality was advanced by the *Description de l'isle des hermaphrodites*, first published around 1605, but republished in Brussels in 1724, with an appendix identifying the isle with the court of Henri.[29] There is no evidence that Pope had read this book or knew about it, but the isle does have eight statues: four of Roman emperors (including Nero) and four of 'ingenieurs du plaisir' (including Sporus). Information about the addressee of this epistle would be helpful, but Butt rejects the traditional identification of 'Colonel Cotterell of Rousham', substituting Anthony Browne of Abscourt, without showing he was a colonel or Cobham's friend. I suspect that Mack is right in suggesting the addressee is General James Dormer, also a colonel, in which case an interest in the male form in Kent's replanning of Dormer's house and garden at Rousham (a striking reproduction of Parmigianino's cupid in the painted parlour, the bronzes, and the garden statuary) might be significant, but there is nothing definite to seize on—which is exactly how Pope would have wanted it to be.[30]

If, as I suspect, one of Pope's aims in *Sober Advice* was to generate electricity for a controversy that would make it easier to attack Hervey in *To Arbuthnot*, he was unsuccessful. With the single exception of *The Poet Finish'd in Prose*, Pope failed to provoke the controversy he had hoped for. Thomas Bentley attacked him (4 March 1735) before publication of the *Works*, but his perceptive criticism of Pope's ridicule of his uncle avoided personal reflections of a helpful sort. *The Poet Finish'd in Prose* did not appear until 26 June 1735, after the publication of the *Works*, though its lengthy discussion of *To Arbuthnot* is confined to the first edition. *Sober Advice* is not mentioned in the dialogue that constitutes the pamphlet's informed and perceptive critique, but it seems likely to have informed its presentation of Pope's sexuality. *The Poet Finish'd in Prose* attributes Pope's nervousness at the start of *To Arbuthnot* to his fear that Sappho would ravish him and to his general aversion to her sex:

for nothing can be imagin'd more terrible than a Rape, to a Gentleman who has not the least Passion for the Sex. whether this want of Inclination proceeds from Constitution or Habit, it is not worth our while to examine at present. Be that as it will, I say, every Man has a Right to pursue his Pleasure by such Ways as his Taste incline him too; and no Body disputes the Elegance of Mr. *Pope*'s. Why, you love a pretty Girl now, and so do I. Very well, Sir, let us follow our Inclinations, I dare say Mr. *Pope* will not interfere with us. Perhaps *He* does not; what is that to us? It would be very hard if we should deny him the same Privileges we pretend to ourselves, and which we may enjoy without the least Interruption from him. 'Tis possible his Fondness for Retirement may have given him this Disrelish for the Sex; but no doubt he

---

[29] The work is attributed to Thomas Artus, sieur d'Embry, but, as with *Le Comte de Gabalis*, interpretation is perplexing. There is a modern edition by Claude-Gilbert Dubois (Geneva: Droz SA, 1996), but the drift of recent French scholarship is not towards simplification. Pierre Bayle identified the work as a satire on the court of Henri III, and his influence was decisive until the reinterpretations of this century.

[30] Mack, *Life*, 680–2. For interesting speculations on Kent's sexuality and his relationship with Burlington (the latter rather rash, in my view), see John Harris, *The Palladians* (London: Trefoil Books, 1981), 18. See also David Nokes's thoughtful consideration of the question of whether Gay was 'gay' (*Gay: A Profession of Friendship*, 43–50).

has found out some other Amusement, equally entertaining to him in his Solitude, and which makes him less sollicitous about losing the Favour of the Ladies. (17–18)[31]

Guerinot remarks that these charges are unparalleled in the attacks on Pope; but I believe they are an entirely appropriate response to *Sober Advice*, raising the issues of sexual orientation and solitude. The Horatian imitations following *Sober Advice* provoked no sexual controversy at all. In this respect the newspaper printing of the version of *Ode 4. 1* with its reference to Patty may have been fortunate, and the beginning of *Epistle 2. 2* clearly baffled Curll, whose *Mr Pope's Literary Correspondence. Volume the Fifth* (1737) contains what it calls a 'Parodie' of the poem, looking for something to attack but saying simply that the Frenchman and his boy are 'ragamuffins' who 'ill befit a Lord'. Any modern reading of Pope's representation of his sexuality must take into account the fact that most hostile contemporaries were indifferent to it.[32] It must also, of course, take into account Pope's attack on Hervey, which has been seen as expressing a revulsion at homosexuality in general. Fuchs rejects the possibility that Pope cast himself as a lover of boys in *Sober Advice* on the grounds that *To Arbuthnot* savagely attacks Lord Hervey for his homosexuality (88), and the same perception is shared by David Nokes:

Pope shared the conventional horror at this vice [homosexual liaisons]. His uncorroborated assertion 'Addison and Steele [were] a couple of h—s' (hermaphrodites) was clearly intended to wound their reputations, while his savage attack on that 'amphibious thing' the 'hermaphroditical' Lord Hervey is full of homophobic rage.[33]

But an examination of the attack on Hervey suggests this interpretation may be based on a misunderstanding.

The earliest sign of a full-scale attack on Hervey is provided by a manuscript draft of the lines that appeared in the *London Evening Post*, 22–5 January 1732, and now form the introduction to the Sporus portrait (283–304). They are partly written in next to the Atticus portrait (428, 22–42), and some of them appear again in the 'But why write Satyr?' page (442. 52–69). I believe Hervey is aimed at from the start, even though the lines offer no direct identification. 'A Master Key to Popery' identifies him as a false friend and a malicious interpreter of *To Burlington*, 'I suspect it was to screen this Author that his gentle Friend Lord Fanny apply'd to this Nobleman [Bathurst] the Character of Villario' (*Prose Works*, 2. 414), and Courthope points to lines in the *State Dunces* in which Hervey is accused of slandering virgin innocence (Elwin–Courthope, 3. 266). Hervey, then, is the fop in the manuscript who invades

---

[31] The pamphlet is well summarized by Guerinot and much of this passage quoted (*Pamphlet Attacks on Alexander Pope*, 254–8).

[32] It is worth noting that Pope's first published love poem was addressed to a boy ('Sapho to Phaon') and that his most famous love poem was addressed to a castrated man (*Eloisa to Abelard*, though Abelard is hardly to be compared with Sporus). Nothing required Pope to assume a female persona in his love poems. In these poems, as in *Rape of the Lock*, Pope seems to have chosen situations in which sex was in the head.

[33] *Gay: A Profession of Friendship*, 45. I think a more serious objection than any of these comes from the treatment of the episode relating to Duckett and Burnet in the *Dunciad*, 3. 179–84, where the note distinguishes between friendship and 'a vice not fit to be nam'd', though the object of attack is John Dennis's crazy defence of Duckett. See *The Dunciad in Four Books*, ed. Valerie Rumbold, 241–3.

the public peace, steals a tear from the virgin, can swear to the identity of Timon's Villa, spreads scandal, and insults fallen worth and beauty in distress. In keeping with *To Arbuthnot*'s emphasis on friendship, the printed version of the poem also has him as the false friend of 'A Master Key to Popery':

> Who has the Vanity to call you Friend,
> Yet wants the Honour injur'd to defend;
> Who tells whate'er you think, whate'er you say,
> And, if he lyes not, must at least betray:
> Who to the *Dean* and *silver Bell* . . . (First edition, 284–8)

If this interpretation is right, and the 'he' of the passage leading up to Sporus is Hervey, it casts an interesting light on Pope's relations with him. The implication is that betrayal is almost as serious as lying. Did Pope actually talk to Hervey about Chandos? It has occurred to me in moments of speculation that the reason Pope gave for his breach with Lady Mary and Lord Hervey, that they 'had pressed him once together . . . to write a satire on some certain persons' (Spence, *Anecdotes*, No. 751), might refer to a discussion about satire of Chandos. If so, it would be in the interest of neither party to admit it, but it would go some way to explain the strength of Pope's resentment against them.

The manuscript lines on Sporus contain nearly all the elements of the full printed sketch, some in exceptionally clear form. Pope appropriates the material of the *Verses* attack on himself and redirects it against Hervey. Whereas in the *Verses* Pope is the impotent outsider, Hervey in *To Arbuthnot* is the treacherous insider. The intimacy of his whispering breath haunts the first draft of the poem. Whereas Pope's bodily form expresses his soul, Hervey's deceives. If Pope is a wasp that stings and dies, Hervey is a 'Bug with gilded wings' a 'painted Child of Dirt that stinks and stings'. Pope even picks up the rhymes of the *Verses* and in reusing them develops the contrast between the straightforward (however loathed) and the deceitful. In the *Verses*:

> 'Tis the gross *Lust* of Hate, that still annoys,
> Without Distinction, as gross Love enjoys . . . (30–1)

In *To Arbuthnot* the failure in penetration, already present in Atticus, is given a new emphasis:

> Whose Buzz the Witty and the Fair annoys,
> Yet Wit ne'er tastes, and Beauty ne'er enjoys . . . (First edition, 300–1)

If in the *Verses* Pope is a porcupine unable to wound with his darts, which fall short of the observers, Hervey in *To Arbuthnot* is an uncertain animal, an amphibious thing, in intimate but deceitful contact, one of those whose slaver kills but not the bite: a spaniel mumbling game, a toad sitting at the Queen's ear, spitting itself abroad 'Half Froth, half Venom'. In his corrupt and perplexing intimacy, Hervey is a type of Satan, 'A Cherub's face, a Reptile all the rest'.

To this repulsive mixture, Pope wished to add his attack on Hervey's sexuality, and to do it in his own name rather than through the anonymity of *Sober Advice*. This strand of attack is present in the manuscript. The name Sporus is used from the outset; the main source is probably Suetonius, *Nero* 28, where the story is told of how Nero castrated the boy Sporus, tried to change him into a woman, and married him. The manuscript Sporus sputters 'unmanly Lyes' (444. 13); is already, in a triplet, 'one poor <u>Antithesis</u>' (444. 32–4); trips a Lady and struts a Lord (444. 38); and is 'This Thing amphibious' (446. 1). It is important to note that it is Hervey's bisexuality and effeminacy that are under attack ('now Master up, now Miss') and that these qualities are allied to political pliability. Sporus' sexuality is of a piece with his political flattery, lies, and insinuations; the portrait may not cast George II as Nero, but the court takes some responsibility for the character of its favourite. Consequently Pope attacks Hervey not from a normative heterosexuality (of which nothing is to be found in *Sober Advice* or, with the exception, I suppose, of Pope's parents, in *To Arbuthnot*) but from the position of independent masculinity. To associate the character of Sporus with homophobic rage, or to see it as the portrayal of an effeminate homosexual, as Rictor Norton does, is to mistake the categories Pope was using.[34] The attack on Sporus is compatible with the distaste of one homosexuality for another, of pederasty for sexual ambiguity in which passivity and effeminacy predominate. Randolph Trumbach attempts a summary of what he sees as 'a profound shift' in 'the conceptualization and practice of male homosexual behavior' at the time Pope was writing:

> It left behind the pattern of homosexual behaviour that had been produced by the emergence of a traditional Western European culture in the twelfth century. In that older pattern, the debauchee or libertine who denied the regulation of sexuality to marriage had been able to find, especially in cities, women and boys with whom he might indifferently, if sometimes dangerously, enact his desires. In the modern pattern, most men conceived first of all that they were male, because they felt attraction to women, and to women alone.[35]

A man who loved both women and boys, or even boys alone, was manly. A man who allowed himself to be loved by men was not. It is not necessary to endorse all aspects of this narrative in order to see the Sporus portrait as drawing strength from such a historical shift. Pope's attack is compatible with the later, and stricter, ordering of identities, and is usually read through it, but it belongs equally comfortably to the older tradition, in which, as we can see from *Sober Advice*, Horace himself was very much at ease. These two possibilities make it much easier for Pope to take his own place in the poem as a contrast to Hervey.

---

[34] *Mother Clap's Molly House: The Gay Subculture in England, 1700–1830* (London: GMP, 1992), 147. Norton's claim that Pope's lines 'had more influence in creating the stereotype of the Effeminate Pouf than any other document in English history' may contain some truth, but I do not think that stereotype encompasses bisexuality as Pope's portrait does.

[35] Trumbach, 'Sodomitical Subcultures, Sodomitical Roles, and the Gender Revolution of the Eighteenth Century: The Recent Historiography', in Robert Purks Maccubbin (ed.), *'Tis Nature's Fault: Unauthorized Sexuality during the Enlightenment* (Cambridge: Cambridge University Press, 1987), 118. The argument is carried forward in Trumbach's *Sex and the Gender Revolution: Heterosexuality and the Third Gender in Enlightenment London* (Chicago: University of Chicago Press, 1998), esp. 3–10. For a view of the Sporus portrait as a Juvenalian rejection of Horatian effeminacy, see Weinbrot, *Alexander Pope and the Traditions of Formal Verse Satire*, 145–6.

The attack on Hervey's sexuality in the *Works* and in subsequent versions of *To Arbuthnot* shifts the centre of gravity of the whole poem by giving a special value to masculinity as a source of personal integrity. It fits in with the concern in the manuscript draft for family, the relation of father to son, and the values Pope espouses as his father's son—the very values imperilled by the public world of writing. The opening lines of the manuscript poem could also have been reworked in the light of the new emphasis

> The Man of Friendship, but no boasting Friend,
> The Man of Courage, but not prone to fight . . . (424. 7–8)

Hervey had, of course, fought a duel with Pulteney, who in *A Proper Reply to a Late Scurrilous Libel* had charged Hervey with effeminacy in terms very similar to Pope's. The man prone to fight was not necessarily a man of courage; an eagerness to quarrel could have been presented as another failure in masculinity. But, in the absence of a response to *Sober Advice* or the first published version of *To Arbuthnot*, Pope was unable to explore such issues in the lines following the portrait of Sporus. When he turns to the account of himself that is necessary as a counterpoise to the attack on Sporus, manliness becomes exclusively metaphorical—the political freedom of Pope's own life in contrast to Hervey's unstable and treacherous commitments.

> Oh keep me what I am! not Fortune's fool,
> Nor Lucre's madman, nor Ambition's tool:
> Not proud, nor servile, be one Poet's praise,
> That, if he pleas'd, he pleas'd by manly ways . . . (*Works*, 326–9)

Pope turns not to his own family life but to the theme of his political independence, his resistance to abuse, and his adherence to the cause of Virtue. His remoteness from the flurries of the public sphere ceases to be a retreat into domestic values and becomes abstracted into a defence of virtue:

> That not in Fancy's Maze he wander'd long,
> But stoop'd to Truth, and moraliz'd his song.
> That not for Fame, but Virtue's better end,
> He stood the furious Foe, the timid Friend . . . (*Works*, 332–5)

Thomas Bentley's hostile remarks on the parallel sentiments in *Satire 2. 1*, 105–22, are equally pertinent here:

Pray, let us know then what you mean by *Virtue*. Is it *Graian* or *Roman*? Or do you mean *Evangelical Graces*? Is it *Charity*, that *suffereth long, and is kind*, that *vaunteth not itself, nor is puffed up*? Is it *Humility, Love of Enemies, &c*? He has nothing of them: You have but little your self. I have sometimes thought, that you put *Virtue* for *Self* and that *Virtue only* is *Self only*; and that *Uni aequus virtuti atque ejus amicis*, means only, *Uni mihi aequus, & mihi amicis*;

> To my self only and my Friends a Friend.
> (Guerinot, *Pamphlet Attacks on Alexander Pope*, 253)

There is an emptiness to this section of the poem that springs from the inability to give a sufficiently powerful account of the self. Pope had seized the opportunity to speak of himself in response to *Verses to the Imitator of Horace* and he saw the possibilities of creating a contrast between his own life and Hervey's, but this section of the poem, providing the direct contrast between the two men, fails in concreteness and detail.

Where *To Arbuthnot* surprisingly regains its strength is in presenting those womanly virtues that Pope acknowledges in himself. He presents himself as 'soft by nature, more a Dupe than Wit' (*Works*, 360) and goes on to picture himself in a maternal role:

> Me let the tender Office long ingage
> To rock the Cradle of reposing Age . . . (450. 5–7)

Pope allows himself these feminine qualities, becoming his mother's mother, because they are asexual and can be bound in to the poem's concern with health. In this respect, the discovery of Arbuthnot as the poem's addressee was invaluable. Arbuthnot was a family man whose occupation involved caring for the sick and weak. Consequently Pope is able to begin his concluding paragraph with a wish appropriate to his central themes:

> O Friend! may each domestick Bliss be thine! (*Works*, 398)

Arbuthnot's role is important because Pope could not have filled the space between the character of Sporus and the poem's ending with an evocation of his own domestic bliss. His clumsy revision of a line early in the poem—*The Poet Finsh'd in Prose* draws attention to it—is relevant. The first draft read:

> The Muse but eas'd some Friend, or Nurse, or Wife . . . (424. 23).

But that was clearly unsatisfactory, so he wrote over the top of 'or Wife', '(not Wife)', and settled for this second thought, even though the expression is clumsy. If Pope had a fully developed view of the contrast between his own masculinity and Hervey's, contemporary conditions did not allow it to be expressed clearly in the poem.

Like *The Dunciad*, the *Epistle to Dr Arbuthnot* is an attempt to engage with eighteenth-century literary culture, to define it and evaluate it. While holding on to specifics of social groupings and personalities, it ranges widely. It is profoundly suspicious of patronage in all its forms, depicting it as an evil for patron and patronized, with pride and servility engaged in mutual support, but it finds no satisfactory alternative in the newer forms of literary power represented by Addison and Pope himself. The power of the gentleman critic may be used unjustly, and public opinion is corrupted quite as easily as court opinion. Wherever literary power is exercised—court, great man's table, bookseller's, or theatre—it may be corrupted by human nature. In the final revision of the poem Pope creates in Sporus the emblem of these corrupted spheres: aristocratic, devious, politically servile, poetically incompetent, sexually ambiguous. Pope's desire to oppose his own life to Sporus' seems not fully realized in this poem, and his appeal to private values remains in some ways paradoxical. The question 'Why then publish?' is always ready to reassert itself. No

'Granville the polite' urged Pope to publish in this instance, still less to publish different versions of his poem to ensure a favourable response. Publication is necessarily a tacit appeal to public opinion: to the significant bourgeois group enabled by the market to buy, read, discuss, and evaluate poetry. But Pope assumed an early role in the degeneration of the public sphere by attempting to manipulate public opinion. If my account of the publication of *Verses Address'd to the Imitator of Horace* and *To Arbuthnot* is right, Pope takes his place as a pioneering exponent of literary propaganda.

8. *The* Works *of 1735–1736: Pope's Notes*

Pope always liked notes, and he used them even in the early manuscripts that have come down to us, but in 1735 and 1736 they started to assume a much greater importance. They became an expression of his direct engagement for the first time with the general public, in an edition that was cheap, elegant, and readily available. They conveyed Pope's evaluation of his poetry, related it to his own life and the work of his predecessors, and invited the public to read him as a historical figure in a historical context. Pope became not just a classic but a popular classic, and he laid down lines of interpretation that have persisted to the present day.

Pope's early notes were part of that formal, book-like quality he aimed at, even in his manuscripts; making a poem usually implied making a book. The manuscript of the *Pastorals*, which Pope dates to 1704, when he was 16, has a note on the first page, with a cue in the text '(a)' and a note '(a) Fontanelle's *Discourse of Pastorals*' at the foot of the page. There are fifteen notes in all in the eight pages of the 'Essay on Pastoral', displaying impressive learning in one so young, but none on the poems themselves. But that did not mean that poems were not to be annotated. The manuscript of *An Essay on Criticism* (1709) has fourteen notes, containing nineteen references, and the *Windsor-Forest* manuscript (1712) has seven, showing that not only critical works merited annotation. The history of his notes proves Pope one of the great literary annotators, with command of a wide range of styles, typographic and literary. Varieties include: notes at the back of the volume, notes at the foot of the page (with or without separating horizontal rules), headnotes called 'Advertisement' or 'Contents', notes separated into columns by vertical rules, notes cued by * and by †, notes led by line number (with or without lemmata), notes in italic and notes in roman, notes with headings (Variations, Remarks, Imitations), notes by fictitious or semi-fictitious commentators, notes by a character in the poem, notes from reference books, references, quotations (and deliberate misquotations), parallel texts, glosses, biographies, botanical and zoological information, compositional notes, textual variants (manuscript or printed), textual corrections, self-justifications, sarcasms, explications, evaluations, and commentary by William Warburton. Except in the *Dunciad Variorum* and its developments, the notes are very disciplined; among the many styles of notes used by Pope, we find no digressions.[1]

---

[1] Gérard Genette in *Paratexts: Thresholds of Interpretation* (Cambridge: Cambridge University Press, 1987), 319–43, has a stimulating discussion. I have drawn on some of his distinctions but not on others. A collection of conference papers on this topic is Stephen A. Barney (ed.), *Annotation and its Texts* (New York: Oxford University Press, 1991). There are several learned and valuable papers in the collection (though some herniation from deconstruction), but by far the best thing in it is the list of questions Barney sent to his contributors before the conference.

Chapter 4 discusses Pope's annotation, part in earnest and part in game, of the *Dunciad Variorum*, modelled on Claude Brossette's edition of Boileau published in Geneva in 1716.[2] This edition was perhaps Pope's first response to Jonathan Richardson, jun.'s proposal for 'making an edition of his works in the manner of Boileau's'.[3] The fully serious response began with the publication of *Works* II in quarto and folio on 23 April 1735, and then expanded during 1735 and 1736 into a four-volume edition of all Pope's works in the much cheaper octavo format, Pope's first edition for the general public. The contents of a table in volume I can be summarized to show the organization of the edition:

I. Consisting of the Author's Original Poems, written under 25 years of age.
II. *An Essay on Man*, being the first Book of Ethic Epistles, to Henry St. John Lord Bolingbroke; Ethic Epistles, the second Book, to several Persons; Epistles, the third Book; Satires of Horace; Satires of Donne; Epitaphs.
III. Temple of Fame, Translations.
IV. The Dunciad, an Heroic Poem, in three books, to Dr Jonathan Swift.

Volume IV in this collection, The *Dunciad*, was extensively, if parodically, annotated already, with its division of the notes into '*REMARKS*' and '*IMITATIONS*'. Two of the other volumes insist on the importance of their annotation. *Works* I (Griffith 413) advertises its notes on its title page: 'WITH Explanatory NOTES and ADDITIONS never before printed', and *Works* II (Griffith 389) has similar wording and typography in its half-title: 'The LAST EDITION CORRECTED, WITH Explanatory NOTES and ADDITIONS never before printed.' Only volume III made no mention of annotation, though in fact it carries a heavy apparatus of Latin and English texts to be compared with Pope's imitations and translations. With this four-volume edition, Pope discovered a way of constantly revising and reissuing his work—all his *Works* became work in progress—but he also launched the project of a full-scale popular annotated edition of his *Works*. The notes were to be textual and historical; they were to be learned and entertaining; and they were to clarify literary and social scandals. New readers were to be guided through the chronology of Pope's literary career and helped to understand the structure and meaning of the poems. Pope revised his plans in 1743 and 1744 when he prepared a new style of edition with William Warburton's commentary, and on his death he left Warburton to complete the project, which finally resulted in the nine-volume octavo edition of 1751.

BROSSETTE'S EDITION OF BOILEAU AND THE HISTORICIZATION OF POPE

Pope learned from Brossette that the way to annotate for the general public was to think of the author as a historical figure. By annotating the text with future genera-

---

[2] Pope's copy, presented to him by James Craggs, is now at Mapledurham House (Mack, *Collected in Himself*, Library, 26).
[3] Several of Pope's manuscripts were given to Richardson 'for the pains I took in collating the whole with the printed editions . . . on my having proposed to him the making an edition of his works in the manner of Boileau's', *Richardsoniana*, 264.

tions in mind, you ensured that you provided the information necessary for a broad contemporary readership. Pope's notes serve a great variety of purposes, but they do so by the implicit construction of Pope himself as a figure distanced in time. The virtues of such note-writing were many: a body of work was established and protected; status as a classic was assured; and a favourable version of social and political history was disseminated. The Geneva Boileau gave the author the ideal combination of presence and absence. Boileau had provided Brossette with his version of his works (text, information, and interpretation), but he had died in 1711 before the production of the edition, which was therefore justified in providing an extensive apparatus explaining the author and his times. The knack of note-writing, Pope concluded, was to think of himself as dead and to interpret himself to future generations. The manœuvre reveals itself more plainly at some times than others. The first edition of *To Arbuthnot* had contained a note on Pope's parents. In *Works* II of 1735 he added the inscription on their monument, which ends 'Parentibus Benemerentibus Filius Fecit, et Sibi', thereby including himself among the memorialized. The general note on the *Pastorals* in the 1736 octavo *Works* says that 'the Author esteem'd these as the most correct ... of all his works', as though the author were no longer alive to make the judgement. A note on the power of gold in *To Bathurst* in the octavo *Works* of 1735 begins 'In our Author's time ...' And in the final note on *To Burlington* in the same collection, Pope accounts for the present and the preceding four years as though they were a remote historical period. The warmth of Pope's friendship with Warburton late in his career may be seen as relief at finally escaping from this self-division by embracing his own Claude Brossette, someone who could endorse the author's self-understanding by merging his authority with the author's own.

Brossette's edition is commendably clear about its claim to distinction, even in its title page: 'Oeuvres de M^R. Boileau Despréaux. Avec des Éclaircissemens Historiques, Donnez par Lui-même.' The explanatory notes that come from the author himself are crucial, but the editor makes even stronger claims for the value of the enterprise in the 'Advertissement de L'Editeur'. The first claim is for completeness. This is the full works, the complete Boileau; each and every piece is presented in its entirety, down to the slightest fragment; even discarded readings have been included. The completeness of Brossette's volume, however, goes beyond a fullness of text; the annotation is also authoritative and copious. The notes, he insists, are based on Boileau's letters or conversation, and very often the words are Boileau's own. But this does not mean that the annotation is confined to information about Boileau himself. The works touch on so many writers and events that the volumes become a history of Boileau's age as well as his works. Through the passing of time, figures once well known at court or in the town have become forgotten, and Brossette assures the reader that he has taken particular pains to give information about obscure figures. Boileau's work is like a dusty painting that has now been cleaned. In a passage that chimes in with Pope's contempt for verbal criticism, Brossette contrasts his own annotation with that of classical scholars, who, lacking his access to private information, resort to a verbal criticism which is dry, forbidding, and of little use. Brossette acknowledges that some of the attraction of his edition will come from the secret information it divulges and the pleasure of being admitted to an unusual intimacy with the author. The reader will become something like the writer's confidant, let into the secret of his thoughts. Brossette concludes with an explanation of the three headings to his notes and a final disclaimer. Under the heading *Changemens* come the readings of the different editions, a textual history. These variants display Boileau's genius. He was not content with writing well; he wanted it to be impossible to write better. His artistic power is revealed in the different ways in which he expressed himself on the same subject, and the list of *Changemens* provides a kind of genealogy of Boileau's thoughts. The *Remarques* are the explanatory notes that have been the major topic of Brossette's discussion. The *Imitations*, directing the reader to parallel passages, show that Boileau was proud of his ingenious use of the works of his predecessors; he himself pointed out some of the more obscure sources. Finally Brossette insists on the moral boundaries of even his extensive annotation; he had adopted a measure of self-censorship. No offensive truths are included, nothing that serves merely to appeal to malignity. The editor has shouldered the responsibility of deciding what the public should know and what it should not: 'Il est de la prudence d'un Ecrivain qui met au jour des faits cachés & des personalités, de distinguer ce que le Public doit savoir, d'avec ce qu'il est bon qu'il ignore.'

Although Pope does not follow Brossette slavishly, this 'Advertissement' supplies many ideas that are vital to his own practice of annotation in 1735 and 1736. Central is the idea that the author is the best commentator on his own poems and that future readers will need his help. I assume throughout this chapter that Pope is the annotator of his poems and that only his sense of decorum prevents him from saying so. The piracy of *Works* II in 1735 surely has it right: *Ethic Epistles, Satires, &c. With the Author's Notes. Written by Mr Pope* (Griffith 391). Both R. H. Griffith and F. W. Bateson want to cast Jonathan Richardson, jun., in the role of editor of the *Works* of this period, edging his involvement ever forward. Griffith cites Richardson's role in collating the manuscripts against the printed works, noting that there are variant readings in the quarto and folio *Works* of 1735. On this basis he says, 'If Richardson was the editor, he as well as or rather than Pope was responsible for the preparation of the notes' (282). Bateson adds:

Richardson may also have been responsible for the elaborate 'Arguments' and analytical notes in the 1735 editions. The 'Arguments' are competently done, but there are one or two slips and clumsinesses . . . that make it virtually certain they were not compiled by Pope. Richardson, or whoever the collator was, may also have contributed some, though certainly not all, of the explanatory notes. (*Twickenham*, 3 (2). 4–5 n.)

Bateson is always interesting, but this is unbridled speculation. What we know is that Richardson did some collating. When we turn to the manuscripts he used, we find that even that collation was under strict supervision. On the title page of the manuscript of *Windsor-Forest*, Pope writes, 'Mem Transcribe what lines are marked *as before in ye pastorals [with Lead *crossed out*] *And what lines differ in ye printed

Copies from the last Edition [*crossed out*]'.[4] Accordingly Pope gives instructions in the margin of this manuscript and that of *An Essay on Criticism* saying what is to be transcribed and what is not. Richardson may have had a freer hand with the *Dunciad*, but from the results Pope would have made his own choice of variants to be printed. For Richardson's writing explanatory notes there is no evidence whatsoever. The information in the notes must come from Pope, and it is inconceivable that, at a time when he was employing his own printer so that he could patrol every capital and comma of his text, he would let someone else compose the notes for him. Bateson rightly draws attention to the 'slips and clumsinesses' in the contents or précis of the ethic epistles, but I am not persuaded that Pope is incapable of slips and clumsinesses, or that in Bateson's examples the contents does not state the topic, as opposed to the complexity of Pope's treatment of it, adequately. Finally, and I think conclusively, the notes and the contents are revised in subsequent editions in a typically Popeian manner (sometimes with old and new texts of the contents only pages apart); Pope treats them exactly like any other part of his text. All the notes, even the *Changemens* or Variations, have been given by *lui-même*; if Jonathan Richardson had been as important as Griffith and Bateson suggest, he would surely have been constructed as a stooge editor like Cleland in the *Variorum*. I suspect he was allowed to do no more than provide copy for the Variations.

The intervention of the poet as annotator is necessary, in Brossette's thinking and in Pope's, to ensure a completeness that defies historical change and enables future readers to approach the poet with confidence, though, as any modern reader would expect, the search for completion may lead to infinite deferral. The first completeness is that of text. As we shall see, Pope felt this requirement acutely, wanting to claim that he had met it, even when his policy was deliberately selective. His notes on this issue grow more sophisticated over time, as does his deployment of Variations (Brossette's *Changemens*) to show his artistry or to present some lines under erasure, partially endorsed, or scrupulously omitted. The second completeness lies in presentation of the author and in the creation of that intimacy with him that Brossette thought would be a pleasurable consequence of his edition. The publication of Pope's *Letters* to coincide with the second volume of *Works* admitted the reader to an even greater intimacy than an annotated edition could, but Pope is none the less careful to fix each poem in its place in the author's career, to provide hints of the author's design, and, when the poem permits it, to give details of the author's life and parentage. The annotation provides a history of the poet's life and career. For his dealings with other writers, providing parallel texts and Imitations, Pope has received too little credit. Like Boileau he believed he transformed what he borrowed and in the octavo edition he specifically draws attention to allusions to particular passages and presents parallel texts so that readers can make their own judgements of his work. The most difficult issues arise from the notes Brossette calls Remarks, where completeness creates its own moral problems. To the scholarly explanatory

---

[4] The asterisk marks the beginning of the section to which an alternative reading or a comment is attached in brackets.

notes of earlier poems, Pope now adds notes that are specifically concerned with historical circumstances that may be unknown to future readers. To his general historical accounts he adds information about notorious contemporary figures, but he is more cautious about persons in power. As he freely confesses, he leaves some satirical attacks for posthumous publication; there are some things that later editors will have to do. This general movement of Pope's annotation towards historical clarification is clear, but in the early stages it was complicated by questions of format. The problem of what to annotate and what to leave unfixed was made more difficult by Pope's sense that he needed to deal with two publics, whose requirements were subtly different.

LARGE- AND SMALL-FORMAT EDITIONS

The year 1735, which R. H. Griffith's humane and readable bibliography calls Pope's *annus mirabilis* (2. 329), saw an important change in Pope's preferred mode of publication, a switch in emphasis from large expensive formats to the much cheaper octavo one. The *Grub-street Journal* of 24 April 1735 announced the publication of *Works* II in quarto, while signalling the end of an era in Pope publishing.

Whereas it hath been the practice of Booksellers to print Editions only in a large Size, which consequently were only to be had at a high price, no greater Number of this Volume is printed in large Folio and Quarto with *expensive Ornaments*, than to answer the like Impressions of *the first Volume* of his Works, and of the *Iliad* and *Odysseys* (so printed and sold off many Years since) at the same price of *one Guinea*: there is also published with it an Edition in a *smaller Folio*, at 12s. and to the End, that whoever has the large Editions in 4to. of the Essay on Man, or Satyres of Horace, may not be obliged to buy them again, they may, on sending them to the abovesaid Booksellers [Gilliver, Brindley, and Dodsley], have all the parts of this Book at 15s.

And whereas Bernard Lintot having the property of the former Volume of Poems, would never be induced to publish them compleat, but only a part of them, to which he tack'd and impos'd on the Buyer a whole additional Volume of other Men's Poems. The present Volume will with all convenient Speed be published in Twelves at 5s. that the Buyer may have it at whatever price he prefers, and be enabled to render compleat any Sett he already has, even that imperfect one printed by Lintot.[5]

Pope was joint proprietor of *Works* II and this is clearly his advertisement, representing his conception of his relation with his readers. He is concerned for the welfare of two possible readerships. The first consists of his loyal following who had subscribed for the *Iliad* and *Odyssey*, and bought *Works* I. Pope feared that readers would not subscribe for the new *Works* unless they had the old one; he imagined a newcomer filling someone else's place rather than just buying volume II. Consequently, an advertisement appeared in the *Daily Courant* of 13 May 1735, saying, 'whoever is willing to part with their first Volumes in large Folio or Quarto, may receive Fifteen Shillings for the same'. The aim was clearly to build up

---
[5] Griffith quotes the advertisement and explains that Lintot's two-volume *Miscellany* of 1720, 1722, 1726–7, and 1732 is being criticized (2. 287–8).

two-volume sets that could be sold to newcomers. It rather looks as though Pope regarded his subscribers/collectors as a diminishing band. William Bowyer had printed 750 quarto copies of the *Works* in 1717, 250 of them on fine paper, and an additional 1,250 folios, making an edition of 2,000 in all, but it appears that in 1735 Pope ordered only 500 copies of *Works* II in quarto, possibly alongside 250 in large folio; he was starting to focus on the production of a popular edition.[6]

The second group Pope had in mind were new readers, casual in their commitment and wanting cheaper books. The recognition of this wider public is a major step in Pope's life as a writer. In the first part of his career he had no financial interest in these readers: his income came from the subscription and high-quality copies; Bernard Lintot dealt with the general public. But now that he had control of his copyrights and his own printer, John Wright, these readers became not only his broader public, an aspect of his fame, but also a potential source of income. The *Grub-street Journal* advertisement first ensures continuity for collectors of the small folios, by offering them for sale at 12*s.*, but then completely rethinks provision for the general public. Two thoughts about these readers go together: that it has been difficult for them to get hold of convenient editions of Pope's early works, and that they would like cheap copies.[7] The advertisement promises a partial remedy 'with all convenient Speed' in the republication of *Works* II in a small format. It says there will be a duodecimo at 5*s.*, but in the event the volume was split into two octavos costing 3*s.* each, bound. The poems of the 1730s, to which Pope held the copyright, became *Works* II, which was published by 31 July 1735 (Griffith 388), while *The Dunciad*, to which Gilliver still held the copyright, was first published under its own name by 31 July (392), and then as *Works* IV in 1736 (Griffith 431). What was not predicted in the advertisement was that Bernard Lintot would prove willing to join the new publishing venture. Whether he responded to the advertisement, or Pope approached him with a characteristic offer of revision, is not clear, but by late August Lintot had a new edition of *Works* I under way, following Pope's example by splitting one volume in quarto or folio into two in octavo. On 14 January 1736, he published *Works* I (Griffith 413), containing Pope's major poems, including the poems belonging to the Tonsons, and then, the following May, *Works* III (417), containing the early fables, translations, and imitations. Although further adjustments had to be made to accommodate the letters and the later Horatian poems, this arrangement set the pattern for the rest of

---

[6] The figures for 1717 are in Keith Maslen and John Lancaster (eds.), *The Bowyer Ledgers* (London: Bibliographical Society, 1991), items 385–6. The figures are similar to those for the *Iliad* and *Odyssey*. When Pope told Ralph Allen about the plans for the edition of the *Letters* on 5 June 1736, he said that 500 quartos would be needed 'to complete the Sets of my Verse Volume' (*Correspondence*, 4. 20). He anticipated that only 100 of those subscribers would come from the final public advertisement as opposed to private solicitation. The advertisement for subscriptions in the *London Gazette*, 8–12 Feb. 1737, implies that only 500 large-format copies will be printed, including folios (*Correspondence*, 4. 41 n.).

[7] The complaint against Lintot, that he made it difficult to buy Pope's early poems on their own, is valid. Pope's works had been republished as part of the two-volume duodecimo *Miscellaneous Poems*, of which the third edition of 1720 (Griffith 124) is representative. In the first volume of 286 pages, Pope's poems occupied pages 5–181 and 223–4; in the second volume of 318 pages there was only one piece by Pope. The *Pastorals* and the other poems owned by the Tonsons that appeared in *Works* I were not reprinted.

Pope's career, with republication roughly every two years.[8] The financial success of the move led Pope to regret the large-format editions: 'indeed I have done with expensive Editions for ever, which are only a Complement to a few curious people at the expence of the Publisher, & to the displeasure of the Many' (*Correspondence*, 4. 350). As a result of the new arrangement, the volumes of Pope's *Works* ceased to be simply the summary of a period of his career that was over and done with. The early works remained of the past, but it was understood that they spoke, alongside new works, to the present. The growing annotation, in a function complementary to that of the *Epistle to Dr Arbuthnot*, read the poems biographically, as the life of the poet's art, and historically, as a record of his times. A body of work, representing the poet, was constantly being developed and readjusted, with new works modifying the reader's sense of the old.

WORKS II IN QUARTO AND FOLIO

The new pattern of annotation starts not with the octavos but with *Works* II in quarto and folio published on 23 April 1735. These large-format books were in a classic style aimed at the 'subscribing' type of reader; the pages were to be kept relatively clean and the Brossette-style Variations and Notes were to appear separately at the end. Unfortunately, the annotation had to be aborted, Pope says, when it was found that the book was getting impossibly large. The large-format *Works* II volumes are presented as obvious companions to the *Works* of 1717. The title is in red and black, like that of Lintot's *Works*, and the same motto appears on the title page. There is no frontispiece—that had to wait for the octavos—but there is an image of Pope on the title page, engraved on a medal, over which two cherubs embrace. Only the inscription on the medal, 'VNI ÆQVVS VIRTVTI ATQ. EIVS AMICIS', the title page motto of the *First Satire of the Second Book of Horace*, 'To virtue only and her friends, a friend', hints at the new satiric gravity that had come over Pope's poetry since 1717. The introductory note to the whole volume, 'The Author to the Reader', deals with the specific issue of comprehensiveness that always seemed to Pope central to editions of his works, but was at odds with his own desire to select the canon of his writings. He takes the opportunity to review the whole of his career, presenting two works, his preface and *To Arbuthnot*, as satisfactory acts of definition: '*ALL* I had to say of my *Writings* is contained in my Preface to the first of these Volumes, printed for *J. Tonson*, and *B. Lintot* in quarto and folio in the year *1717:* And all I have to say of *Myself* will be found in my last Epistle [*To Dr Arbuthnot*].'[9] Although Pope

---

[8] Griffith notes that there are two editions of each volume dated 1735 or 1736, undistinguished by edition number or date (Griffith 388, 389; 413, 414; 417, 418; 431, 432). He identifies 414 with an item in Woodfall's accounts dated 9 Sept. 1737 (*Notes and Queries*, 11 (1855), 377–8). I suspect the first editions (of 3,000 copies) sold in around two years, and that, because there were no new poems to add, they were then reprinted (in editions of 2,000?) without alteration. When the Horatian poems necessitated an alteration in the volume numbering (in 1738–9), they were again reprinted with adjustments and new dates.

[9] Quotations are from the folio *Works* (Griffith 370). As usual in dealing with extracts removed from their context, whole passages that would be in italic are presented in roman, and italic and roman are reversed where the preponderance would be italic.

distinguishes 'my *Writings*' and '*Myself*', the two accounts are both regarded as appropriate to—perhaps required by—the *Works*, which in some way fuses two Popes by binding the separate poems into a form of personal history. The 'Author to the Reader' goes on to say that the two volumes contain 'whatsoever I have written and design'd for the press' (the second verb narrowing the scope of the first), except for the *Iliad* and twelve books of the *Odyssey*, the preface to the *Shakespeare*, and a few *Spectator*s and *Guardian*s. The material in the four volumes of *Miscellanies* he thinks 'too inconsiderable to be separated and reprinted here', though he identifies his own contributions. He concludes, 'It will be but justice to me to believe, that nothing more is mine, notwithstanding all that hath been publish'd in my Name, or added to my Miscellanies since *1717*, by any Bookseller whatsoever.' In settling his canon and fencing off unauthorized pieces, he uses the same equivocation as in the 1717 preface. There he argued that, because he had burned some pieces, the world owed him 'the justice in return' to accept his definition of his works. Now he claims that it will be 'but justice to me to believe, that nothing more is mine'. In neither case does he explicitly deny writing other works. I take it that Pope's position is that it is fair—given all the trouble he has gone to in trying to be a good writer and in compiling these volumes of works—that they alone should count as his writings. In the scales of justice conscientious artistry outweighs, even cancels, the occasional impropriety.

Works II ought to have seen the first full publication of Pope's *opus magnum*, consisting of two books of ethic epistles. *An Essay on Man* was the first book; the second was to consist of nine epistles, among which *To Cobham*, *To a Lady*, *To Bathurst*, and *To Burlington* would have had some place.[10] The project was unfinished and Pope was ambivalent about its representation in the volume. At first the presentation is confident. In quarto the four epistles of *An Essay on Man* are grouped with a section title as 'Ethic Epistles, to Henry St. John, L. Bolingbroke. Written in the Year 1732'. The four further epistles he had completed were then presented in a section title as the second book: 'Ethic Epistles, The Second book. To Several Persons.' 'Several Persons' at this stage made a neat contrast with the first book, which had four epistles to one person, Bolingbroke. But this adherence to the *opus magnum* plan was not sustained. After *To Burlington*, which should have been the fourth and final epistle of the second book, comes *To Addison*, which was numbered 'V', 'To Oxford' 'VI', and *To Arbuthnot* 'VII'. These poems are not thematic and cannot properly be regarded as ethic epistles. A new arrangement, in which *An Essay on Man* alone was regarded as consisting of ethic epistles, was coming into being. The Satires of Horace, the Satires of Donne, and Epitaphs, which follow, made clearer groupings, but their

---

[10] Leranbaum, *Alexander Pope's 'Opus Magnum'*, gives a good account. The second book was to consist of the following (subsidiary works in a group are in parentheses): 'Of the Limits of Human Reason' ('Of the Use of Learning'; 'Of the Use of Wit'); 'Of the Knowledge and Characters of Men'; 'Of the Particular characters of Women'; 'Of the Principles and Use of Civil and Ecclesiastical Polity' ('Of the Use of Education'); 'A View of the Equality of Happiness in the Several Conditions of Men' ('Of the Uses of Riches, &c.').

annotation is much less important than that of the two planned books of ethic epistles or of *To Arbuthnot*.[11]

The notes in the large-format copies of *Works* II are of various sorts: preliminary notes, footnotes, and endnotes. The most important preliminary notes are the general 'Author to the Reader', which has already been discussed, and the advertisement to the imitations of Horace and Donne. The advertisement intensifies the note of self-justification that was present in the 'Author to the Reader' and reappears in some of the endnotes. Such notes are contextual because they give the reader the circumstances that generated the work, but they are also evaluative because they exonerate the author and convict his opponents. These exculpatory notes were dangerous for Pope: in trying to protect the text, they introduce an accusation that can be fully pursued only outside it. They have had a significant influence on the criticism of Pope, and we shall find that on occasion he decided to delete them. In the advertisement Pope argues that his imitations are defensive in nature; they follow distinguished predecessors in showing contempt for vice in high places and they are satires not libels. This defence is, of course, one that Pope elaborated in the Horatian poems themselves. The other form of preliminary note found in *Works* II is the précis of each epistle, called the contents, that Pope had already provided for the collected edition of *An Essay on Man* in imitation of Michael Maittaire's edition of *De Rerum Natura*. In a surprising move, he decided to parallel the second book of ethic epistles with the first, by providing separate contents for *Cobham*, *A Lady*, *Bathurst*, and *Burlington*, and printing them together at the start of the second book of ethic epistles. The contents find it difficult to cope with poems as unsystematic as these, but their presence strongly reinforces the section titles in representing the rump *opus magnum* in these *Works*. Later, in the octavos, the contents also became footnotes, and I shall discuss them later in relation to the octavos.

The footnotes to the quartos and folios reproduce the scholarly and entertaining information that characterized individual editions of the poems; elements of the more systematic and historical annotation that was to mark the octavos are found in the endnotes. One of the purposes of the footnotes is to add variety and colour to the text. Some briefly clear up difficulties; others provide curious and amusing information. The few footnotes to *An Essay on Man*, which derive from the first collected edition of the four epistles, lighten the gravity of the work. They have a role in reinforcing the authority of the poetry with information and references, but they also show a simple delight in the natural world—even when they are wrong. For example, a note supporting the claim that Providence balances its gifts is splendidly eccentric:

[I] VER. 205.—*the headlong Lioness*—] The manner of the Lions hunting their Prey in the Deserts of Africa is this; at their first going out in the night-time they set up a loud Roar, and then listen to the Noise made by the Beasts in their Flight, pursuing them by the Ear, and not

---

[11] For a valuable discussion of textual changes in these epistles, see Julian Ferraro, 'From Text to Work: The Presentation and Re-presentation of Epistles to Several Persons', in Howard Erskine-Hill (ed.), *Alexander Pope: World and Word*, Proceedings of the British Academy 91 (Oxford: Oxford University Press for the British Academy, 1998), 111–34.

by the Nostril. It is probable, the story of the Jackall's hunting for the Lion was occasion'd by observation of the Defect of Scent in that terrible Animal.

I think Mack is right in suggesting that this information came from Pope's friend and surgeon William Cheselden. 'Cheseld' is written at the appropriate point in the Houghton manuscript, and, before it, Pope has placed a † indicating that this information is to go in a note.[12] Pope is passing on the benefits of lively conversation. The account of the Nautilus has a similar charm, though this time Pope gives a reference:

[III] VER. 178.] Oppian. Halieut. Lib. I. describes this Fish in the following manner. They swim on the surface of the Sea, on the back of their Shells, which exactly resemble the Hulk of a Ship; they raise two Feet like Masts, and extend a Membrane between which serves as a Sail; the other two Feet they employ as Oars at the side. They are usually seen in the Mediterranean.

Other notes have similar curiosity value, while doing little more than restate the material of the poem (1. 174, 3. 72).

[III] VER. 72] Several of the Ancients, and many of the Orientals at this day, esteem'd those who were struck by Lightning as sacred Persons, and the particular Favourites of Heaven.

The interest in natural history shown in *An Essay on Man* persists into the second book of ethic epistles, where we are told in *To Cobham* that 'There are above 300 Sorts of Moss observed by Naturalists' (18), a figure Pope forgot, or pretended to forget, in annotating the *Dunciad* (4. 450). *To a Lady* has no footnotes, but *To Bathurst* has two, one of them serious. To the couplet that reads,

> Where *London's Column pointing at the skies
> Like a tall bully, lifts the head, and lyes . . . (339–40),

Pope adds the note, '*The Monument built in Memory of the Fire of London, with an Inscription importing that City to have been burn'd by the Papists.' This is already Brossette-style historical clarification, making the object of the poem's critique explicit; the endnotes to this poem do similar work, even more boldly. Pope also reprints contextual notes on the Earl of Burlington's publication of Inigo Jones's designs and Palladio's work on the antiquities of Rome. The note saves the poem the labour of orientation, allowing it to begin conversationally. In *To Arbuthnot*, where the issues of context are more complex, he merely gives a footnote on Midas, and explains that 'Burnets, Oldmixons, and Cooks' (140) were 'Authors of secret and scandalous History', reserving the difficult material for endnotes. In the quarto edition of the imitations of Horace and Donne, but not the folio, there are more workaday notes: praise for Pope's friend Peterborough, whose 'lightning pierc'd th'Iberian Lines' (*Satire 2. 1*, 131); explanation of 'barbecu'd' (*Satire 2. 2*, 26); and, in the Satires of Donne, clarification of references to waxworks, the tapestries of Hampton Court, and the giant Ascapart. In short, the footnoting of these works is

---

[12] See *Twickenham*, 3 (1). 42. In *Last and Greatest Art*, 328, Mack has not transcribed the marginal note, but I doubt that is because he questions its authenticity.

charming, casual, intermittent. Pope pauses to give curious information that has caught his attention, or to advance the reputation of his friends. These notes are not exculpatory and they are not evaluative; interpretation of the poems is not seriously affected by them, and no new material is developed. The same, however, is not true of the endnotes in this edition.

There are two sorts of endnotes to *Works* II: Variations and Notes. The Notes deal in corruption, scandal, and literary in-fighting. They supplement the poems by delving into the dark side of eighteenth-century society in a fascinated, journalistic way that the poetry avoids. They were compiled in an attempt at full annotation, combining information about the author, his times, and the text, but as a postscript explains they were incomplete:

> It was intended in this Edition, to have added *Notes* to the *Ethic Epistles* as well as to the *Dunciad*, but the book swelling to too great a bulk, we are oblig'd to defer them till another Volume may come out, of such as the Author may hereafter write, with several Pieces in Prose relating to the same subjects.
>
> In the mean time, that nothing contained in the former Editions may be wanting in this, we have here collected all the *Variations* of the separate Impressions, and the *Notes* which have been annexed to them, with the addition of a few more which have been judg'd the most necessary.

This note carries the concern with completeness to a new level, going beyond inclusion of all the works to inclusion of all the versions of all the works. As in Brossette's edition of Boileau, nothing is to be lost. The reason Pope gives for not completing the endnotes is valid—the Bodleian's two copies show the quarto 'swelling to too great a bulk' and splitting in two—but the task of completing this annotation was a heavy one and probably unrealizable in the time Pope and Jonathan Richardson had available. These quartos and folios have Variations (*Changemens*) for *An Essay on Man* and *To Arbuthnot* only, and Notes (*Remarques*) on *To Bathurst* and *To Arbuthnot* only. There are no Imitations, and even later editions found little scope for annotating these poems in that way; the earlier poems provided more scope.

The endnotes of *To Bathurst* are socio-historical in the manner of Brossette, telling future readers about the author's times and thereby informing contemporary readers as well. They contextualize the poem and explain some of its allusions, but they provide something of an ethical holiday, indulging in curiosity for its own sake. They treat current affairs laced with scandal, adopting a style of news-reporting that is intensely curious about individuals. In a kind of prurience of corruption, individuals serve as representatives of general trends, but they are also of interest in themselves for their personal disgrace. The notes take Pope's practice of naming a step further than it had reached in the *Dunciad Variorum*, providing biographical information that is indisputably serious and treating figures from the higher classes of society.[13] The first endnote to *To Bathurst* is representative. It is too long to quote in full, but the following extract represents it in freely abridged form.

---

[13] Walter Harte had included a translation of Boileau's 'Discourse on Satires, Arraigning Persons by Name' with his *Essay on Satire*, but the neoclassical treatment of the issue in Boileau's essay is of little assistance in understanding Pope's practice.

[Ver. 20.] JOHN WARD . . . convicted of Forgery . . . expelled the House [of Commons] . . . stood in the Pillory . . . join'd in a Conveyance with Sir John Blunt to secrete fifty thousand pounds . . . conceal'd all his personal Estate, which was computed to be one hundred and fifty thousand pounds . . . imprisoned . . . his amusement was to give Poyson to Dogs and Cats, and see them expire by slower or quicker torments . . . *worth* above two hundred thousand pounds . . .

FR. CHARTRES . . . infamous for all manner of Vices . . . drumm'd out of the Regiment for a Cheat . . . banish'd Brussels, and drumm'd out of Ghent . . . took to lending of Money at exorbitant Interest . . . seizing to a minute when the Payments became due . . . House was a perpetual Bawdy-house . . . twice condemn'd for Rapes . . . The Populace at his Funeral rais'd a great Riot, almost tore the Body out of the Coffin, and cast dead Dogs, *&c.* into the Grave with it. [Epitaph by Arbuthnot] . . . *worth* seven thousand pounds a year Estate in Land, and about one hundred thousand in Mony.

Mr. WATERS, the third of these Worthies . . . But this Gentleman's History must be deferr'd till his Death, when his *Worth* may be known more certainly.

The note specifies the poem's claim that riches are not given to Heaven's favourites, but 'To Ward, to Chartres, and the Devil', but the detail exceeds the demands of specification. To the stories of financial misdeeds, Pope adds sensational details: Ward's poisoning of cats and dogs; Chartres's house being a bawdy-house; and the throwing of dead dogs into his coffin. In a clever typographical play, Pope italicizes '*worth*' contrasting different systems of value, but the annotation itself seems to be very interested in money, detailing the wealth of the three 'worthies'; in later editions it becomes even more excited by these sums and puts them in italic, thus weakening the play on 'worth'. Several of the other notes show precisely the same qualities, enlivening the financial details with a sense of paradox. Sir William Colepeper, 'a Person of an ancient Family and ample Fortune, without one other quality of a Gentleman' (53), ruined himself at the gaming table and then spent his time watching the ruin of others. Richard Turner, 'possessed of three hundred thousand pounds' (84), having lost £70,000, took to his chamber, where he could save the cost of clothes and other expenses. Denis Bond (102) was one of the directors of the Charitable Corporation but fond of the maxims '*God hates the Poor*' and '*every man in want is Knave or Fool*'. Peter Walter (125), referred to in the note on line 20 and still living, gets a more circumspect entry:

PETER WALTER, a Person not only eminent in the Wisdom of his Profession, as a dextrous Attorney, but allow'd to be a good, if not a safe, Conveyancer; extremely respected by the Nobility of this Land, tho' free from all manner of Luxury and Ostentation: His Wealth was never seen, and his Bounty never heard of; except to his own Son, for whom he procur'd an Employment of considerable Profit, of which he gave him as much as was *necessary*. Therefore the taxing this Gentleman with any *Ambition*, is certainly a great wrong to him.

Walter outlived Pope and so the satire has to rely on the slipperiness of the adjectives: 'dextrous' suggests skill but not probity, 'good, if not safe' implies that the interests of the client were neglected by this money scrivener. The remainder of the note picks up again the moral blindness of the first note, unable to distinguish generosity and ostentation, with the introduction of the word 'bounty' setting up a critique of what might otherwise be ordinary family relations.[14]

The sole truly contrasting note is on the MAN of ROSS (269):

This Person who with no greater Estate actually performed all these good Works, and whose true Name was almost lost (partly by the Title of the *Man of Ross* given him by way of Eminence, and partly by being buried without so much as an Inscription) was called Mr. *John Kyrle*. He died in the year 1724 aged 90, and lies interr'd in the Chancel of the Church of Ross in Herefordshire.

This note represents the positive side of naming and overcomes a difficulty. Pope's villains are named in the poem and their biographies appear in the notes. But because he transcended his social limitations, Pope's hero loses his personal identity and becomes the MAN of ROSS, in honorific caps. and smalls. The title is an honour, but it represents the loss of his own name. In honouring him Pope wishes to restore that personal identity for the historical record, and this he succeeds in doing in his note. Through the combination of poem and note John Kyrle gives up his identity to his bounty without losing it. The annotating poet is a researcher with a responsibility to social realities, and the note succeeds, rather like some notes in the *Dunciad Variorum*, in merging imagination and the everyday.

The endnotes to *To Arbuthnot* are also concerned with persons, but in a subtly different way. The persons admitted to these notes are qualified not by public distinction or notoriety, but by their relation to the poet. Readers are no longer being informed or reminded of material already in the public domain; they are admitted instead to a shady territory surrounding the poet. The problem with such notes is that they rarely give the reader the sense of completion: they simply reinforce the poem, without supplying references, explanation, or specific information, or they introduce new problems, tempting the reader to pursue them elsewhere. The full note on line 346 is only a partial exception to the general imprecision:

VER. 346. *Abuse on all he lov'd, or lov'd him, spread.*] Namely on the Duke of Buckingham, the Earl of Burlington, Lord Bathurst, Lord Bolingbroke, Bishop Atterbury, Dr. Swift, Dr. Arbuthnot, Mr. Gay, his Friends, his Parents, and his very Nurse, aspersed in printed papers: by James Moore and G. Ducket, Esquires, Welsted, Tho. Bentley, and other obscure persons, *&c.*

The note contrasts the large number of distinguished friends with Pope's undistinguished enemies, but it supplies no information about those enemies or about the nature of their abuse. It is, however, more specific than this poem's first endnote, which points up the different stages in the evolution of the text.

THIS Epistle contains an Apology for the Author and his Writings. It was drawn up at several times, as the several Occasions offer'd. He had no thought of publishing it, till it pleas'd some Persons of Rank and Fortune to attack in a very extraordinary manner, not only his *Writings*,

---

[14] For Walter, see Howard Erskine-Hill, *The Social Milieu of Alexander Pope: Lives, Example, and the Poetic Response* (New Haven: Yale University Press, 1975), 243–59.

but his *Morals, Person*, and *Family*: of which he therefore thought himself obliged to give some account.

Neither the 'Persons of Rank and Fortune' whose attack precipitated the poem, nor the several occasions of its composition, are specified. What Pope wants to convey is his own stance in relation to publication and not any further detail. Pope's notes never provide information about those 'Persons', Lord Hervey and Lady Mary Wortley Montagu; they are figures too powerful to name, let alone attack directly. Consequently, the note insists rather than clarifies. The same is true of the positive note on 'Granville the polite' (129):

These are the persons to whose account the Author charges the publication of his first Writings. The Catalogue might have been extended very much to his honour, but that he confin'd it to Friends of that early date.

The note does little more than repeat what had been said; it insists on the poet's popularity. Much the same character of insistence belongs also to the two notes on Dryden (at 135 and 241) and on the *Dunciad* (366). A note that goes beyond these in imparting information but makes a virtue of its lack of reference is that on Atticus (207):

It was a great Falshood which some of the Libels reported, that this Character was written after the Gentleman's death, which see refuted in the Testimonies prefix'd to the Dunciad. But the occasion of writing it was such, as he would not make publick in regard to his memory; and all that could further be done was to omit the Name in the Edition of his Works.

The note begins by repeating a charge and denying it, but then it becomes entangled in the issue of naming. I think this is the only time Pope extends his self-conscious sensitivity about naming to his notes. From the point of view adopted here, naming becomes quasi-magical and divorced from issues of reference; to name someone is analogous to cursing. After all, the lead-in to the portrait of Atticus had been altered from the manuscript version to read 'not Addison himself was safe'. There is no doubt that Atticus 'is' Addison, yet a great show is made of protecting him. Under the weight of such scrupulousness, the second sentence of the note falls into incoherence: Pope fails to explain why the lines are being published at all. Non-publication would have saved him the labour of trying to achieve the combination of non-naming, naming, and self-praise for non-naming, all at the same time. There is no doubt that Pope found this note troublesome. There is another attempt at it, more successful but not completely so, in his handwriting on the Earl of Oxford's copy of the individual folio, now in the Bodleian.[15]

The only three endnotes to *Arbuthnot* that do conventional note-work come towards the end of the poem. Even here the notes are only partially successful, as indignation and self-justification jostle for space with information. The first of the three, on 'The Tale Reviv'd, the Lye so oft o'erthrown' (342) has a slightly complex history. In the first folio separate edition, it appeared like this:

---

[15] See McLaverty, 'Pope in the Private and Public Sphere: Annotations in the Second Earl of Oxford's Volume of Folio Poems, 1731–1736', 51, 57–8.

Lies so oft o'erthrown.] Such as those in relation to Mr. *A*—, that Mr. *P.* writ his Character after his death, *&c.* that he set his Name to Mr. *Broom*'s Verses, that he receiv'd Subscriptions for *Shakespear*, &c. which tho' publickly disprov'd by the *Testimonies* prefix'd to the *Dunciad*, were nevertheless shamelessly repeated in the Libels, and even in the Paper call'd, *The Nobleman's Epistle.*

Because the charges relating to 'Mr. A—' (again, easily identifiable in this note) were given a note of their own, they were omitted from this note in the *Works*, which reordered the other allegations against Pope. The next major note is of the same sort, though even more particular.

VER. 367. *Welsted's Lye.*] This Man had the Impudence to tell in print, that Mr. P. had occasion'd a *Lady's death*, and to name a person he never heard of. He also publish'd that he had libel'd the Duke of Chandos; with whom (it was added) that he had liv'd in familiarity, and receiv'd from him a Present of *five hundred pounds:* The Falsehood of which is known to his Grace, whom Mr. P. never had the honour to see but twice, and never receiv'd any Present, farther than the Subscription for Homer, from him, or from Any Great Man whatsoever.

Budgel in a Weekly Pamphlet call'd the *Bee*, bestow'd much abuse on him, in the imagination that he writ some things about the *Last Will* of Dr. *Tindal*, in the *Grubstreet Journal*; a *Paper* wherein he never had the least Hand, Direction, or Supervisal, nor the least knowledge of its Authors. He took no notice of so frantick an Abuse; expecting that any man who knew himself Author of what he was slander'd for, would have justify'd him on that Article.

It is noticeable that Pope ceases to be referred to as 'the Author' in these notes, in order to become 'Mr. P.' If anything, this style, rather like that of a legal document or a public statement, implies strongly that this is Pope's authorized utterance. The information must, of course, come from him, and the change in designation marks the new absorption in biographical material. The note is, right from the start ('the Impudence'), a protest, but it commits Pope to giving a brief account of the charges against him as well as his defence. The difficulty is that neither account can be adequate in the space available. The end of the note sidetracks us into a dispute with whoever wrote the *Grub-street Journal* article (possibly Richard Russel), giving the whole a centrifugal force. Pope counteracts this danger in the final note. It repeats the charge that his birth is obscure, and then attempts a detailed refutation. As noted earlier in the chapter, the *Works* adds an air of finality to the account published in the first edition by adding the inscription to Pope's family monument, concluding with a line to bring the work to a close:

PARENTIBUS BENEMERENTIBUS FILIUS FECIT, ET SIBI.

The anticipation of his own death is a neat way to close down a poem which included 'all I have to say of *Myself*'.

The notes of *Works* II in quarto and folio are followed by Variations, a textual history of the selected poem. The collations are highly selective—they omit, for example, the embarrassing 'A mighty maze of walks without a plan' from the first edition of *An Essay on Man*—but they include some of the major textual revisions of the two books of ethic epistles and *To Arbuthnot*. *An Essay on Man* had been subjected to several layers of revision since its first publication, and one of Pope's aims was to illustrate that striving for perfection, and fertility of mind, that he believed characterized the great artist. Consequently the notes draw attention to new material as well as to variants. For example, in the note on the *Essay*, 1. 251–6, he explains, rather proudly, 'These six Lines are added since the first Edition', whereas in other cases he prints lines that have been deleted since the first edition, with the sense that they are still worthy of the reader's attention.

An important feature of the *Essay on Man* Variations is a focus on lines dealing with difficult theological material, especially the issue of immortality. The degree to which it was appropriate to discuss immortality remained a problem for Pope, and some of the decisions recorded in these editions were reversed when Pope came to revise the poem for the death-bed edition in 1744. The history of his own thinking recorded in this edition possibly influenced their later reinstatement. The second Variation is an example:

> Ed. I. Ver. 95 [1. 98]
> If to be perfect in a certain State,
> What matter, here or there, or soon or late?
> He that is blest to day, as fully so,
> As who began ten thousand years ago.
> *Omitted in the subsequent Editions.*[16]

Mack points out that in 1744 these became 1. 73–4, 287–8, and 75–6, with some verbal changes. They are lines that have been subjected to close critical attention in the light of Pope's letter to Caryll of 8 March 1733, in which he suggests that the anonymous author 'quits his proper subject, *this present world*, to insert his belief of *a future state* and yet there is an *If* instead of a *Since* that would overthrow his meaning'. I suspect it was to escape this middle state—committed too far for the restricted scope of his essay and yet not far enough to please Christian thinkers—that Pope omitted the lines. In two other cases (2. 21–2 and 35–8) he restored or semi-restored the earlier reading in 1744, again suggesting he consulted this apparatus. In one case, the concluding address to Bolingbroke, he used the 1735 endnotes to supplement the text that had appeared earlier in the volume:

> Ver. 380 [4. 387–8]
> When Statesmen, Heroes, Kings, in dust repose,
> Whose Sons shall blush their Fathers were thy Foes.
> *Omitted by mistake in the Folio Edition.*

These lines appeared in the quarto four-epistle edition of *An Essay on Man* in 1734, but not in the equivalent folio. That meant they did not appear in the Folio *Works*, which used the same setting of type as the four-epistle edition. But, contrary to the impression given in the note, they did not appear in the quarto *Works* either, which was from a fresh setting of type. The omission may have been an oversight, but it

---

[16] As usual, I have given the original number, with the relevant *Twickenham* number in brackets, and reversed italic and roman when the preponderance would be italic.

might alternatively have sprung from a timidity Pope later regretted. As detailed analysis of the text of *Works* II has shown, Pope seems to have got cold feet about other passages with dangerous personal and political implications. However, this reminder to himself in the Variations worked, and the lines were restored in the octavo copies of the *Works*.

An inverse difficulty arose with four lines in *To Cobham*. This time the lines were in the poem, though they should not have been, at least according to the note.

Triumphant Leaders, &c. These four Verses having been misconstrued, contrary to the Author's meaning, they are suppressed in as many Copies as he cou'd recall.

The lines are:

> Triumphant Leaders at an Army's head,
> Hemm'd round with Glories, pilfer Cloth or Bread,
> As meanly plunder as they bravely fought,
> Now save a People, and now save a Groat. [146–9]

The reference is to the Duke of Marlborough, with a characteristic allegation of avarice; Bateson identifies the allusion to Sir Solomon Medina's paying Marlborough £6,000 a year to provide bread for the army. These lines are a trace of the attacks on the Duke and Duchess of Marlborough Pope was planning for these *Works* but abandoned, perhaps in acknowledgement of the Duchess's support for the anti-Walpole, Patriot campaign. Courthope reproduces two pages from the quarto collected edition of *An Essay on Man* with manuscript revision of the lines on the hero (4. 291–308) which turn them into a savage attack on Marlborough. Courthope concludes, and he is surely correct, that the revision was prepared for the quarto edition of the 1735 *Works*. Similarly the line-numbering of *To a Lady* in the 1735 *Works* suggests that Pope had originally intended to include a version of the Atossa portrait in that poem, but had later decided to exclude it.[17] The stories that Pope removed these attacks in response to pressure from the Duchess are given greater credibility by this endnote. The implication is that Pope's attention was drawn to the possible interpretation of the line after publication, hence the attempt to suppress the passage in 'as many Copies as he cou'd recall'. Bateson says he has seen no copy in which the lines are suppressed, and I cannot understand how the note can appear, as it does in the Bodleian copies, without the lines being removed by a cancel. The note is disingenuous; Pope was using it as a substitute for withdrawing the lines. They were included in the subsequent octavos and had to be cancelled in the death-bed edition, perhaps at Warburton's suggestion.[18]

---

[17] For a short account of the bibliographical puzzles surrounding *To a Lady*, see Foxon, *Pope and the Book Trade*, 127. Bateson's interesting arguments against the identification of the Duchess of Marlborough as a target of the Atossa passage are based on the assumption, profoundly mistaken in my view, that such portraits are directed at one person (*Twickenham*, 3 (2) appendix A). The trick was to fuse two targets.

[18] The cancellandum does not survive, but Bateson's argument in *Twickenham*, 3 (2). 25–6 n., is persuasive. Warburton did not include the lines in his edition.

The most important remaining notes concern *To Arbuthnot*, whose text was discussed in the previous chapter. Others primarily stress Pope's artistry and diligence, with particular emphasis on the fulness of the *Works* text. The four last lines of *To Bathurst* were originally only a couplet. The twelve verses following 'Who then shall grace, or who improve the Soil?' in *To Burlington* (169 [177]) 'are added since the separate Editions'. The four on Dryden in *To Arbuthnot* are 'not in the separate Editions', and neither is the couplet on vengeance (340–1 [348–9]). Of these, the additional lines in *To Burlington* are significant because, in making an explicit tribute to Bathurst and Burlington, Pope gives his first picture of a successful landlord:

> Whose chearful tenants bless their yearly toil,
> Yet to their Lord owe more than to the soil . . . (175–6 [183–4])

These lines do something to counterbalance the general satirical treatment of the aristocracy in this controversial poem, but only through praise of Pope's particular friends.

OCTAVO *WORKS* II

The octavo edition of *Works* II that appeared in the summer of 1735 was built on the basis of the socio-historical annotation of the quarto and folio, but it showed a different understanding of its readership. The most striking changes are the absence of Variations and the use of the contents as a form of running commentary on *An Essay on Man* and the four linked epistles. The absence of Variations from *Works* II tells us something about Pope's conception of his readers. Whereas the readers of the folios and quartos were conceptualized as collectors, interested in every word he wrote and anxious to have a full record, such information was thought unnecessary for the casual purchasers of the octavos. What they needed was an up-to-date version of Pope's work that was accessible and comprehensible. The elimination of historical and social obstacles to understanding, an objective emphasized by Brossette, therefore assumed a central place. But to this Pope added another: the clarification of the structure and argument of the poetry. Readers were to be guided through the philosophical poetry by a prose companion. This guidance had already been offered in the contents to the *Essay* and *Epistles*, fulfilling a double function of index and explanation, but now the emphasis shifted to explanation, as the contents moved alongside the poems, summarizing and clarifying. The reader would move through the material with the poet at his elbow; topics coming up for discussion would be declared and complexities cleared away. A change in typography supported the new importance of running commentary. In the quartos and folios footnotes were introduced casually with an asterisk or a dagger, and a residue of this system is to be found on some pages of octavo *Works* II. But in the octavos the pages are compartmentalized, with a rule at the foot of the verse separating it from the notes, and underneath the rule, each note is usually introduced by a line number (originally an elaborate 'VER. 1.]' etc., later simply 'V.1.' etc.), sometimes with a lemma. The implication is that the poems are to be subject to regular commentary and that this is a thoroughgoing annotated edition. Different sorts of notes (contents and Remarks) are distinguished by different type (roman and italic), but not in a consistent pattern from poem to poem. Of course, using the contents twice, at the beginning and alongside the footnotes, took up more space, but as Pope already had some unused Variations and could easily be provided with others by Richardson, filling the volume was unlikely to be his motive. It is true that when the octavo *Works* I was published in 1736, Variations were provided, but by that time the project had changed a little—this was to be a full edition of the *Works* rather than just a popular version of one volume—and Pope had had no previous opportunity of providing Variations for *Works* I. By then the prospect of producing a full-scale edition of the *Works* in octavo had opened up.

The footnotes derived from the contents take on a structural importance. They are used to give the topic of each epistle ('Of the Nature and State of MAN, with respect to the UNIVERSE'), and then each item of the contents declares a subsection ('*The* PASSIONS, *and their* Use').[19] This exercise in signposting was built on by Warburton, who assigned numbers to the sections of *An Essay on Man* created by the précis. At this stage Pope also pointed up the unity of his work by cross-references. *To a Lady* is declared 'a Corollary to the former Epistle [*To Cobham*]' and *To Burlington* is 'a Corollary to the preceding [*To Bathurst*]', while the notes to *An Essay on Man* frequently give cross-references from epistle to epistle in pursuit of a topic, with six footnotes referring to a total of twenty lines. The contents in the footnotes are full of small revisions, both from the preliminary contents and from previous versions, which makes me confident they are the work of Pope himself, in spite of some difficulties of fit with the complex arguments of the epistles. The alterations are very characteristic, almost inconsistent enough to be called tinkering but frequently charged with purpose. Sometimes they are typographical, with significant additions of caps. and smalls. Sometimes they correct slackness of expression, changing, for example, an indistinct 'these' to 'sensible *and* mental *Faculties*' (*Essay on Man*, 1. 200). But quite often, at least in *An Essay on Man*, they subtly modify meanings. For example, in the collected edition of the *Essay*, the summary of 1. 69 etc. is 'Man is not therefore to be deem'd *Imperfect*, but a Being suited to his *Place* and *Rank* in the Creation, agreeable to the *General Order* of Things, and conformable to *Ends* and *Relations* to him unknown.' In the octavo *Works* of 1735 the 'therefore' is, quite correctly, taken out of the précis. But then in the footnotes to the octavo the whole is rephrased to avoid illogicality and maximize assertion: '*He is not therefore a Judge of his own perfection, or imperfection, but is certainly such a Being as is suited to his* Place *and* Rank *in the Creation*' (1. 36). The 'therefore' has come back, but it now relates to the limitations of human judgement, and not to a logical conclusion about man's place in the creation. However, the 'certainly' then comes in to provide confidence, without claiming to have achieved it by ratiocination. There are a number of other cases where Pope combines philosophical caution with assertiveness and a twist to a

---

[19] Quotations are from the octavo *Works* (Griffith 389). I have retained Pope's pattern of roman and italic in quoting notes.

human perspective. 'Extravagance, Madness, and Pride' becomes in the footnotes '*Extravagance, Impiety, and Pride*' (1. 250); 'Of that which is called the *State* of *Nature*' becomes '*Of the* STATE *of* NATURE; *That it was* SOCIAL' (3. 148); and 'the *true End of all*' becomes '*the* TRUE USE OF ALL' (3. 284). There are similar revisions in *To Cobham* (101, 136, 162), *To a Lady* (45, 69, 81), *To Bathurst* (20, 161), but Pope's work on *To Burlington* seems to have gone furthest.

To Burlington was the oldest of the four epistles and Pope took advantage of the octavo printing to change the account of the poem provided by the preliminary contents into something more accessible to the reader. In one section the note gives greater authority to human judgement. The original 'The chief proof of it [good sense] is to *follow Nature*, even in Works of mere Luxury and Elegance. Instanced in *Architecture* and *Gardening*' becomes 'The chief proof of good Sense in this, as in every thing else, is to *follow Nature*, but with Judgment, and Choice' (47). At other times, the footnote expands on the original contents vividly and colloquially. So 'A Description of the *False Taste of Magnificence*; the first grand Error of which is to imagine that *Greatness* consists in the *Size* and *Dimension*, instead of the *Proportion* and *Harmony*, of the *Whole*' becomes 'The first wrong Principle, is to imagine true Greatness consists in *size* and *dimension:* whereas, let the work be ever so vast, unless the parts cohere in one harmony, it will be but a great many Littlenesses put together' (100) There is an obvious attempt to speak to the reader more directly. The second error becomes simply '*Disproportion*, small things joined to large ones' (109). Later the rather philosophical 'since it [wealth] is dispersed to the Poor and Laborious part of mankind' is changed to the everyday 'A bad Taste employs more hands, and diffuses Expence, more than a good one' (167). The footnote commentary is much more expansive than the preliminary one, with summaries being broken from their blocks and dispersed through a section. At one point the annotation is surprisingly revealing: 'Ornaments of building or sculpture, either too *much multiplied*, or *ill-placed*, or where *Nature* does not favour 'em. All the Examples are taken from some known Gardens' (119). It is, of course, this mingling of examples from different sources that criticism has found it so difficult to accept.

In addition to the contents, there are some new notes in the octavo *Works* and some notes left out. Some of these additions and omissions are significant. *To Cobham* has an important conclusion to a new note explaining 'sober *Lanesb'row*, dancing in the Gout' (247 [251]):

An ancient Nobleman, who continued this practise long after his Legs were disabled by the Gout. Upon the death of Prince George *of* Denmark, *he demanded an Audience of the* Queen, *to advise her to preserve her Health, and dispell her Grief by* Dancing.
The rest of these Instances are strictly true, tho' the Persons are not named.

The note provides a gossipy anecdote to back up an allusion the reader might not grasp, but then Pope insists that his other illustrations are references which may be even more difficult to pick up. The note insists on the purchase on reality that his no-names policy cannot demonstrate to a remote readership. A new note on *To a Lady* goes further by pointing to not merely the unnamed but the unreferred-to:

[103] *Between this and the former lines, and also in some following parts, a want of Connection may be perceived, occasioned by the omission of certain* Examples *and* Illustrations *of the Maxims laid down, which may put the reader in mind of what the Author has said in his Imitation of* Horace,

> Publish the present age, but where the text
> Is Vice too high, reserve it for the next.

This is a striking use of the footnote to supplement the poem. It presents the poem as incomplete and unfree, suggesting there are women who should be satirized but are not because they are too powerful. It encourages the reader to speculate on the fuller text that is not yet realized but is implicit in the present one. *To Bathurst* presents a poet with more freedom and daring, though the stakes are not so high. Its notes are more specifically political, and, as in quarto and folio *Works*, they risk naming or alluding to personalities. An example is the new note on 'And *Worldly* crying Coals from street to street' (50):

Some Misers of great Wealth, Proprietors of the Coal-mines, had enter'd at this time into an Association to keep up Coals to an extravagant price, whereby the Poor were reduced almost to starve, till one of them taking the advantage of underselling the rest, defeated the design. One of these Misers was *worth ten thousand*, another *seven thousand* a year.

The note avoids mentioning Wortley Montagu and concentrates on giving its readers an account of a historical episode, but the detail makes identification a possibility. The typography is now excited by sums of money as well as by '*worth*'.

Two other new notes, one on *To Bathurst* and the other on *To Burlington*, show Pope taking political risks; they may have been designed to balance one another. The first, on the suggestion that paper credit can 'fetch or carry *Kings*' (72), shows Pope loyal to the house of Hanover in a picture of the age:

In our Author's time, many Princes had been sent about the world, and great Changes of King's projected in Europe. The Partition-Treaty had dispos'd of *Spain*, *France* had set up a King for *England*, who was sent to *Scotland*, and back again; King *Stanislaus* was sent to *Poland*, and back again; the Duke of *Anjou* was sent to *Spain* and *Don Carlos* to *Italy*.

The attitude to kingship in this note is irreverent, and the attitude to the Stuarts markedly so. Their threat to the throne is presented as the product of French money; they are made to look like French puppets. The note is gratuitous: Pope does not need to comment on Jacobite rebellion, but he goes out of his way to disown it. In the major addition to *To Burlington*, however, he shows himself equally critical of the house of Hanover. The final lines of the poem create a positive effect with their praise of public works:

> These Honours, Peace to happy *Britain* brings,
> These are *Imperial Works*, and worthy *Kings*.

But a final footnote in the octavo edition turns this passage into an implicit critique of contemporary Britain:

The Poet after having touched upon the proper objects of Magnificence and expence, in the private Works of Great Men, comes to those great and publick Works which become a Prince.

This Poem was published in the year 1732: when some of the new built Churches, by the Act of Q. Anne, were ready to fall, being founded in boggy land, and others vilely executed, thro' fraudulent cabals between Undertakers, Officers, &c. when Dagenham Breach had done very great mischefs; when the Proposal of building a Bridge at Westminster had been petitioned against, and rejected; when many of the High ways throughout England were hardly passable, and most of those which were repaired by Turnpikes, made Jobbs for private Lucre, and infamously executed, even to the Entrances of London itself. There had, at this time, been an uninterrupted Peace in Europe for above twenty years.

This state of England clearly does not become its Prince. The poem's positive vision is transformed by the note into an attack on the incompetence and timidity of the government. The attack on the Hanoverians is quite as harsh as that on Jacobite rebellion.

The remaining poem with significant changes to its annotation in the octavo *Works* II is *To Arbuthnot*, which took on a settled shape only in this edition. The imprecise note on '*Granville* the polite' (139) is now merged with the first note on Dryden, and the whole is revised to define the group of Pope's first admirers and build up the parallel between Pope and Dryden. Other problems in the annotation of *To Arbuthnot* are solved by excision. The two very defensive notes, on Atticus and on 'The lye so oft o'erthrown', were simply left out. From the note on Welsted's lie Pope dropped the claim that he had seen Chandos only twice and omitted the closing reference to the writer of the paper on Budgell. These notes raised charges against Pope without the opportunity to dispose of them adequately. They started arguments that could not be completed; the danger was that the charge rather than the exoneration would be left in the reader's mind. Pope's early instinct had been to include such material; it made the account of his life more comprehensive, and many of his readers would already have been aware of these controversies. But, once again, his conception of the readers of the octavos was different: they might have been unaware of the charges represented in the notes and it was unnecessary and undesirable to enmesh them in these controversies.

OCTAVO *WORKS* I AND III

When, later in the summer of 1735, Pope turned to preparing the octavo edition of his early poems, he brought to it considerable enthusiasm. Lintot was amused and impressed, as his letter to Broome of 26 August 1735 makes clear: 'I am again printing for Mr. Pope,—the first volume of his miscellaneous works, with notes, remarks, imitations, &c.,—I know not what' (*Correspondence*, 3. 489). The 'notes, remarks, imitations' once again conjure up Brossette (*Remarques, Changemens, Imitations*), but the nature of these poems required a different sort of annotation from *Works* II. Pope is still conscious of the historical resonance of his poetry, and at times the annotation draws this out or attempts to bring it under control, but a more regular form of annotation slots each poem into the history of Pope's career and draws attention to his artistry. In compiling the notes for the new edition Pope was not tilling virgin soil. The 1717 *Works* already had a light body of annotation on which the 1736 edition could build, largely in the form of footnotes from the separate editions of the poems. The most elaborately annotated poem was 'Messiah', whose combination of footnotes and endnotes was discussed in Chapter 3, but other poems had references to sources and occasional limited explanations of allusions or meaning.

Most of the annotation in 1717 drew attention to parallel passages in authoritative texts or identified particular figures. As might be expected, *An Essay on Criticism* provided a series of notes identifying sources. The commonest were Quintilian with seven references and Cicero with three, but there were seven additional single references. These notes had appeared in the first edition of the poem, reinforcing its status as a learned poem but ignoring its relation to contemporary debate. The 1717 notes to *The Rape of the Lock* were also adopted from the individual edition of the poem, and were directed at explaining the poem by drawing attention to parallels with epic. As Howard Erskine-Hill notes, nothing was done to draw attention to the major textual development of 1717, Clarissa's speech urging realism and accommodation at the start of canto 5.[20] *Windsor-Forest* is the most interesting of the poems which retained identifying notes in 1717. There were five in total. Two (on London and on Cowley) were innocuous, but the other three played with the political and historical dimensions of the poem and the potential melding of William of Orange with the Norman conquerors. It is not unduly suspicious to see the footnotes' insistence that these are Norman figures as pointing up the original ambiguity. So the 'savage laws' (45) to which the lands are a prey (with 'Kings more furious and severe than they') were identified as '*The forest Laws*'; the fields which are 'ravish'd from th'industrious swains' (65) were said to allude to the New Forest '*and the tyrannies exercis'd there by* William *the first*'; and the man who died hunting ('At once the chaser and at once the prey' (82)) was identified as 'Richard, *second son of* William *the Conqueror*', and not, of course, as William III. *The Temple of Fame*, a learned poem like *An Essay on Criticism*, retained notes identifying, and sometimes explaining the mode of representation of, Sesostris, Zamolxis, Odin, Alexander, Aristides, and Horace. *Eloisa to Abelard* went a little further than this, supplementing the poem with missing biographical information. This poem, and consequently its annotation, were new, and I think this was the first work where Pope realized that he could use his annotation to free his poem of certain responsibilities to reveal information. Finally, it is worth noting that two footnotes to the *Thebais* (lines 173 and 282–3) are rare examples of the verbal criticism Pope attacked so sharply in *The Dunciad*. Although he later amended his text of Horace, Pope never discussed his emendations in the notes.

For the 1736 octavos, the 1717 volume was split in two and this division created more room for annotation. The notes become more significant typographically, adopting the pattern of *Works* II, with compartmentalized pages, line numbers, and lemmata, though *An Essay on Criticism* and *The Rape of the Lock* adopt a mixed system, with asterisks and daggers from the earlier editions running alongside the

---

[20] Erskine-Hill, *Poetry of Opposition and Revolution*, 89–93, a particularly subtle and persuasive analysis. Quotations from the 1717 *Works* are from Griffith 80.

new classical style of footnote, and *Eloisa to Abelard*'s few notes are linked to the text by asterisks. The first two pages of verse carry the compartmentalization further, with Brossette's divisions into REMARKS and IMITATIONS, as though Pope's original aim was to give even greater dignity to these notes, but the pattern is not sustained. There are also minimal preliminary notes on the section titles of the poems, giving information about when the poems were written. The description of this volume in the introductory table is 'the Author's Original Poems, written under 25 years of age', and though little is made of the claim to originality (there is more interest in debts in volume III), there is a recurrent concern with the poet's age. Pope makes good use of perspectival annotation, designed to give the user a sense of his developing career. This had already been present in the 1717 section titles, but now it spreads to the initial notes to the poems. The emphasis is on Pope's precocity and the rapid development of his talent. The *Pastorals* were 'Written in the Year 1704', 'Ode on St Cecilia's Day' in 1708, *Essay on Criticism* in 1709, *Rape of the Lock* in 1712 (though the subsequent note claims 1711 and full publication a year later). *Eloisa to Abelard* has no date—a good thing, because it was composed after the poet was 25.

The *Pastorals* are the most heavily annotated section of octavo *Works* I, possibly because they came first and took the full force of revisionary energy. The first note is concerned with the poet's career and prompts admiration:

> These Pastorals were written at the age of sixteen, and then past thro' the hands of Mr. *Walsh*, Mr. *Wycherley*, G. *Granville*, afterwards Lord *Lansdown*, Sir *William Trumbal*, Dr. *Garth*, Lord *Halifax*, Lord *Somers*, Mr. *Mainwaring*, and others. Notwithstanding the early time of their production, the Author esteem'd these as the most correct in the versification, and musical in the numbers, of all his works. The reason for his labouring them into so much softness, was, that this sort of poetry derives almost its whole beauty from a natural ease of thought and smoothness of verse; whereas that of most other kinds consists in the Strength and fulness of both. In a Letter of his to Mr. *Walsh* about this time, we find an enumeration of several Niceties in Versification, which perhaps have never been strictly observ'd in any *English* poem, except these Pastorals. They were not printed till 1709.[21]

The influence of Brossette is apparent throughout this note. The use of the third person is, of course, required by decorum, but the past tense 'esteem'd' suggests that Pope, like Boileau in 1716, is already dead and that we are being posthumously allowed into the secret of his opinions. The tone aimed at combines an intimate knowledge of the author's ideas with distant and respectful admiration for his genius. The reference to the letter to Walsh on versification points up the role of the *Letters*, which Pope tricked Curll into publishing in 1735, as companions to the *Works*. The letter alluded to had not yet, of course, been published in an edition authorized by Pope, but it had appeared in the edition foisted on Curll. Dated 22 October 1706, it is a fabrication based on a letter to Cromwell of 25 November 1710. Pope elaborates the discussion of versification, while presenting it as the work of

---

[21] Quotations are from Griffith 413. When displaying a note, as here, I have followed Pope's pattern of roman and italics, but in other cases I have changed all italic to all roman.

four years earlier, thus fitting it to be a companion to the *Pastorals*. The note and letter go hand in hand as a representation of Pope's relation to Walsh, who was called by Dryden 'the best critic of our nation'.[22] These are indeed ' éclaircissemens donnez par lui-même', based on evidence Pope has fabricated himself.

Pope's detailed annotation of the *Pastorals* draws attention to his revision of them for this edition. The revisions seem to have been prompted by a rereading of his *Guardian* essay of 27 April 1713, 'On the Subject of Pastorals', in preparation for the edition of his prose works.[23] In this essay, which ironically compares his own poems with Ambrose Philips's, Pope says particularly:

> When I remarked it as a principal Fault to introduce Fruits and Flowers of a Foreign Growth, in Descriptions where the Scene lies in our Country, I did not design that Observation should extend also to Animals, or the Sensitive Life; for *Philips* hath with great Judgement described *Wolves* in *England* in his first Pastoral. Nor would I have a Poet slavishly confine himself (as Mr. *Pope* hath done) to one particular Season of the Year, one certain time of the Day, and one unbroken Scene in each Eclogue.[24]

The point about wolves is reflected in the decision to print a variant reading at 2. 79, 'And list'ning wolves grow milder as they hear', which Warburton later annotated: 'the author, young as he was, soon found the absurdity which *Spenser* himself overlooked, of introducing wolves into England.' The point is picked up again in *Windsor-Forest*. The other element in the passage from the *Guardian*, the implicit praise for unity of season, hour, and scene, also generated revision, and annotation to underline it, in 1736. Pope pays particular attention to the beginning of each pastoral, highlighting the poem's unity, relation to tradition, and place in the group. The note to line 1 of the first pastoral is typical: 'This is the general Exordium and opening of the Pastorals, in imitation of the 6th of *Virgil*, which some have ... thought to have been the first originally.' The beginnings of the other three pastorals consequently imitate lines of the 'three chief Poets in this kind'. So the beginning of 'Summer' imitates Spenser; 'Autumn' Virgil; and 'Winter' Theocritus. The patterning continues with the time-setting of the first pastoral, with Pope providing a stage-direction at line 17: 'The Scene of this Pastoral a Vally, the Time the Morning', which goes on to point out that the text has been revised, from 1717, to strengthen this temporal setting. The second pastoral, 'Summer' similarly has a new note to underline the setting: 'The Scene of this Pastoral by the River's side; suitable to the heat of the season; the Time, Noon', while the second note draws attention to the revision of the 1717 text to create the allusion to Spenser and the appropriate time of day. The third pastoral has a similar note, 'The Scene, a Hill; the Time, at Sun-set', but this time there is no need of revision, with 'setting Phœbus' evoked in line 13. The fourth pastoral also has an appropriate note on time and place: 'The Scene of

---

[22] For the letter, see *Correspondence*, 1. 22–4. For Dryden and Walsh, see the admirable, if unreadable, note in *Twickenham*, 1. 59.

[23] For a discussion of these plans, often frustrated, see Foxon, *Pope and the Book Trade*, 124–38.

[24] *Prose Works*, 1. 99–100. This essay was reprinted as an appendix in the *Dunciad Variorum* in 1729.

this Pastoral lies in a grove, the Time at midnight'; again Pope prints four lines of variants to show that he has tightened up the presentation of time, but in this case the lines had already been revised for the first printing. At the very end of the poem Pope directs our attention to a final touch of artistry: 'These four last lines allude to the several *Subjects* of the four Pastorals, and to the several *Scenes* of them, particularized before in each.' The threads of the scheme are thus neatly drawn together, and the reader's attention directed to the poet's artistry.

These framing notes to the *Pastorals* can be seen to unite Brossette's three types: drawing attention to Pope's design (*Remarques*), clarifying his relations to his predecessors (*Imitations*), and revealing developments in Pope's artistry (*Variations*). All three are continued in the historical annotation that runs along the bottom of the pages of the *Pastorals*. We are told who Sir William Trumbull, George Granville, Samuel Garth, and Mrs Tempest were; we are given the meaning of 'wond'rous Tree' (1. 86, Charles II's oak) and the explanation of the riddle of the thistle and the lily (1. 90). These last two notes tint the poem with Stuart politics, but this form of annotation is not sustained. The majority of the notes are Imitations, pointing confidently to allusions to Virgil. For example, at 1. 41, Pope says 'Literally from *Virgil*' and quotes three relevant lines. In all there are twenty-six references to Virgil in the notes. Variations are also well represented, with a total of thirteen, giving full evidence of Pope's youthful creativity. Sometimes the superseded lines clearly demonstrate Pope's ability to improve on his first thoughts. Contrast, for example, 'And the fleet shades fly gliding o'er the green', where 'fleet', 'fly', and 'gliding' seem competing for the same work, with the replacement, 'And the fleet shades glide o'er the dusky green' (3. 64). But sometimes the emphasis seems to be on the fertility of Pope's invention, as in the manuscript alternatives he presented to Walsh. So 'And swelling clusters bend the curling vines' (1. 36) is accompanied by its rejected alternative, 'And clusters lurk beneath the curling vines.' I suspect Pope liked both lines, and the new extensive annotation gave him the opportunity of including them both.

The notes to *Windsor-Forest* share with those to the *Pastorals* the aim of giving an account of Pope's early development. They emphasize the poem's broken textual history, enabling the reader to reconstruct the different stages of composition and the unification of the whole. Most of the new notes are concerned with the early stages of the poem's life. The first note, expressing a customary concern for dating, breaks up the stability of the text: 'This Poem was written at two different times: the first part of it which relates to the country, in the year 1704, at the same time with the Pastorals: the latter part was not added till the year 1710, in which it was publish'd.' The information given in outline here is supplied in detail later, in a note on line 288. As well as being allowed to reconstruct the original poem, the reader is given various earlier states of the text. There are fourteen notes giving manuscript readings, ranging in length from two to eight lines. Nearly all are examples of stylistic revision, achieving greater energy and concentration. On two occasions Pope not only gives his reading but justifies his choice. In the 1736 text, lines 55 to 60 read:

> The swain with tears his frustrate labour yields,
> And famish'd dies amidst his ripen'd fields.
> What wonder then, a beast or subject slain
> Were equal crimes in a despotick reign?
> Both doom'd alike, for sportive Tyrants bled,
> But that the subject starv'd, the beast was fed.

To this he adds a note, giving a Variation:

> VER. 57, &c.   *No wonder savages or subjects slain—*
> [60]   *But subjects starv'd while savages were fed.*

It was originally thus, but the word Savages is not so properly apply'd to beasts as to men; which occasion'd the alteration.

The Twickenham editors point out that 'savages' was repeatedly used for beasts in the *Iliad*, and that Macaulay used it in that sense as late as 1831, but that is *OED*'s last instance of a meaning declared obsolete. This is an unusual case of Pope's modernizing his poem by bringing his language up to date. Line 72 is similarly corrected, but this time in order to bring it into line with Pope's own criticism of Ambrose Philips.

> VER. 72. *And wolves with howling fill,* &c.] The Author thought this an error, wolves not being common in *England* at the time of the Conqueror.

In both these cases there is an impressive willingness not only to correct mistakes but draw attention to them. To correct is not to fall below the highest standards of artistry but to sustain them. However, no explanation is provided for the most interesting omission, nor for the decision to semi-restore it in a footnote at line 91:

> Oh may no more a foreign master's rage
> With wrongs yet legal, curse a future age!
> Still spread, fair Liberty! thy heav'nly wings,
> Breath plenty on the fields, and fragrance on the springs.

These lines originally had a clear reference to William III, especially as they followed on from the account of the deaths of William I's sons, Richard and William Rufus, who, like William III, met their deaths through hunting. Brought back in a footnote during the reign of George II, they seem inevitably to point to the house of Hanover, and yet Pope can claim credit for his original decision to exclude the lines. This is another case, like the lines on 'Triumphant leaders' in *To Cobham*, where Pope uses his note to have his cake and eat it.

In the notes to the *Pastorals* and *Windsor-Forest*, we see a strong interest in the history of the poem. The reader is taken behind the scenes in order to appreciate the growing unity and art of the *Pastorals* and encouraged to trace the growth and development of *Windsor-Forest*. The annotation of *An Essay on Criticism* is less interested in its development than its design, with a complementary purpose of guiding readers through the argument of the poem. The chief means of doing this is through a prefatory contents list like those to *An Essay on Man* and the related epistles. Pope saw the *Essay on Criticism* as a poem that would be enhanced by the provision of a précis to guide and instruct the reader. The consequence, as with the *Essay on Man*, was to open up the possibility of claiming that the poem offered a more systematic treatment of the topic than Pope had originally intended, a systematic treatment that Warburton eventually supplied in his commentary. Warburton and Pope both aimed to rebut Addison's account of the poem's organization in his generally favourable review in *Spectator*, 253, of 20 December 1711: 'The Observations follow one another like those in *Horace's Art of Poetry*, without that methodological Regularity which would have been requisite in a prose author.' Pope believed that the poem was more carefully organized than that and used the contents to demonstrate the point.[25]

The most important structural aspect of the *Essay on Criticism* is its tripartite division. The 1736 edition makes that clear in the contents. 'PART I' has no general topic, but in 1744 Warburton helpfully suggests it 'gives rules for the Study of the Art of Criticism'; 1736 lists eleven numbered topics, each beginning on a separate line. 'PART II' has the general topic 'Causes hind'ring a true Judgment', followed by ten numbered topics run on. 'PART III', 'Rules for the Conduct of Manners in a Critic', has perhaps seven topics, though it is difficult to judge how Pope might have counted them, because they are not numbered; it was possible to précis the *Essay*, even in the third section, but not in so orderly a manner. The contents is not an analysis of the argument of the *Essay*; but it is successful in grouping related series of topics. In presenting the tripartite structure of the poem in the contents, the 1736 edition was clarifying a structure that had been advertised all along. In the first edition of 1711 Pope had intended to mark the sections by leaving a blank line and using full caps to start the new section. Unfortunately disagreements with the printer resulted in only partial realization of the plan: the first blank line is wasted, almost unnoticeable at the top of page 14, though there is unusual capitalization, 'OF all the Causes' (201–2).[26] The beginning of the third section reverses the pattern of success: the blank line works but the capitals are not present (560).[27] The arrangements in 1736 were much more successful in making their point: the *Essay* may not be systematic, but it is ordered.

The majority of new notes to the *Essay* in 1736 are textual. Twenty-three Variations are added to the notes present in 1717. The assignment to place this time is particularly careful, 'Between Verse 25 and 26, were these lines' and the like. As was the case with the *Pastorals*, some lines must have seemed simply too good to waste. The longest note gives ten lines satirizing inferior modes of wit—clowns, crambo, anagrams, acrostics, puns—another example of Pope's refining the texture of his poem. It is characteristic of Elwin's scholarship and animus against Pope that

---

[25] Warburton quotes Addison's review in the first note to the 1743/4 edition, in order to refute it. For a contrary view of the *Essay*, see *Richardsoniana* (1776), 264, quoted in *Twickenham*, I. 223 n. 3.

[26] I have used the line-numbering of the *Twickenham* edition. Because the *Twickenham* editors have dutifully followed the policy of combining a late text with earliest accidentals, they have preserved this bizarre 'OF', unparalleled, I suspect, in Pope's typography.

[27] David Foxon has a fascinating discussion of this section of the poem, without a concern with the issue of the poem's structure, in *Pope and the Book Trade*, 167–80.

he says that this passage 'was probably written after the poem was first published' because he senses a debt to Addison's *Spectator*, 63, of 12 May 1711. This conjecture is rightly rejected by the *Twickenham* editors on the ground that the lines are in the manuscript that went to the printer, but there is something of a puzzling coincidence here. Pope's poem (Foxon P806) was entered in the Stationers' Register on 11 May 1711, the day before Addison's essay appeared; it was published, probably, on 15 May 1711. The likelihood must be that Pope's lines influenced Addison and that he was given access to them in advance of the publication of the poem.

The most interesting note on *An Essay on Criticism* returns us to questions of history by drawing attention to a case of self-censorship. It is significant because it seems to say something about Pope's politics. Towards the end of the second part of the essay, Pope comments on the vices of the reign of Charles II, and then comments on those of William III:

> The following licence of a Foreign reign
> Did all the dregs of bold *Socinus* drain;
> Then unbelieving Priests reform'd the nation,
> And taught more pleasant methods of salvation . . . (546–9, *Twickenham*, 544–7)

The edition of 1736 supplied a note to line 548:

VER. 548. *The Author has omitted two lines which stood here, containing a National Reflection, which in his stricter judgment he could not but disapprove, on any People whatever.*

The omitted two lines are

> Then first the *Belgian* morals were extoll'd;
> We their religion had, and they our gold . . .

Pope really does seem to have repressed these lines for good. The contrast with the lines on the 'foreign master's rage' retrieved from the manuscript into the footnotes of *Windsor-Forest* is striking. One possible interpretation of the omission is that the arrival of the house of Hanover so altered Pope's view of William III and his Dutch followers, and that he was no longer prepared to make the gibe about low countries, Socinianism, or the draining of English resources to Holland. But the semi-restoration of the attack on the 'foreign master's rage' in *Windsor-Forest* suggests the omission had more to do with tone. My suspicion is that Pope is here deliberately showing himself capable of the delicacy recommended by Brossette. There was something politically crude in the lines, which, after all, appear in a passage mainly concerned with theological corruption. A follower of the Church of Rome would, on consideration, be reluctant to entangle questions of religious orthodoxy with questions of nationality. Pope was scrupulous enough to drop the attack, rather than merely relegate it to a footnote—but not scrupulous enough to avoid advertising his good deed.

The only other poem with extensive notes in octavo *Works* I, is the *Rape of the Lock*. *Eloisa to Abelard* is heavily revised (sixty-seven changes in accidentals), but not annotated. Pope seems in popularizing mode again in his notes to the *Rape*. He gives the customary information about publication informally, 'The first sketch of this Poem was written in less than a fortnight's time, in 1711', and goes on to introduce readers to the operation of mock-epic and explain passages they might find difficult. So the line, '*So spoke the Dame, but no Applause ensu'd' (5. 35) is annotated:

*It is a verse frequently repeated in *Homer* after any speech,
> So spoke—and all the Heroes applauded.

This is a long way from the bare classical references of the early editions of *An Essay on Criticism*. A more particular but similarly explanatory note is given on Umbriel's clapping his wings at the fight below (5. 53):

Minerva in like manner, during the Battle of *Ulysses* with the Suitors in *Odyss.* perches on a beam of the roof to behold it.

These notes develop a style set by the 1714 explanation that the history of Belinda's bodkin is 'In imitation of the progress of *Agamemnon*'s sceptre in Homer, *Il.* 2.' Additionally we find a reassuring note on the woman who imagined herself a goose pie, and a helpful and learned note on the sylphs:

[1.] VER. 145. *The busy* Sylphs, *&c.*] Ancient Traditions of the *Rabbi's* relate, that several of the fallen Angels became amorous of Women, and particularize some; among the rest *Asael*, who lay with *Naamah*, the wife of *Noah*, or of *Ham*; and who continuing impenitent, still presides over the Women's Toilets. *Bereshi Rabbi* in *Genes.* 6. 2.

Other notes are references only, to Homer (7), Virgil (4), Ovid (2), and Milton and Ariosto (1 each), and aimed at a different kind of reader, but they run alongside a persistent attempt to make the *Rape of the Lock* accessible to the non-learned, while demonstrating its author's own learning.

In comparison with the detailed re-presentation of the major poems of volume I, the annotation of the remaining material from 1717 in the new volume III (Griffith 417), 'Consisting of Fables, Translations, and Imitations', was relatively straightforward. The extra space created by the division of the volume enabled Pope to restore some of the annotation of the first editions, which had been dropped in 1717 and subsequently, and to provide extensive quotations from Chaucer to accompany the *Temple of Fame*. The advertisement to the volume shows the familiar care in sketching the details of Pope's career, and provides publication details. Because the poems are genuine juvenilia, the pride in the author's precocity mingles with a sense that they are hardly worthy of publication in a volume of their own; the tone is more apologetic than boastful. The extra space allowed Pope to put aside his worries about his debt to Chaucer's *House of Fame*. His original note expressed some unease:

The Hint of the following Piece was taken from *Chaucer*'s *House of Fame*. The Design is in a manner entirely alter'd, the Descriptions and most of the particular Thoughts my own: Yet I could not suffer it to be printed without this Acknowledgement, or think a Concealment of this Nature the less unfair for being common. The Reader who would compare this with *Chaucer*, may begin with his Third Book of *Fame*, there being nothing in the Two first Books that answers to their Title.

Like several of Pope's notes, this one is strangulated by his insistence on his own probity; it gives the impression that he felt a strong temptation to hide his debt, whereas the aim is to point up his superiority to those who hid theirs. In 1736 this was cleared up. Pope put a full stop after 'Acknowledgement' and left out 'or think a Concealment of this Nature the less unfair for being common'. He then added at the end, 'Wherever any hint is taken from him, the passage itself is set down in the marginal notes.' The ability to print the Chaucer in the footnotes eliminates the problem. Readers can judge for themselves, and the risks of denying a debt or exaggerating it are removed.

The Temple of Fame, with its passages from Chaucer and learned notes deriving from the first edition, set a standard for annotation that is not sustained in the rest of the volume, but the translations that followed hardly required it. Pope took the opportunity to print some of the Latin texts in full at the bottom of the page: 'Sapho to Phaon', 'Vertumnus and Pomona', 'The Fable of Dryope', 'The First Book of Statius'. But 'January and May' and 'The Wife of Bath' were printed without their Chaucerian equivalents. Conceived as a whole, the re-editing of the 1717 *Works* was a distinguished achievement, setting the poems in their biographical context, explaining literary traditions and influences, and attempting to remove difficulties for inexperienced readers.

THE ACHIEVEMENT OF THE OCTAVO ANNOTATION

Looked at from the perspective of Pope's entire career, the initiation of the octavo *Works* was a golden moment. The breach with Bernard Lintot was healed, and Pope could gather his whole literary career into his own keeping. Although the plans of his period of independence had not yet been fully realized, there was still time in which the gaps might be filled; and in the meantime it was possible to use the apparatus of the *Works* to show the range and coherence of what had been achieved so far, to place it in its historical context, and to assist the general reader in interpretation. In a letter to Swift of 16 February 1733, anticipating the second volume of the *Works*, Pope had said that in one respect they would be 'like the works of Nature, much more to be liked and understood when consider'd in the relation they bear with each other, than when ignorantly look'd upon one by one' (*Correspondence*, 3. 348). The grouping of the poems in the new books (with the early epistles moved to join the later in volume II, for example), the careful placing of the poems in a chronology of publication, and the demonstration of the unity of the *Essay on Man*, the four ethic epistles of the second book, and the *Pastorals*, all served to bind the works together into a coherent whole, impressive for its learning, integrity, and artistry. And the notes were both interesting and fun. They gave a historical context, but spiced it with curious facts, old anecdotes, and scandals, or they varied the canonical lines with alternatives, or hinted at mysterious and dangerous references; they pleased collectors by discussing textual variants, and they helped new readers by paraphrase and explanations. Only the self-defensive notes, giving the oxygen of publicity to stale quarrels and dreary charges, were a mistake, and Pope had the wisdom to eliminate some of them. The

Éclaircissemens were not complete—they never could be—but they were close to being just right. Although there are some lines missing in Pope's octavo *Works*, they are still probably one of the very best ways to read him.

What is striking over 250 years later is how successful Pope was in steering criticism of his work. Just as Scriblerus anticipated Aubrey Williams's interpretation of the *Dunciad*, so Pope's annotation leads to Warburton's systematization of his essays, with a whole variety of nineteenth- and twentieth-century responses, to Elwin–Courthope and then the *Twickenham* edition, to Reuben Brower's *The Poetry of Allusion* (1959), to the sections on social corruption in Howard Erskine-Hill's *The Social Milieu of Alexander Pope* (1975), and to the quarrels that have fed various biographical accounts of Pope, culminating in Maynard Mack's *Life*. Parts of this critical tradition might on reflection have worried Pope: Warburton's perverse, ingenious commentaries and the hopelessly unfunny hypercritics of Aristarchus, the venom of Elwin, the continuing controversy over *An Essay on Man*, perhaps even the overpowering generosity of the *Twickenham*'s industrious editors. But on the whole he would surely have found that the octavo *Works* had succeeded in their task of mesmerizing posterity.

Works Cited

Primary material, especially the work of Pope, is not listed here. Publication details for works published before 1900 are confined to place and date.

ADDISON, JOSEPH, and STEELE, Sir RICHARD, *The Spectator*, ed. Donald F. Bond, 5 vols. (Oxford: Clarendon Press, 1965).
ARTUS, THOMAS, sieur D'EMBRY, *Description de l'isle des hermaphrodites*, ed. Claude-Gilbert Dubois (Geneva: Droz SA, 1996).
ATIYAH, P. S., *The Rise and Fall of Freedom of Contract* (Oxford: Clarendon Press, 1979).
ATKINS, G. DOUGLAS, 'Pope and Deism: A New Analysis', *Huntington Library Quarterly*, 35 (1972), 257–78.
AULT, NORMAN, *New Light on Pope* (London: Methuen, 1949; repr. Hamden, Conn.: Archon, 1967).
—— *Pope's Own Miscellany* (London: Nonesuch Press, 1935).
AUSTIN, J. L., *How to Do Things with Words*, 2nd edn. (Oxford: Oxford University Press, 1975).
BAKHTIN, M. M., *The Dialogic Imagination: Four Essays*, ed. Michael Holquist, trans. Caryl Emerson and Michael Holquist (Austin: University of Texas Press, 1981).
—— *Problems of Dostoevsky's Poetics*, ed. and trans. Caryl Emerson (Manchester: Manchester University Press, 1984).
BALLASTER, ROSALIND, *Seductive Forms: Women's Amatory Fiction, 1684–1740* (Oxford: Clarendon Press, 1992).
BARNEY, STEPHEN A. (ed.), *Annotation and its Texts* (New York: Oxford University Press, 1991).
BLOOM, LILLIAN D., 'Pope as Textual Critic: A Bibliographical Study of his Horatian Text', *Journal of English and Germanic Philology*, 47 (1948), 150–5.
BOILEAU-DESPRÉAUX, NICOLAS, *Œuvres*, 1, ed. Jérôme Vercruysse (Paris: Garnier-Flammarion, 1969).
BOLINGBROKE, HENRY ST JOHN, Viscount, *The Works of Henry St. John, Lord Viscount Bolingbroke*, ed. David Mallet, 5 vols. (London, 1754).
BROWER, REUBEN ARTHUR, *Alexander Pope: The Poetry of Allusion* (London: Oxford University Press, 1959).
BROWNELL, MAURICE R., *Alexander Pope and the Arts of Georgian England* (Oxford: Clarendon Press, 1978).
BRÜCKMANN, PATRICIA, 'Virgins Visited by Angel Powers', in G. S. Rousseau and Pat Rogers (eds.), *The Enduring Legacy: Alexander Pope Tercentenary Essays* (Cambridge: Cambridge University Press, 1988), 3–20.
BUCHOLZ, R. O., *The Augustan Court: Queen Anne and the Decline of Court Culture* (Stanford, Calif.: Stanford University Press, 1993).
BURNET, GILBERT, *History of his Own Time*, 6 vols. (Oxford: Oxford University Press, 1833).
BUTT, JOHN, 'Pope's Poetical Manuscripts', Warton Lecture on English Poetry, *Proceedings of the British Academy*, 40 (1954), 23–39.

CARRETTA, VINCENT, ' "Images Reflect from Art to Art": Alexander Pope's Collected Works of 1717', in Neil Fraistat (ed.), *Poems in their Place: The Intertextuality and Order of Poetic Collections* (Chapel Hill: University of North Carolina Press, 1986), 195–233.
CASE, A. E., *A Bibliography of English Poetical Miscellanies* (Oxford: Bibliographical Society, 1935).
CIBBER, COLLEY, *A Letter from Mr Cibber to Mr Pope*, introd. Helene Koon, Augustan Reprint Society 159 (Los Angeles: William Andrews Clark Library, 1973).
CLAUSEN, WENDELL, *A Commentary on Virgil's Eclogues* (Oxford: Clarendon Press, 1994).
COLEIRO, EDWARD, *An Introduction to Vergil's Bucolics with a Critical Edition of the Text* (Amsterdam: B. R. Grüner, 1979).
COLOMB, GREGORY C., *Designs on Truth: The Poetics of the Augustan Mock-Epic* (Philadelphia: Pennsylvania State University Press, 1992).
DENNIS, JOHN, *The Critical Works of John Dennis*, ed. Edward Niles Hooker, 2 vols. (Baltimore: Johns Hopkins University Press, 1939–43).
DENTITH, SIMON (ed.), *Bakhtinian Thought: An Introductory Reader* (London: Routledge, 1995).
DERRIDA, JACQUES, 'Living on. Border Lines', in *Deconstruction and Criticism*, ed. Harold Bloom et al. (New York: Seabury Press, 1979), 75–176.
——*A Derrida Reader: Between the Blinds*, ed. Peggy Kamuf (Hemel Hempstead: Harvester Wheatsheaf, 1991).
DEUTSCH, HELEN, *Resemblance & Disgrace: Alexander Pope and the Deformation of Culture* (Cambridge, Mass.: Harvard University Press, 1996).
DIXON, PETER, 'Pope and James Miller', *Notes and Queries*, 215 (1970), 91–2.
DOWNES, KERRY, *The Architecture of Wren*, 2nd edn. (Reading: Redhedge, 1988).
EISENSTEIN, ELIZABETH L., *The Printing Press as an Agent of Social Change*, 2 vols. (Cambridge: Cambridge University Press, 1979).
ELLIOTT, ROBERT C., *The Power of Satire: Magic, Ritual, Art* (Princeton: Princeton University Press, 1960).
EMPSON, WILLIAM, *Seven Types of Ambiguity*, 3rd edn. repr. (Harmondsworth: Penguin, 1965).
ERSKINE-HILL, HOWARD, *The Augustan Idea in English Literature* (London: Edward Arnold, 1983).
——'Literature and the Jacobite Cause: Was There a Rhetoric of Jacobitism?', in Eveline Cruickshanks (ed.), *Ideology and Conspiracy: Aspects of Jacobitism, 1689–1759* (Edinburgh: John Donald, 1982), 49–69.
——*Poetry of Opposition and Revolution: Dryden to Wordsworth* (Oxford: Clarendon Press, 1996).
——'The Satirical Game at Cards in Pope and Wordsworth', in Claude Rawson and Jenny Mezciems (eds.), *English Satire and the Satiric Tradition* (Oxford: Basil Blackwell, 1984), 183–95.
——*The Social Milieu of Alexander Pope: Lives, Example, and the Poetic Response* (New Haven: Yale University Press, 1975).
FERBER, MICHAEL (ed.), *A Dictionary of Literary Symbols* (Cambridge: Cambridge University Press, 1999).
FERRARO, JULIAN, 'From Text to Work: The Presentation and Re-presentation of *Epistles to Several Persons*', in Howard Erskine-Hill (ed.), *Alexander Pope: World and Word*, Proceedings of the British Academy 91 (Oxford: Oxford University Press for the British Academy, 1998), 111–34.

FERRARO, JULIAN, '"Rising into Light": The Evolution of Pope's Poems in Manuscript and Print' (unpublished doctoral thesis, University of Cambridge, 1993).
—— 'The Satirist, the Text and "The World Beside": Pope's *First Satire of the Second Book of Horace Imitated*', *Translation and Literature*, 2 (1993), 37–63.
FLETCHER, F. T. H., *Pascal and the Mystical Tradition* (Oxford: Blackwell, 1954).
FORCE, JAMES E., *William Whiston: Honest Newtonian* (Cambridge: Cambridge University Press, 1985).
FOUCAULT, MICHEL, 'What Is an Author?', in David Lodge (ed.), *Modern Criticism and Theory* (London: Longman, 1988), 196–210.
FOXON, DAVID F., *English Verse 1701–1750*, 2 vols. (Cambridge: Cambridge University Press, 1975).
—— *Pope and the Early Eighteenth-Century Book Trade* (Oxford: Clarendon Press, 1991).
FRAENKEL, EDUARD, *Horace* (Oxford: Clarendon Press, 1957).
FUCHS, JACOB, *Reading Pope's 'Imitations of Horace'* (Lewisburg, Pa.: Bucknell University Press, 1989).
GALLAGHER, CATHERINE, *Nobody's Story: The Vanishing Acts of Women Writers in the Market Place, 1670–1820* (Oxford: Clarendon Press, 1994).
GAY, JOHN, *Poetry and Prose*, ed. Vinton A. Dearing, 2 vols. (Oxford: Clarendon Press, 1974).
GENETTE, GÉRARD, *Paratexts: Thresholds of Interpretation* (Cambridge: Cambridge University Press, 1987).
GREETHAM, DAVID C., '[Textual] Criticism and Deconstruction', *Studies in Bibliography*, 44 (1991), 1–30.
GREGG, EDWARD, *Queen Anne* (London: Routledge, 1980).
GRICE, H. P., 'Meaning', *Philosophical Review*, 66 (1957), 377–88.
GRIFFIN, DUSTIN, *Literary Patronage in England, 1650–1800* (Cambridge: Cambridge University Press, 1996).
GRIFFITH, REGINALD H., *Alexander Pope: A Bibliography*, 2 vols. (Austin: University of Texas Press, 1922–7).
GRUNDY, ISOBEL, *Lady Mary Wortley Montagu: Comet of the Enlightenment* (Oxford: Oxford University Press, 1999).
—— '*Verses Address'd to the Imitator of Horace*: A Skirmish between Pope and Some Persons of Rank and Fortune', *Studies in Bibliography*, 30 (1977), 96–119.
GUERINOT, J. V., *Pamphlet Attacks on Alexander Pope, 1711–1744* (London: Methuen, 1969).
HABERMAS, JÜRGEN, *The Structural Transformation of the Public Sphere: An Inquiry into a Category of Bourgeois Society*, trans. Thomas Burger, with Frederick Lawrence (Cambridge: Polity Press, 1989).
HALSBAND, ROBERT, *Lord Hervey: Eighteenth-Century Courtier* (Oxford: Clarendon Press, 1973).
—— *'The Rape of the Lock' and its Illustrations, 1714–1896* (Oxford: Clarendon Press, 1980).
HAMMELMANN, HANS, *Book Illustrators in Eighteenth-Century England*, ed. and completed by T. S. R. Boase (New Haven: Yale University Press, 1975).
HAMMOND, BREAN, *Professional Imaginative Writing in England, 1670–1740: Hackney for Bread* (Oxford: Clarendon Press, 1997).
HARRIS, JOHN, *The Palladians* (London: Trefoil Books, 1981).
HOLMES, GEOFFREY, *The Trial of Dr Sacheverell* (London: Eyre Methuen, 1973).
HOLQUIST, MICHAEL, *Dialogism: Bakhtin and his World* (London: Routledge, 1990).
The Holy Bible: A Facsimile in a Reduced Size of the Authorized Version Published in the Year 1611, introd. A. W. Pollard (Oxford: Oxford University Press, 1911).

HORACE (Quintus Horatius Flaccus), *Horace: Satires and Epistles; Persius: Satires*, trans. N. Rudd (London: Penguin Books, 1979).
—— *Horace: Satires, Epistles, and Ars Poetica*, trans, H. Rushton Fairclough (London: Heinemann, 1926, rev. 1929).
HUME, DAVID, *An Enquiry Concerning Human Understanding*, ed. Tom Beauchamp (Oxford: Oxford University Press, 1999).
HUNTER, G. K., 'The "Romanticism" of Pope's Horace', *Essays in Criticism* (1960), 390–4.
ILCHESTER, Earl of, *Lord Hervey and his Friends, 1726–1738* (London: John Murray, 1950).
JACK, IAN, *The Poet and his Audience* (Cambridge: Cambridge University Press, 1984).
JARDINE, LISA, *Erasmus, Man of Letters: The Construction of Charisma in Print* (Princeton: Princeton University Press, 1993).
JOHNSON, SAMUEL, *Lives of the English Poets*, ed. George Birkbeck Hill, 3 vols. (Oxford: Clarendon Press, 1905).
JONSEN, ALBERT R., and TOULMIN, STEPHEN, *The Abuse of Casuistry: A History of Moral Reasoning* (Berkeley and Los Angeles: University of California Press, 1988).
JONSON, BEN, *Ben Jonson*, ed. C. H. Herford, and Percy and Evelyn Simpson, 11 vols. (Oxford: Clarendon Press, 1925–50).
KEENER, FREDERICK M., 'Pope, *The Dunciad*, Virgil, and the New Historicism of Le Bossu', *Eighteenth Century Life*, 15 (1991), 35–57.
KENNEY, E. J., *The Classical Text: Aspects of Editing in the Age of the Printed Book* (Berkeley and Los Angeles: University of California Press, 1974).
KERNAN, ALVIN, *Printing Technology, Letters & Samuel Johnson* (Princeton: Princeton University Press, 1987).
KNYFF, L. (artist), and KYP, I. (engraver), *Britannia Illustrata or Views of Several of the Queens Palaces as Also of the Principal Seats of the Nobility and Gentry of Great Britain* (London, 1709).
LANGFORD, PAUL, *A Polite and Commercial People: England, 1727–1783* (Oxford: Clarendon Press, 1989).
LAW, ERNEST, *The History of Hampton Court Palace*, 3 vols. (London, 1885–91).
LEE, HERMIONE, *Virginia Woolf* (London: Chatto & Windus, 1996).
LERANBAUM, MIRIAM, *Alexander Pope's 'Opus Magnum'* (Oxford: Oxford University Press, 1977).
LORD, G. DE F., et al. (eds.), *Poems on Affairs of State*, 7 vols. (New Haven: Yale University Press, 1963–75).
MACAULAY, THOMAS BABINGTON, Lord, *The History of England from the Accession of James the Second*, ed. Charles Harding Firth, 6 vols. (London: Macmillan, 1913).
MCGANN, JEROME, 'Modernism and the Renaissance of Printing', in his *Black Riders: The Visible Language of Modernism* (Princeton: Princeton University Press, 1993), 3–41.
MACK, MAYNARD, *Alexander Pope: A Life* (New Haven: Yale University Press, 1985).
—— *Collected in Himself* (Newark: University of Delaware Press, 1982).
—— *The Garden and the City: Retirement and Politics in the Later Poetry of Pope* (Toronto: University of Toronto Press, 1969).
—— *The Last and Greatest Art: Some Unpublished Poetical Manuscripts of Alexander Pope* (Newark: University of Delaware Press, 1984).
MCKEON, MICHAEL, *The Origins of the English Novel, 1600–1740* (Baltimore: Johns Hopkins University Press, 1987).
MCLAVERTY, JAMES, 'The Contract for Pope's Translation of Homer's *Iliad*: An Introduction and Transcription', *Library*, 6th ser. 15 (1993), 206–25.

McLaverty, James, 'The First Printing and Publication of Pope's Letters', *Library*, 6th ser. 2 (1980), 264–80.
—— 'Issues of Identity and Utterance', in Philip Cohen (ed.), *Devils and Angels: Textual Editing and Literary Theory* (Charlottesville: University Press of Virginia, 1991), 134–51.
—— 'Lawton Gilliver: Pope's Bookseller', *Studies in Bibliography*, 32 (1979), 101–24.
—— 'The Mode of Existence of Literary Works of Art: The Case of the *Dunciad Variorum*', *Studies in Bibliography*, 37 (1984), 82–105.
—— ' "Of Which Being Publick the Publick Judge": Pope and the Publication of *Verses Address'd to the Imitator of Horace*', *Studies in Bibliography*, 51 (1998), 183–204.
—— 'Pope in the Private and Public Spheres: Annotations in the Second Earl of Oxford's Volume of Folio Poems, 1731–1736', *Studies in Bibliography*, 48 (1995), 33–59.
—— 'Pope and Giles Jacob's *Lives of the Poets*: The Dunciad as Alternative Literary History', *Modern Philology*, 83 (1985–6), 22–32.
—— *Pope's Printer, John Wright*, Oxford Bibliographical Society Occasional Publication 11 (Oxford: Bibliographical Society, 1977).
McLuhan, Marshall, *The Gutenberg Galaxy: The Making of Typographic Man* (London: Routledge & Kegan Paul, 1962).
Manley, Delarivier, *The New Atalantis*, ed. Rosalind Ballaster (London: Penguin Books, 1991).
Maslen, Keith I. D., and Lancaster, John (eds.), *The Bowyer Ledgers* (London: Bibliographical Society, 1991).
Mengel, Elias, jun., 'The *Dunciad* Illustrations', *Eighteenth-Century Studies*, 7 (1973–4), 161–78.
Merrill, Walter McIntosh, *From Statesman to Philosopher: A Study in Bolingbroke's Deism* (New York: Philosophical Library, 1949).
Monod, Paul Kléber, *Jacobitism and the English People, 1688–1788* (Cambridge: Cambridge University Press, 1989).
Montagu, Lady Mary Wortley, *Essays and Poems*, ed. Robert Halsband and Isobel Grundy (Oxford: Clarendon Press, 1993).
Newey, Vincent, *Cowper's Poetry: A Critical Study and Reassessment* (Liverpool: Liverpool University Press, 1982).
Nichols, John, *Literary Anecdotes of the Eighteenth Century*, 9 vols. (London, 1812–15).
Nokes, David, *John Gay: A Profession of Friendship* (Oxford: Oxford University Press, 1995).
Norton, Rictor, *Mother Clap's Molly House: The Gay Subculture in England, 1700–1830* (London: GMP, 1992).
Nussbaum, Felicity, *The Brink of All We Hate: English Satires on Women, 1660–1750* (Lexington: University Press of Kentucky, 1984).
Nuttall, A. D., *Pope's 'Essay on Man'* (London: George Allen & Unwin, 1984).
O'Sullivan, Maurice J., jun., 'Ex Alieno Ingenio Poeta: Johnson's Translation of Pope's *Messiah*', *Philological Quarterly*, 54 (1975), 579–91.
Papali, G. F., *Jacob Tonson, Publisher* (Auckland: Tonson, 1968).
Pascal, Blaise, *Pensées sur la religion et sur quelques autres sujets*, introd. Louis Lafuma, 3 vols. (Paris: Éditions du Luxembourg, 1951).
—— *Pensées*, trans. and introd. A. J. Krailsheimer (Harmondsworth: Penguin Books, 1966).
Piper, David, *The Image of the Poet: British Poets and their Portraits* (Oxford: Clarendon Press, 1982).

POLLAK, ELLEN, *The Poetics of Sexual Myth: Gender and Ideology in the Verse of Swift and Pope* (Chicago: University of Chicago Press, 1985).
POPE, ALEXANDER, *The Correspondence of Alexander Pope*, ed. George Sherburn, 5 vols. (Oxford: Clarendon Press, 1956).
—— *Dunciad of 1728: A History and Facsimile*, ed. David L. Vander Meulen (Charlottesville: University Press of Virginia for Bibliographical Society of University of Virginia and New York Public Library, 1991).
—— *The Dunciad Variorum 1729* [facsimile] (Menston: Scolar Press, 1968).
—— *The Dunciad in Four Books*, ed. Valerie Rumbold (Harlow: Longman, 1999).
—— *An Epistle from Mr Pope to Dr Arbuthnot* [facsimile], ed. David F. Foxon (Menston: Scolar Press, 1970).
—— *An Essay on Criticism (1711)* [facsimile], ed. David F. Foxon (Menston: Scolar Press, 1970).
—— *An Essay on Man* [facsimile], ed. David F. Foxon (Menston: Scolar Press, 1969).
—— et al., *Memoirs of the Extraordinary Life, Works, and Discoveries of Martinus Scriblerus*, ed. Charles Kerby-Miller (New York: Russell & Russell, 1966).
—— *The Prose Works of Alexander Pope*, 1, ed. Norman Ault (Oxford: Basil Blackwell, 1936).
—— *The Prose Works of Alexander Pope*, 2, ed. Rosemary Cowler (Oxford: Shakespeare Head for Basil Blackwell, 1986).
—— *The Rape of the Lock* [facsimile], ed. David F. Foxon (Menston: Scolar Press, 1969).
—— *The Twickenham Edition of the Poems of Alexander Pope*, ed. John Butt et al., 11 vols. (London: Methuen, 1939–69).
—— *The Works of Alexander Pope*, ed. W. Elwin and W. J. Courthope, 10 vols. (London, 1871–89).
—— *The Works of Alexander Pope*, ed. Joseph Warton, 9 vols. (London, 1797).
PUTTENHAM, GEORGE, *Arte of English Poesie*, ed. Gladys D. Willcock and Alice Walker (Cambridge: Cambridge University Press, 1936).
RAWSON, CLAUDE, 'Heroic Notes: Epic Idiom, Revision and the Mock-Footnote from the *Rape of the Lock* to the *Dunciad*', in Howard Erskine-Hill (ed.), *Alexander Pope: World and Word*, Proceedings of the British Academy 91 (Oxford: Oxford University Press for the British Academy, 1998), 69–110.
REID, GEORGE WILLIAM, *Catalogue of Prints and Drawings in the British Museum*, division 1, volume 2: *June 1698–1733* (London: Printed by Order of the Trustees, 1873).
RICHARDSON, JONATHAN, jun., *Richardsoniana* (London, 1776).
RIDEOUT, TANIA, 'The Reasoning Eye: Alexander Pope's Typographic Vision' (unpublished doctoral thesis, University of Cambridge, 1994).
—— 'The Reasoning Eye: Alexander Pope's Typographic Vision in the *Essay on Man*', *Journal of the Warburg and Courtauld Institutes*, 55 (1992), 249–62.
RIGGS, DAVID, *Ben Jonson: A Life* (Cambridge, Mass.: Harvard University Press, 1989).
ROBB, NESCA A., *William of Orange*, 2 vols. (London: Heinemann, 1962–6).
ROGERS, ROBERT W., *The Major Satires of Alexander Pope* (Urbana: University of Illinois Press, 1955).
—— Review of Lilian D. Bloom's article, listed above, *Philological Quarterly*, 28 (1949), 397–8.
ROSE, MARK, *Authors and Owners: The Invention of Copyright* (Cambridge, Mass.: Harvard University Press, 1993).

RUDD, NIALL, *The Satires of Horace* (Cambridge: Cambridge University Press, 1966).
RUMBOLD, VALERIE, *Women's Place in Pope's World* (Cambridge: Cambridge University Press, 1989).
SCARBORO, DONNA, ' "Thy Own Importance Know": The Influence of *Le Comte de Gabalis* on *The Rape of the Lock*', *Studies in Eighteenth-Century Culture*, 14 (1985), 231–41.
SEARLE, JOHN R., *Expression and Meaning: Studies in the Theory of Speech Acts* (Cambridge; Cambridge University Press, 1979).
—— *Intentionality: An Essay in the Philosophy of Mind* (Cambridge: Cambridge University Press, 1983).
—— *Speech Acts: An Essay in the Philosophy of Language* (Cambridge: Cambridge University Press, 1969).
SHERBURN, GEORGE, *The Early Career of Alexander Pope* (Oxford: Clarendon Press, 1934).
SOLOMON, HARRY M., *The Rape of the Text: Reading and Misreading Pope's 'Essay on Man'* (Tuscaloosa: University of Alabama Press, 1993).
SOMMERVILLE, JOHANN P., 'The "New Art of Lying": Equivocation, Mental Reservation, and Casuistry', in Edmund Leites (ed.), *Conscience and Casuistry in Early Modern Europe* (Cambridge: Cambridge University Press, 1988), 159–84.
SPENCE, JOSEPH, *Observations, Anecdotes, and Characters of Books and Men*, ed. James M. Osborn, 2 vols. (Oxford: Clarendon Press, 1966).
SPURGEON, CAROLINE F. E., *Five Hundred Years of Chaucer Criticism and Allusion, 1357–1900*, 3 vols. (Cambridge: Cambridge University Press, 1925).
STACK, FRANK, *Pope and Horace: Studies in Imitation* (Cambridge: Cambridge University Press, 1985).
STALLYBRASS, PETER and WHITE, ALLON, *The Politics and Poetics of Transgression* (London: Methuen, 1986).
STEIGER, RICHARD, *The English and Latin Texts of Pope's 'Imitations of Horace'* (New York: Garland, 1988).
STEPHEN, Sir LESLIE, *History of English Thought in the Eighteenth Century*, 3rd edn., 2 vols. (London: Smith, Elder, 1902).
THEOBALD, LEWIS, *Shakespeare Restored* [facsimile] (London: Frank Cass, 1971).
TRACY, CLARENCE, *The Rape Observ'd* (Toronto: University of Toronto Press, 1974).
TREVELYAN, GEORGE MACAULAY, *England under Queen Anne*, 3 vols. (London: Longmans Green, 1930–4).
TRUMBACH, RANDOLPH, *Sex and the Gender Revolution: Heterosexuality and the Third Gender in Enlightenment London* (Chicago: University of Chicago Press, 1998).
—— 'Sodomitical Subcultures, Sodomitical Roles, and the Gender Revolution of the Eighteenth Century: The Recent Historiography', in Robert Purks Maccubbin (ed.), *'Tis Nature's Fault: Unauthorized Sexuality during the Enlightenment* (Cambridge: Cambridge University Press, 1987), 109–21.
VILLARS, MONTFAUCON DE, *Le Comte de Gabalis ou Entretien sur les sciences secretes. La critique de Bérénice*, ed. Roger Laufer (Paris: A. G. Nizet, 1963).
—— *Comte de Gabalis, Rendered in English*, 'Published by the Brothers' (London: Old Bourne Press, 1913).
—— *The Diverting History of the Count de Gabalis*, 2nd edn. (London, 1714).
—— *La Première Critique des 'Pensées'*, ed. Dominique Descotes (Lyons: Centre National de la Recherche Scientifique, 1982).
WANLEY, HUMFREY, *The Diary of Humfrey Wanley, 1715–1726*, ed. C. E. Wright and Ruth C. Wright, 2 vols. (London: Bibliographical Society, 1966).

WARNER, WILLIAM B., *Licensing Entertainment: The Elevation of Novel Reading in Britain, 1684–1750* (Berkeley and Los Angeles: University of California Press, 1998).
WARTON, JOSEPH, *An Essay on the Genius and Writings of Pope*, 5th edn., 2 vols. (London, 1806).
WATT, IAN, *The Rise of the Novel* (London: Chatto, 1957).
WEINBROT, HOWARD D., *Alexander Pope and the Traditions of Formal Verse Satire* (Princeton: Princeton University Press, 1982).
—— *Augustus Caesar in 'Augustan' England* (Princeton: Princeton University Press, 1978).
—— *Eighteenth-Century Satire: Essays on Text and Context from Dryden to Peter Pindar* (Cambridge: Cambridge University Press, 1988).
—— *The Formal Strain: Studies in Augustan Imitation and Satire* (Chicago: University of Chicago Press, 1969).
WHITE, DOUGLAS H., *Pope and the Context of Controversy: The Manipulation of Ideas in 'An Essay on Man'* (Chicago: University of Chicago Press, 1970).
—— and TIERNEY, THOMAS P., '*An Essay on Man* and the Tradition of Satires on Mankind', *Modern Philology*, 85 (1987), 27–41.
WILLIAMS, AUBREY L., 'Pope and Horace: *The Second Epistle of the Second Book*', in Caroll Camden (ed.), *Restoration and Eighteenth Century Literature* (Chicago: Chicago University Press, 1965).
—— *Pope's 'Dunciad': A Study of its Meaning* (London: Methuen, 1955).
WIMSATT, W. K., *The Portraits of Alexander Pope* (New Haven: Yale University Press, 1965).
WOLLASTON, WILLIAM, *The Religion of Nature Delineated* (New York: Garland, 1978).
WOODMANSEE, MARTHA, 'The Genius and the Copyright: Economic and Legal Conditions of the Emergence of the "Author"', *Eighteenth-Century Studies*, 17 (1984), 425–48.
—— and JASZI, PETER (eds.), *The Construction of Authorship: Textual Appropriation in Law and Literature* (Durham, NC: Duke University Press, 1994).

Index

There is no entry for Pope; his poems, including his translations, are entered independently under their titles. Peers are entered under their titles.

Act for laying several duties upon all sope and paper 3
Addison, Joseph 14, 21, 58, 60, 76–7, 183, 187–8, 203, 207, 217, 223–4, 231, 237, 238
advertisement: preliminary 73, 76, 77, 88, 90, 105, 149, 176–7, 190, 218, 239; for sale 25, 102, 113, 214, 215
Allgemeines Oeconomisches Lexicon (1753) 61
Anacreon 48
Anne, Queen 30, 32, 33, 37, 40–1, 172, 229, 231
Apollo 64–5, 67–8, 70–1, 110, 188, fig. 4
Arbuthnot, John 185, 207, 222; see also *Epistle from Mr Pope to Dr Arbuthnot*
Archias 66
Ariosto, Ludovico 239
Aristophanes 46
Aristotle 52, 74, 101
Arnall, William 191
Atiyah, P. S. 155
Atterbury, Francis, Bishop of Rochester 162, 222
Ault, Norman 17, 69
Austin, J[ohn] L[angshaw] 8
'Author to the Reader' (*Works* II) 216–17

Bakhtin, Mikhail M. 84–7, 88, 90, 91, 93
Ballaster, Rosalind 34
Barber, John 35, 173
Barret, J. 17
Bate, Mr 17
Bateson, F. W. 212–13, 226
Bathurst, Allen, 1st Earl 116, 146, 203, 222, 227, see also *Epistle to . . . Bathurst*
Baxter, William 158
Beaufort, Henry Somerset, 2nd Duke of 36
Beaumont, Francis 53
Bentinck, William, 1st Earl of Portland 34
Bentley, Richard 5, 55, 87, 88, 90, 157, 158, 168, 195, 196, 199
Bentley, Thomas 149, 202, 206, 222
Bethel, Hugh 165, 169
Betterton, Thomas 17, 18
Blackmore, Sir Richard 4, 101, 151, 163
Blome, Richard 95
Bloom, Lillian D. 157–8
Blount, Martha (Patty) 18, 22, 201
Blount, Teresa 18, 22
Boethius, Anicius Manlius Severinus 131

Boileau-Despréaux, Nicolas 11, 48, 64, 70, 87–9, 90, 91–2, 151, 166–7, 210–14, 220, 233
Bolingbroke, Henry St John, 1st Viscount 66, 109, 110, 116–17, 120–1, 130, 137, 143–4, 157, 162, 167, 200, 217, 222, 225
Bond, Denis 221
Bond, William 103
Boreman, Thomas 113, 195, 196
Bosc, Claude du 26
Bossu, René Le 99–105
Boswell, James 52
Bowyer, William, sen. 5–6, 7, 47, 61, 215
Bowyer, William, jun. 5–6
Bramston, James 151, 189
Breval, John Durant 103
Bridgman, Charles 146, 174
Brindley, John 195, 214
Brooke, Henry 189
Broome, William 16, 17, 19, 111, 113, 144, 189, 224, 231
Brossette, Claude 88–90, 167, 210–14, 216, 219, 220, 227, 231, 233, 235, 238
Brower, Reuben 241
Browne, Anthony 202
Brutus, Marcus Junius 162
Buckingham, John Sheffield, 1st Duke of 7, 50, 61, 63, 66, 67, 70–1, 222
Budgell, Eustace 96, 163, 224, 231
Burlington, Dorothy Savile, Countess of 146
Burlington, Richard Boyle, 3rd Earl of 111, 145, 146, 219, 222, 227; see also *Epistle to . . . Burlington*
Burnet, Gilbert, Bishop of Salisbury 30
Burnet, Sir Thomas 219
Butt, John 179, 182–3, 185, 201–2
Byron, George Gordon, 6th Baron 85

Caesar, Augustus (Gaius Julius Caesar Octavianus) 157, 162–4, 168
Callimachus 68, 70
capitals 8, 10, 13, 24, 25, 87, 89, 125, 152, 159, 163, 166, 237; caps. and smalls 9, 135–6, 159, 162, 163, 166, 195, 222, 228
Caroline (of Brandenburg-Ansbach), Queen 145, 179–80, 204
Carretta, Vincent 62, 64
Caryll, John 17, 18, 113–14, 121, 149, 201, 225

catchwords 89
Caxton, William 4, 82
Celebrated Beauties, The (1709) 19
Centlivre, Susannah 96
Chandler, Edward 77–8
Chandos, James Brydges, 1st Duke of 144, 145, 181, 204, 224, 231
Charles II, King 235, 238
Chartres [or Charteris], Francis 221
Chaucer, Geoffrey 18, 48, 73, 74–5, 239–40
Cheselden, William 219
Chetwood, William Rufus 4
Chubb, Thomas 131, 138, 140
Cibber, Colley 84, 96, 97, 185, 191
Cicero, Marcus Tullius 66, 74, 127, 232
class 11–12, 105, 143, 144–7, 153–4, 172, 198, 199, 220
Cleland, William 105–6, 164, 213
Cobham, Richard Temple, Viscount 146, 202
Colepepper, Sir William 221
Collins, Anthony 77–8
commendatory poems 16, 50, 66–8, 70, 71
Comte de Gabalis, Le 37–41
Concanen, Matthew 90, 95
Congreve, William 53
contents 66, 68–70, 82, 117–19, 122, 124, 127, 209, 213, 218, 227–9, 236–7
Cook[e], Thomas 219
copyright 49–52, 53–6, 144, 189, 195, 215
Copyright Act (1709) 51, 55, 56
Corbet[t], Charles 195
Cotton, Charles 53
court, the 3, 5, 11, 14, 22, 30–1, 34–6, 37, 40–1, 100, 103, 109–10, 143–7, 161, 162–3, 166, 175, 177, 178–80
Courthope, William John 203, 226, 241
Cowley, Abraham 232
Cowper, John 145
Craggs, James 88, 111
Creech, Thomas 9
Cromwell, Henry 17, 55, 233
Crousaz, Jean-Pierre de 55, 114, 116, 137
cues, *see* indices
Cunningham, Alexander 157, 158
Cupid 73, 202
Curll, Edmund 4, 37, 82, 90, 98, 104, 113, 178, 195, 203, 233

Dacier, André 28, 151, 155, 161, 162, 163, 166, 169–70, 195, 198–9
Dearing, Vinton 16
dedication 5, 14, 18, 22–4, 37, 41, 61, 96, 101, 188, 257
Defoe, Daniel 53, 84
Dennis, John 6, 38–9, 55, 63, 82, 90, 91, 94, 95, 96–7, 104, 187
Descartes, René 128

design, as prelim 117
Desprez, Ludovicus 157
dialogism 82–106
dialogue, formal 148, 152, 154, 155–7, 161–2, 164, 173, 202
'Discourse on Pastoral Poetry' 49, 71, 75
Dixon, Peter 151
Dodd, Anne 178, 180
Dod[d]ington, George Bubb, Baron Melcombe 188, 193
Dodsley, Robert 214
Donaldson v. Becket (copyright case, 1774) 51
Donne, John 147, 217, 218, 219
Dormer, James 202
Downes, Kerry 30
Dryden, John 15, 17, 20, 46–7, 48, 53, 73, 82, 87, 93, 105, 149, 192–3, 194, 223, 227, 231, 234, fig. 9
Duckett, George 222
Dunciad, The 1, 2, 3–5, 26, 34, 44, 144, 149, 178, 186, 187, 207, 213, 215, 219, 220, 223, 224, 232, 241; *Dunciad* (1728) 5, 23, 84, 92; *Dunciad Variorum* 3, 6, 10, 11, 47, 75, 82–106, 109, 111, 113, 117, 149, 157, 164, 168, 209, 210, 213, 220, 222, figs. 6, 9; *Dunciad in Four Books* 5–6, 84
duodecimo 25, 215
dropped head 2, 44, 149, 155, 196

Eisenstein, Elizabeth 84
'Elegy to the Memory of an Unfortunate Lady' 49
Elliott, Robert C. 28–9
Eloisa to Abelard 50, 67, 73–4, 232, 233, 238
Elwin, Whitwell 237–8, 241
Empson, Sir William 161, 196
endnotes 78, 183, 218, 219, 220–7
Ennius, Quintus 163
'Epilogue to *Jane Shore*' 50
'Episode of Sarpedon, The' 15, 121
Epistle from Mr Pope to Dr Arbuthnot, An 10, 11, 12, 52, 59, 113, 146, 147–8, 164, 174, 175–208, 211, 216, 217–18, 219, 220, 222–4, 227, 231, fig. 8
Epistle to a Lady, An 113, 201, 217, 218, 219, 226, 228, 229–30, fig. 7
Epistle to . . . Allen, Lord Bathurst, An 108, 109, 113, 116, 147, 151, 152, 211, 217, 218, 219, 220–2, 227, 228, 229, 230
'Epistle to Miss Blount, on her Leaving the Town, after the Coronation' 21, 49–50
'Epistle to Miss Blount, with the Works of Voiture' 17
Epistle to Mr Jervas 26
Epistle to . . . Richard Earl of Burlington, An 12, 36, 108, 113, 143–7, 151, 152, 155, 165, 172, 174, 186, 190, 193, 203, 211, 217, 218, 227, 228, 229, 230–1

Epistle to . . . Richard Lord Visct. Cobham, An 217, 218, 219, 226, 228, 229, 236
'Epitaph. On Sir William Trumbull' 50
epitaphs 217
Erasmus, Desiderius 46, 103
Erskine-Hill, Howard 33–4, 232, 241
Eschenbach, Wolfram von 86
essay, as poetic form 116–25
Essay on Criticism, An 2, 7, 8–9, 21, 49–50, 61, 69–70, 71, 89, 149, 186, 209, 213, 232, 233, 236–8, 239, fig. 4
Essay on Man, An 11, 12, 55, 107–41, 151, 159, 214, 217, 218–19, 220, 224–5, 226, 227, 228–9, 236–7, 240, 241
Etherege, Sir George 53
Eusden, Laurence 102, 104

F., R. 20
'Fable of Dryope, The' 49, 240
'Fair Nun, The' 20
Fairclough, H. Rushton 174, 198, 199
fame 16, 21, 22, 56, 60, 62, 63, 64–5, 66–8, 70, 71, 87, 96, 147, 188, 206, 215; see also *Temple of Fame*
Fenton, Elijah 16, 17, 19, 20, 48, 66, 67–8, 70–1, 111, 113, 144, 189
Fermor, Arabella 11, 14, 15, 18, 19, 22–4, 29, 34, 35, 41–4
Ferraro, Julian 185
Fichte, Johann Gotlieb 51
Fielding, Henry 84, 86
fine paper 25, 47, 48, 215
First Book of Statius his Thebais, The 17, 18, 73, 232, 240
First Epistle of the Second Book of Horace, The 149, 192, 200, 259
First Satire of the Second Book of Horace, The 3, 108, 109–10, 142, 143–4, 146, 147–68, 169, 171, 174, 178–9, 181–2, 187, 191, 197, 200, 206, 216, 219, fig. 10
Fletcher, John 53
folio 17, 25, 44, 46, 47, 48, 50, 95, 142, 148, 151–2, 166, 168, 176, 183, 184; see also *Works* II quarto and folio
Fontenelle, Bernard le Bovier de 75, 209
footnotes 2, 10, 13, 75, 82, 86, 89, 90, 93, 94, 95–9, 102, 103–4, 106, 149, 152, 180, 183, 218–20, 227–8, 232
format 11, 44, 47, 48, 75, 87, 142, 151–2, 214–16; see also duodecimo; folio; octavo; quarto
Fortescue, William 154, 157, 168, 179, 180, 191
Foucault, Michel 50–3
Fourth Satire of the First Book of Horace 147–8, 161, 164
Fox, Henry 164, 191
Fox, Stephen 145, 164
Foxon, David Fairweather 1, 6, 25, 47, 48, 144, 182–3, 195

Fracastoro, Girolamo 88
Frederick Louis, Prince of Wales 154
Frege, Gottlob 52
Froben, Johann 46
frontispiece 26–9, 31, 33, 36, 47, 49, 62–4, 73, 89, 95, 151, 176, 216
Fuchs, Jacob 166, 200, 203

Gallagher, Catherine 34–5
Garth, Sir Samuel 233, 235
Gay, John 16, 17, 18, 48, 120, 190, 192, 193–4, 222
genius 50, 53–4, 59, 60, 67, 68, 71, 76, 90, 101, 146, 193, 212, 233
George I, King 172
George II, King 100, 145, 157, 162–4, 168, 179, 192, 205, 236
gilding 4, 47, 95
Gildon, Charles 90, 94, 187
Gilliver, Lawton 109–10, 111–12, 113, 140, 144, 149–51, 152, 154, 175, 188, 195, 214, 215
Godolphin, Sidney, 1st Earl of 34, 35
Goethe, Johann Wolfgang von 64
Granville, George, Baron Lansdowne 96, 187, 208, 223, 231, 233, 235
Gribelin, Simon 47, 73
Grice, H. P. 7–8, 9
Griffith, R. H. 212, 214
Grub-street Journal, The 2, 110, 159, 164, 214–15, 224
Guardian, The 2, 82, 100, 105, 217, 234
Guerinot, J. V. 181, 203
Guernier, Louis du 26

Habermas, Jürgen 55, 64
Halifax, Charles Montagu, 1st Earl of 193, 233
Halsband, Robert 26, 28
Hampton Court 21, 26, 29–31, 34–6, 40–1, 43, 44, 72, 176, 219, 246, figs. 1, 2, 5
Hampton Court, Richmond, and Kensington Miscellany (1733) 30
Hanmer, Sir Thomas 5
Harcourt, Simon 66–7, 68, 70, 74
Harte, Walter 11, 48, 107–41, 151, 189, figs. 7–8
Haywood, Eliza 84, 95
headpieces 14, 25–6, 28, 44, 47, 49, 64–6, 70–4
Hearne, Thomas 5, 153
Heinsius, Daniel 157, 158, 196
Henley, John ('Orator') 4–5, 191
Henri III, King of France 201–2
Hervey, John, Baron 11, 12, 142, 145, 146, 148, 161, 164, 175–82, 190, 191, 194–5, 198, 199–200, 202, 203–7, 223, 224
Het Loo 30
Hill, Aaron 145, 189
Hills, Henry 16
Hillsborough, Trevor and Mary Hill, Viscount and Viscountess 197–8

Hobbes, Thomas 92–3
Homer 47, 48, 51, 63, 64, 66, 67–8, 69, 71, 74, 76, 84, 92–3, 98, 109, 239, fig. 4; see also 'Episode of Sarpedon'; *Iliad*; *Odyssey*
Hooghe, Romeyn de 36
Hooker, Richard 53, 55
Horace (Quintus Horatius Flaccus) 9, 11, 68, 70, 74, 75, 89, 94, 101, 142–74, 181, 195, 217, 219, 232, 237; see also *First Epistle of the Second Book; First Satire of the Second Book; Fourth Satire of the First Book; Ode to Venus; Second Epistle of the Second Book; Second Satire of the Second Book; Sixth Satire of the Second Book; Sober Advice*
Horace his Ode to Venus 200–1, 203
'Horace, Satyr 4. Lib 1. Paraphrased' 147–8, 161
Huggonson, John 110, 159
Hughs, John 157, 159, 195
Hume, David 137–8

Iliad 1, 14, 25, 26, 47, 48, 49, 51, 68, 69, 70, 71, 74, 100, 101, 104, 121, 154, 193, 214, 217, 236, 239
illustrations 11, 12, 14, 22, 25–34, 36, 37, 38, 47, 49, 61, 64–6, 70–4, 80–1, 188
indices (cues) 142, 151, 160, 169, 171, 195
initials 5, 14, 25–6, 44, 47, 61, 70, 71, 80–1
intentions 7–10, 11, 12, 13, 23, 41, 42, 93, 113–14, 124, 154, 181
Invalides, Les 36
Isaiah 75–81
italic and roman 2, 7, 8–9, 13, 24, 38, 61, 89, 97, 135, 139, 142, 151, 160–9, 173, 183, 184, 185, 197, 221, 228

J., W. 100–1
Jack, Ian 14
Jacob, Giles 52, 95–8, 104, 106
Jansenism 37–8
'January and May' 15, 71, 73, 240
Jeffreys, George 189
Jervas, Charles 26, 47, 49, 63, fig. 3
Johnson, Samuel 23, 51–2, 79, 81
Jones, Inigo 219
Jonson, Ben 46
Juvenal (Decius Junius Juvenalis) 29, 87, 89, 166

Keener, Frederick M. 99–105
Kensington Palace 36
Kernan, Alvin 1, 3–4
Key to the Lock, A 36, 41, 55, 89, 145, 177
Knapp, Francis 66, 67, 69
Knight, Ann 111, 117
Knight, John 111, 117
Kyrle, John 222

Laelius, Caius 162, 167
Lanesborough, James Lane, Viscount 229

Langford, Paul 11–12
Lansdowne, George Granville, Baron 96, 187, 208, 223, 231, 233, 235
Law, Ernest 30
Lee, Nathaniel 53
Leranbaum, Miriam 144
Letter to a Noble Lord 179–80
Letters (Pope's) 2–3, 12, 52, 143, 164–5, 178, 188, 195, 213, 233
Lewis, William 49
line spacing 152, 159–60, 168–9, 195
Lintot[t], Bernard 4, 6, 15–18, 25, 26, 37, 44, 47, 48–50, 53, 56, 86, 111, 153, 189, 214, 215, 216, 231, 240
Lintot, Henry 153
Lucian 92
Lucilius, Gaius 163, 164–7, 174
Lucretius (Titus Lucretius Carus) 118, 127
Lucullus, Lucius Licinius 66
Lyttelton, George, 1st Baron 189

Macaulay, Thomas Babington, Baron 236
Mack, Maynard 1, 11, 48, 56, 80, 108, 116, 120, 125, 127, 133, 165, 185, 195, 201, 202, 219, 225, 241
McLuhan, Marshall 1, 3
Maecenas, Gaius Cilnius 162
Mainwaring [or Maynwaring], Arthur 40–1, 233
Maittaire, Michael 118, 158, 218
Mall: or, the Reigning Beauties, The (1709) 19
Mallet, David 111–12, 189
Manilius, Marcus 127
Manley, Delarivier 34–6
Mansfield, William Murray, 1st Earl of 201
manuscripts, Pope's 1, 2, 18, 22–3, 56, 57, 58–9, 68–9, 89, 154, 161, 164, 182–3, 185–94, 195, 203–5, 206, 209, 212–13, 219, 223, 226, 235–6, 238
Mar, Frances Erskine, Countess of 197
Marlborough, John Churchill, 1st Duke of 33, 35, 36, 226
Marlborough, Sarah Churchill, Duchess of 35, 40–1, 172, 193, 226
Martinus Scriblerus 3–5, 82, 88, 90, 91, 95, 99–105, 106
Mary, Queen 30, 31, 36
Masham, Abigail, Lady 40–1
'Master Key to Popery, A' 145–6, 147, 161, 203–4
Maynwaring [or Mainwaring], Arthur 40–1, 233
meaning 7–10, 13, 15, 52–3, 75, 85, 90, 134–5, 160, 198, 199, 225, 226
Medina, Sir Solomon 226
Memoirs of Mrs Manley 36
Menander 91
Mengel, Elias F., jun. 26
Mercury 73
Messiah 10, 69, 70, 71, 75–81, 232
Miller, James 109–10, 151, 189

Index

Milton, John 79, 88, 120, 187, 195, 239
Miscellaneous Poems and Translations (Lintot's, 1712) 14, 15–20, 21–2, 25
Miscellanies (Pope and Swift's) 100, 217
Montagu, Edward Wortley, sen. 170–1, 230
Montagu, Edward Wortley, jun. 171, 197
Montagu, Lady Mary Wortley 12, 58, 142, 170–1, 175, 176–82, 191, 195, 197, 198, 200, 204, 223
Moore, A. (pseud bookseller) 95
Morphew, John 16
Morrice, Bezaleel 103
motto 59, 66, 110, 127, 152, 216
Murray, William, 1st Earl Mansfield 201
Muses 59, 70–1, 79, 94

Narrative of Dr. R. Norris 104
Nero 127, 202, 205
New Atalantis 34–6
'New Ballad to the Tune of Fair Rosemond, A' 41
Newcastle, Margaret Cavendish, Duchess of 95
Newton, Sir Isaac 128
Nokes, David 193, 203
Norton, Rictor 205
notes 2, 4, 7, 10, 11, 12, 56, 60, 61, 70, 75, 78, 81, 82, 87, 88–9, 90, 92, 94, 95–9, 102, 103–4, 106, 112, 117, 123, 157, 175, 195, 197–8, 209–41
Nuttall, A. D. 137

'Occasion'd by Some Verses of his Grace the Duke of Buckingham' 50, 67
octavo 2, 14, 44, 47, 84, 95; see also *Works* octavo series
Ode for Musick, on St Cecilia's Day 18, 71, 233
Ode to Venus 200–1, 203
Odyssey 1, 6, 16, 26, 86, 100, 113, 121, 154, 189, 214, 217, 239
Of False Taste: An Epistle to . . . Richard Earl of Burlington 12, 36, 108, 113, 143–7, 151, 152, 155, 165, 172, 174, 186, 190, 193, 203, 211, 217, 218, 227, 228, 229, 230–1
Of the Characters of Women: An Epistle to a Lady 113, 201, 217, 218, 219, 226, 228, 229–30, fig. 7
Of the Knowledge and Characters of Men. . . . An Epistle to . . . Richard Lord Visct. Cobham 217, 218, 219, 226, 228, 229, 236
Of the Use of Riches, an Epistle to . . . Allen, Lord Bathurst 108, 109, 113, 116, 147, 151, 152, 211, 217, 218, 219, 220–2, 227, 228, 229, 230
offprints 18, 25
Oldfield, Anne 170
Oldfield, Richard 170
Oldmixon, John 96, 98, 219
'On a Fan' 50
'On a Flower which Belinda gave me from her Bosom' 19
'On Silence' 17

'On the Birth-Day of Mr *Robert Trefusis*; Being Three Years Old' 20
ornaments, printer's 5, 61, 113, 195
Osborne, Thomas 4
O'Sullivan, Maurice 81
Otway, Thomas 53
Ovid (Publius Ovidius Naso) 17, 68, 239
Oxford, Edward Harley, 2nd Earl of 113, 192–4, 217, 223
Oxford, Robert Harley, 1st Earl of 24, 66
Ozell, John 96

pagination 18
Palladio, Andrea 219
paper 3, 61, 142; fine 25, 47, 48, 215; large 4, 111; writing royal 47, 48
parallel texts 10, 11, 75, 78, 105, 142–74, 181, 195, 213, figs. 9, 10
Parmigianino (Girolamo Francesco Maria Mazzola) 202
Parnell, Thomas 48, 66, 68, 70, 71–2, 73, 74
'Part of the Thirteenth Book of Homer's Odysses' 73
Pascal, Blaise 37
Pastorals, The 2, 15, 19, 49, 66, 69, 71, 209, 211, 212, 233–5, 236, 237, 240; see also *Messiah*
Paulus Maximus 201
Payne, Olive 195
Perrault, Charles 87
Persius (Aulus Persius Flaccus) 29, 89
Peterborough, Charles Mordaunt, 3rd Earl of 167, 179, 219
Petre, Robert, 7th Baron 18, 22, 34, 35
Philemon 91
Philips, Ambrose 2, 191, 234, 236
Philips, John 20
Phillips, Constantia 170
Phoebus Apollo 64–5, 67–8, 70–1, 110, 188, fig. 4
Pindar 74
Pine, John 158
Piper, David 56, 63
plates 14, 25–34, 36–7, 44
Plautus, Titus Maccius 46
Poems on Several Occasions: as edited by Pope 64, 69; as title 47–8
Poet Finish'd in Prose, The 202–3, 207
Poetical Miscellanies Tonson's (1709) 15, (1714) 49
Pollio, Gaius Asinius 79; *see also* Virgil, *Eclogue* 4
Pompey (Gnaeus Pompeius Magnus) 66
'Pop upon Pope, The' 104
Portland, William Bentinck, 1st Earl of 34
Preface to *Works* (1717) 2, 10, 23–4, 47, 49, 56–60, 64, 66, 67, 68, 69, 70–1, 176, 216, 217
presswork 2, 6
print, definitions 1–2
Prior, Matthew 17, 48

public, the 2, 10, 12, 23, 25, 47, 53, 55–60, 67, 68, 111–12, 146, 209–10, 212, 214–16
'Publisher to the Reader' 23
Pulteney, William 206
Pushkin, Alexander Sergeyevich 85, 86
Puttenham, George 28–9

quarto 25, 26, 47–8, 50, 63, 84, 153, 155, 168; see also *Works* II quarto and folio
Queensberry, Charles and Catherine Douglas, 3rd Duke and Duchess of 193
Quinault, Philippe 87
Quintilian (Marcus Fabius Quintilianus) 232
quotation marks 122–5, 155

Radcliffe, Ann 85
Rahner, Karl 53
Rape of the Lock, The 5, 7, 10, 11, 14–45, 61, 69, 70, 71, 72, 106, 156, 232, 233, 238–9, figs. 1, 2, 5; *Rape of the Locke* 18–22
Rapin, René 75
'Rapin Imitated, in a Pastoral Sent to *Belinda* upon her leaving *Hattley*' 19
'Receipt to Make an Epic Poem, A' 100–1
revision 7, 10, 11, 14, 22, 24, 37–45, 116, 128, 182–94, 207, 215, 224–7, 228–9, 233–6, 240
Richard, son of King William I 232, 236
Richardson, Jonathan, jun. 89, 167, 210, 212–13, 220, 228
Richardson, Samuel 84, 85
Rival Dutchess, The 41
Roberts, James 178
Rochester, John Wilmot, 2nd Earl of 53
Rogers, Robert W. 157–8
roman, *see* italic and roman
Rose, Mark 50, 51, 52, 54–5
Rosicrucianism 37–8
Rowe, Nicholas 16, 25, 48, 53, 55
Rudd, Niall 148, 164
Rumbold, Valerie 22, 35
Russel, Richard 224

Sacheverell, Henry 33, 35
Saint-Gemain 36
St James's Park: A Satyr (1708) 19
Sallust (Gaius Sallustius Crispus) 127
'Sapho to Phaon' 55, 71, 73, 240
Sappho, Greek poet 48
satyrs 28–9, 38, 71–2, 73
Scipio Aemilianus Africanus 162, 167
Scriblerus 3–5, 82, 88, 90, 91, 95, 99–105, 106, 241
Searle, John R. 7–10, 52–3
Second Epistle of the Second Book of Horace, The 201–2, 203
Second Satire of the Second Book of Horace, The 152, 153, 155, 156, 157–8, 159–60, 165, 168–74
section titles 18, 70, 73, 217, 218, 233

Sedley, Sir Charles 53
Seneca, Lucius Annaeus 127
Serle, John 190
Settle, Elkanah 94
Shakespeare, William 5, 48, 52–3, 55, 88, 91, 95, 98–9, 122, 187, 217, 224
Sidney, Sir Philip 53
Sixth Satire of the Second Book of Horace 168
Smith, Edmund 17
Smythe, James Moore 90, 103, 106, 191, 222
Sober Advice from Horace 2, 8, 10, 12, 113, 149, 157, 159, 170, 171, 174, 175–6, 182, 194–203, 204, 205, 206
Socrates 127
Solomon, Harry M. 127
Somers, John, Baron 233
Somerset House 36
Southcott [or Southcote], Thomas 17
Souza, D. Emmanueli Caietano de 158
Spectator, The 14, 21, 38, 55, 58, 75–7, 78, 217, 237
spelling 20, 28, 158
Spence, Joseph 15, 22, 92, 111, 112, 143, 157
Spenser, Edmund 48, 234
Sporus 148, 176, 185, 188, 190–1, 194–5, 200, 202, 203, 204–5, 206, 207
Stack, Frank 151, 160, 163
Statius, Publius Papinius 17–18, 74, 232
Steele, Sir Richard 58, 75–6, 77, 153, 203
Suarez, Michael 53
subscription 2, 14, 25, 47, 112, 153, 215
Suckling, Sir John 53
Suetonius (Gaius Suetonius Tranquillus) 205
superior numbers, *see* indices
Sutherland, James 92–3
Swift, Jonathan 35, 36, 37, 55, 58, 68, 91, 168, 173, 200, 222, 240

tailpieces 25, 49, 70 n. 31
Talbot, James 158
Tassoni, Alessandro 92
Taylor, John 97
Tempest, Althea or Henrietta 235
Temple of Fame: A Vision, The 17, 66–7, 68, 69, 70, 71, 74–5, 232, 239–40
Terence (Publius Terentius Afer) 46
Theobald, Lewis 47, 86, 88, 90, 91, 94–5, 96, 97, 98–9, 102, 104, 106
Theocritus 71, 94, 234
Tibullus, Albius 48
Tillotson, Geoffrey 19, 22, 23, 29
Tindal, Matthew 140, 224
Titian (Tiziano Vecellio) 67, 73
title page 11, 46, 54, 66, 70, 82, 107, 108–10, 112, 113, 127, 152–4, 158, 174, 180, 195–6, 210, 211, 212, 216, fig. 8
To a Lady (Of the Characters of Women) 113, 201, 217, 218, 219, 226, 228, 229–30, fig. 7

'To a Lady Sitting before her Glass' 19
'To a Young Lady on Leaving the Town after the Coronation' 21, 49–50
To Addison 217
To Arbuthnot 10, 11, 12, 52, 59, 113, 146, 147–8, 164, 174, 175–208, 211, 216, 217–18, 219, 220, 222–4, 227, 231, fig. 8
To Bathurst (Of the Use of Riches) 108, 109, 113, 116, 147, 151, 152, 211, 217, 218, 219, 220–2, 227, 228, 229, 230
To Burlington (Of False Taste) 12, 36, 108, 113, 143–7, 151, 152, 155, 165, 172, 174, 186, 190, 193, 203, 211, 217, 218, 227, 228, 229, 230–1
To Cobham (Of the Knowledge and Characters of Men) 217, 218, 219, 226, 228, 229, 236
'To Mr Pope on his Translation of Homer' 69
'To Oxford' 217
'To the Author of a Poem, Intitled, Successio 17
To the Imitator of the Satire of the Second Book of Horace 178; see also *Verses [Address'd] to the Imitator of Horace*
Tonson, Jacob, sen. 15, 16, 17, 19, 46, 49, 53, 56, 88, 215, 216
Tonson, Jacob, jun. 6, 49, 50, 88, 215
town, the 6, 21, 49–50, 57, 73, 100, 179–80, 196, 211
Trapp, Joseph, *Prælectiones Poeticæ* 5, 25, 61, 70
Trebatius (C. Trebatius Testa) 148, 154, 155, 157, 162, 167
Trefusis, Robert 20
Trevelyan, George Macaulay 35
Trumbach, Randolph 205
Trumbull, Sir William 233, 235
Turner, Richard 221
'Two Choruses to the Tragedy of Brutus' 49

variants (in notes) 191, 192, 220, 224–8, 234–5, 236
Velázquez, Diego Rodriguez de Silva y 73
Venn, Richard 140–1
Venus 28, 73, 200–1
Verdi, Giuseppe Fortunino Francesco 53
Vernon, Thomas 153, 173
Versailles 30
'Verses to be Prefix'd before Bernard Lintot's New Miscellany' 16–17, 18
Verses [Address'd] to the Imitator of Horace 12, 142, 164, 174, 175, 176–82, 185, 189–91, 195, 204, 207, 208
'Verses to the Memory of an Unfortunate Lady' 49
Vertue, George 47, 49, 63, fig. 3
'Vertumnus and Pomona' 17, 240
Vida, Marcus Girolamo 88
Villars, Abbé Montfaucon de 37–41
Virgil (Publius Vergilius Maro) 48, 63, 64, 66, 67, 68, 69, 71, 73, 74, 82, 93, 94, 96, 234, 235, 239, fig. 4; Eclogue 4 (Pollio) 75–81

Walpole, Sir Robert 144, 154, 157, 162, 168, 173, 178, 179, 194, 226
Walsh, William 187, 233–4, 235
Walter, Peter 173, 221–2
Wanley, Humfrey 153
Warburton, William, Bishop of Gloucester 6, 12, 42, 55, 56, 59, 89, 93, 112, 116, 117, 123–4, 137, 154, 180, 201, 209, 210, 211, 226, 228, 234, 237, 241
Ward, Edward 33, 95, 96, 97, 98
Ward, John 221
Warmsley, Catherine 22
Warner, William B. 35
Warton, Joseph 107–8, 112
Watts, John 5, 6
Weinbrot, Howard 29
Welsted, Leonard 97, 222, 224, 231
West, Gilbert 189
Whiston, William 77, 78
White, Douglas H. 134
'Wife of Bath her Prologue, The' 73, 240
Wilford, John 110
William I, King 232, 236
William II (Rufus), King 30–1, 236
William III, King 30–1, 33, 34, 36, 232, 236, 238
Williams, Aubrey L. 1, 3, 99, 241
Wimsatt, W. K. 63
Winchilsea, Anne Finch, Countess of 66, 67, 69
Windsor-Forest 17, 21, 30–1, 44, 66, 68, 69, 71, 80, 209, 212, 232, 234, 235–6, 238
Wollaston, William 140
Woodfall, Henry 47, 178, 180
Woodmansee, Martha 50–1, 52, 53–4, 61
Worde, Wynkyn de 4
Wordsworth, William 165
Works, Pope's: *Works* I (1717) 2, 4, 5, 7, 9, 10, 11, 28–9, 46–81, 88, 153, 154, 176, 188, 214–15, 216, 231–2, fig. 3; *Works* II quarto and folio (1735) 26, 68, 107,113, 124, 143, 144, 146, 153, 156, 159, 168, 170, 175, 176, 182–5, 191, 192–3, 194, 195, 200, 202, 206, 210, 211, 212, 214–15, 216–27, 240, fig. 7; *Works* octavo series (1735/6) 2, 10, 47, 89, 153, 183–4, 191, 195, 200, 210, 211, 213, 214–16, 217, 227–41
world, the (the public) 23, 24, 57–8, 59, 60, 68, 91, 100, 166, 217
Wren Sir Christopher 30
Wright, John 6–7, 91, 109, 110, 149, 159, 168, 173, 175, 182, 188, 213
Wycherley, William 15, 46, 53, 66, 67, 69, 233

Yates, Edward 29
Young, Edward 51, 108